# Introduction to
# *Artificial*
# *Intelligence*

## THIRD EDITION

### Philip C. Jackson, Jr.

Dover Publications, Inc.
Mineola, New York

*Bibliographical Note*

*Introduction to Artificial Intelligence: Third Edition,* first published by Dover Publications, Inc., in 2019, is a new edition of the work originally published in 1974 by Petrocelli/Charter, New York, and in an enlarged Second Edition by Dover in 1985. This 2019 edition includes a new Preface and Acknowledgments section which replaces the 1985 Preface, and a new section, "Artificial Intelligence in the 21st Century," which replaces Developments, 1975–1984.

*Library of Congress Cataloging-in-Publication Data*

Names: Jackson, Philip C., 1949- author.
Title: Introduction to artificial intelligence / Philip C. Jackson, Jr.
Description: Third edition. | Mineola, New York : Dover Publications, Inc.,
 [2019] | Originally published: New York : Petrocelli Books, 1974. |
 Includes bibliographical references and index.
Identifiers: LCCN 2019002273 | ISBN 9780486832869 | ISBN 0486832864
Subjects: LCSH: Artificial intelligence.
Classification: LCC Q335 .J27 2019 | DDC 006.3—dc23
LC record available at https://lccn.loc.gov/2019002273

Manufactured in the United States by LSC Communications
83286401    2019
www.doverpublications.com

*This book is dedicated to*

The memory of my parents, Philip and Wanda Jackson

My wife Christine

# CONTENTS

## 2. MATHEMATICS, PHENOMENA, MACHINES     33

## 3. PROBLEM SOLVING     67

# 8. PARALLEL PROCESSING AND EVOLUTIONARY SYSTEMS

# 9. THE HARVEST OF ARTIFICIAL INTELLIGENCE

# PREFACE (2019)

Are we intelligent enough to understand intelligence? One approach to answering this question is "artificial intelligence," the field of computer science that studies how machines can be made to act intelligently. In general this book is addressed to all persons interested in studying the nature of thought, and hopefully much of it can be read without previous formal exposure to computers.

Much progress has been made in research on artificial intelligence since the First Edition of this book was published in 1974. The book as originally written remains a general introduction to the foundations of the field, which have long been called "good old-fashioned AI" (GOFAI).[1] Hopefully, this Third Edition will be more useful because of material added to summarize progress over the decades and guide the reader into topics especially relevant for AI in the 21st century. For simplicity, this supplementary material, including its own bibliography, is added as a separate section immediately following this Preface.

Artificial intelligence can and should be studied in ways that are not strictly technical. It is important for us to realize how this science is related to the hopes (and fears) of humanity. To do this we must try to understand people, not just machines. If artificial intelligence is to be developed beneficially, it will have to become one of our most humanistic sciences.

◊

It is important to thank everyone who helped make this work possible, though time and space would make any list incomplete, and regretfully these words are written too late for some to read.

I gratefully acknowledge the help and guidance of the late Dr. Ned Chapin, Editor for the First Edition of this textbook, and the help and guidance of John Grafton, Editor for the Second and Third Editions.

---

[1] The term was coined by Haugeland (1985).

I am grateful to all who have contributed directly or indirectly to my studies and research on artificial intelligence and computer science, in particular:

Harry Bunt, Walter Daelemans, John McCarthy, Arthur Samuel, Patrick Suppes, C. Denson Hill, Sharon Sickel, Michael Cunningham, Ira Pohl, Filip A. I. Buekens, H. Jaap ven den Herik, Paul Mc Kevitt, Carl Vogel, Paul A. Vogt, Edward Feigenbaum, Bertram Raphael, William McKeeman, David Huffman, Michael Tanner, Frank DeRemer, James Q. Miller, Bryan Bruns, David Adam, Noah Hart, Marvin Minsky, Donald Knuth, Nils Nilsson, Faye Duchin, Douglas Lenat, Robert Tuggle, Henrietta Mangrum, Warren Conrad, Edmund Deaton, Bernard Nadel, Thomas Kaczmarek, Carolyn Talcott, Richard Weyhrauch, Stuart Russell, Igor Aleksander, Helen Morton, Richard Hudson, Vyv Frederick Evans, Michael Brunnbauer, Jerry Hobbs, Laurence Horn, Brian C. Smith, Philip N. Johnson-Laird, Charles Fernyhough, Antonio Chella, Robert Rolfe, Brian Haugh, K. Brent Venable, Jerald Kralik, Alexei Samsonovich, Peter Lindes, William G. Kennedy, Arthur Charlesworth, Joscha Bach, Patrick Langley, John Laird, Christian Lebiere, Paul Rosenbloom, John Sowa.

They contributed in different ways, such as teaching, questions, guidance, discussions, reviews of writings, permissions for quotations, collaboration, and/or correspondence. They contributed in varying degrees, from sponsorship to encouragement, lectures, comments, conversations, objective criticisms, disagreements, or warnings that I was overly ambitious. I profoundly appreciate all these contributions. To be clear, in thanking these people it is not claimed they would agree with everything I've written or anything in particular.

◊

In general, my employment until retirement in 2010 was in software development and information technology. This was not theoretical research, though in some cases it involved working with other AI specialists on AI applications. I was fortunate to work with many of the best managers and engineers in industry, including Phil Applegate, Karen Barber, Doug Barnhart, Barbara Bartley, Ty Beltramo, Pete Berg, Dan Bertrand, Charles Bess, William Bone, Sam Brewster, Michelle Broadworth, Mark Bryant, Gregory Burnett, Tom Caiati, Pam Chappell, David Clark, David Coles, Bill Corpus, Justin Coven, Doug Crenshaw, Fred Cummins, Robert Diamond, Tom Finstein, Geoff Gerling, Dujuan Hair, Phil Hanses, Steve Harper, Kathy Jenkins, Chandra Kamalakantha, Kas Kasravi, Phil Klahr, Rita Lauer, Maureen Lawson, Kevin Livingston, David Loo, Steve Lundberg, Babak Makkinejad, Mark Maletz, Bill Malinak, Arvid Martin, Glenda Matson, Stephen Mayes, Stuart McAlpin,

Eileen McGinnis, Frank McPherson, Doug Mutart, Bruce Pedersen, Tyakal Ramachandraprabhu, Fred Reichert, Paul Richards, Anne Riley, Saverio Rinaldi, Marie Risov, Patrick Robinson, Mike Robinson, Nancy Rupert, Bob Rupp, Bhargavi Sarma, Mike Sarokin, Rudy Schuet, Dan Scott, Ross Scroggs, Pradip Sengupta, Scott Sharpe, Cheryl Sharpe, Christopher Sherman, Patrick Smith, Michael K. Smith, Scott Spangler, Kevin Sudy, Saeid Tehrani, Zane Teslik, Kathy Tetreault, Lakshmi Vora, Rochelle Welsch, Robert White, Terry White, Richard Woodhead, Scott Woyak, Glenn Yoshimoto, Ruth Zarger. Again, any list would be incomplete and in thanking these people it is not claimed they would agree with everything I've written or anything in particular.

◊

It should be expressly noted that I alone am responsible for the content of this book. Naturally, I hope the reader will find that its value greatly outweighs its errors, and I apologize for any errors it contains.

I will always be grateful to my late parents, whose faith and encouragement made this effort possible. Heartfelt thanks also to other family and friends for encouragement over the years.

I'm especially grateful to my wife Christine, for her love, encouragement and patience with this endeavor.

PHILIP C. JACKSON, JR.

# ARTIFICIAL INTELLIGENCE
# IN THE 21ST CENTURY

This supplementary section gives a brief introduction to the current state and future prospects of artificial intelligence, and suggestions for how to learn more about the field. These pages focus on major topics that appear likely to be important for AI in the 21st century.

Perhaps no introduction to the field can be complete, at this point. A vast amount of research has been performed over the decades, and is being conducted around the world regarding artificial intelligence and cognitive science. These pages are just the author's perspective.

So, this new material is only an introduction to AI research in the 21st century, just as the original text of this book is only an introduction to AI research up to 1974. This section summarizes and gives pointers to research. Hopefully, the reader will follow these pointers to gain greater knowledge of the entire field, including research not cited.

*References and § Notation*

In some cases, references are made to the following textbook by just using Chapter numbers. Almost all citations are to entries in the Supplementary Bibliography at the end of this new section. A few of the citations before 1970 are to entries in the original Bibliography at the end of the book. Throughout these pages the § notation is used to refer to chapters and sections in *Toward Human-Level Artificial Intelligence* (Jackson, 2019). For example, §2.1 refers to the first section in Chapter 2 there. Its first subsection is §2.1.1.

## 1. How to Define Human-Level Intelligence?

As discussed in Chapter 1, Turing's (1950) paper *Computing Machinery and Intelligence* challenged scientists to achieve human-level artificial

xvii

intelligence. However, the term 'artificial intelligence' was not officially coined until the Dartmouth summer research project proposal by McCarthy, Minsky, Rochester, and Shannon (1955), who conjectured that every aspect of learning and intelligence could be simulated by a computer. Newell and Simon (1976) further formalized the conjecture, stating the Physical Symbol System Hypothesis that symbolic processing systems are necessary and sufficient for achieving general intelligence. Newell (1973) challenged us to achieve a science of human cognition commensurate with its power and complexity. Newell (1990) advocated developing 'unified theories of cognition' which would include language, learning, motivation, imagination, and self-awareness – the complete scope of human intelligence.

Turing suggested scientists could say a computer thinks if it cannot be reliably distinguished from a human being in an "imitation game" now called a Turing Test. He did not attempt to define 'thinking', though he countered several arguments that a computer cannot think. He did not set any limits to the questions people could ask a computer in an imitation game – scientists could be as rigorous as possible, but the computer could also not answer questions. For example a computer pretending to be a person might say it was not very good at writing poetry, or intentionally make occasional errors in arithmetic. Turing predicted that in about 50 years: 1) a computer would not be identified as a computer more than 70% of the time by 'average' interrogators after five minutes playing the imitation game; 2) people would commonly say machines think and this would be an educated opinion.

While people do informally speak of machines thinking, it is widely understood that computers do not yet really think or learn with the generality and flexibility of humans. While an average person might confuse a computer with a human in a typewritten Turing Test lasting only five minutes there is no doubt that within five to ten minutes of dialog using speech recognition and generation (successes of AI research), it would be clear that computers do not yet have human-level intelligence. We are still a very long way from achieving human-level AI – there is much research and development to be done.

Over the decades there have been cycles of optimism and pessimism about the prospects for achieving human-level AI. A survey in 2012 and 2013 of about 550 AI experts found almost 18% believed no research approach would ever achieve human-level machine intelligence (Müller & Bostrom, 2016). The open issue of how to define human-level intelligence may contribute to such doubts. It has been suggested that intelligence may not be a concept which can be analyzed and duplicated (Kaplan, 2016). As we shall see in the next section, some philosophers, mathematicians, and scientists have argued human-level AI is impossible.

While a Turing Test may help recognize human-level AI if it is created, the test does not define intelligence nor indicate how to design, implement, and achieve human-level AI. Therefore, a different approach was proposed in (Jackson, 2014), to define human-level intelligence by identifying capabilities achieved by humans and not yet achieved by any AI system, and to inspect the internal design and operation of any proposed system to see if it can in principle support these capabilities, which I call *higher-level mentalities*. They include human-level natural language understanding, higher-level learning, metacognition, imagination, and artificial consciousness. These topics are further discussed throughout (Jackson, 2019), and in the following pages as needed to address AI in the 21st century. Section 8.1 focuses on higher-level mentalities.

## 2. Theoretical Objections to the Possibility of Human-Level AI

The goal of Chapter 2 in this book is to present some of the mathematical theory underlying artificial intelligence and computer science in general. In particular it discusses whether there is any way in theory of proving mathematically that machines could or could not be intelligent. In addition, it presents some practical limitations that affect computers because they are real-world machines subject to the laws of physics. These results from mathematics and physics are useful in reasoning about computers and the limitations of artificial intelligence, but not in themselves sufficient to prove or disprove the attainability of human-level artificial intelligence.

Many scientists have discussed this question, arguing both for and against the ultimate achievability of human-level intelligence by computers. And into this debate they have introduced ideas from other sciences.

Regarding the general theoretical limitations of artificial intelligence, Haugeland (MD, 1981) included several papers arguing against the possibility of artificial intelligence which could duplicate or surpass human thought, as well as other papers that discuss AI methodology but are not skeptical of its ultimate success. The arguments against AI (by Dreyfus, Haugeland, Searle, Davidson, and others) draw on issues in the fields of psychology, philosophy, and biology.

They argue that computers cannot duplicate the biochemistry of the human brain, which prevents AI from duplicating moods, emotions, awareness, feelings, and other phenomena important to human thought. Also, they argue that "understanding" concepts is fundamentally different from symbol manipulation; that sensorimotor (and other) skills are not developed by thought processes such as those studied by AI and cognitive science; that human thought is "holistic" and cannot be divided into sub-processes in the way that AI approaches it; that human thought deals with infinite exceptions and ambiguities and thus is too complex for computers. (I do not say that each of the authors listed above subscribes to all of these claims.)

I alluded to some of these concerns in Chapter 2, for example by noting that the universe might contain phenomena which are not finitely describable, and that the human brain is architecturally different from digital computers. I concluded it is an open question whether computers could ever duplicate all the abilities of human intelligence, though it seems clear they can emulate some.

## 2.1 Searle's 'Chinese Room' Argument

Searle (1980) gave an argument called the "Chinese Room", that symbol manipulation cannot be equivalent to human understanding. He used a variation of the Turing Test: Imagine a person placed in a room, who understands English but does not understand Chinese. The person has instructions written in English saying how to process sentences written in Chinese. Pieces of paper with Chinese sentences on them are pushed through a slot into the room. The person follows the instructions in English to process the Chinese sentences and to write sentences in Chinese that are pushed through the slot out of the room.

Searle asks us to consider the person in the Chinese Room as equivalent to a computer running a software program, and to agree that neither the Chinese Room nor the person inside it using English instructions understands Chinese. From this, Searle argues that no computer running a program can truly understand a natural language like English or Chinese.

Searle's argument contradicts AI research based on the foundational hypothesis that symbolic processing systems can support human-level AI (per section 1 above). The Chinese Room argument has been the subject of unresolved debate since 1980, though the philosophical issues are complex enough that people on both sides may believe they resolved it in their favor, long ago. Cole (2009) provides a survey of this debate.

The most frequent reply to Searle's argument (which he does not accept) is called the 'systems reply', and says the Chinese Room as a whole really would understand Chinese, even though none of its components would: A system can have a property and an ability not possessed by each of its components – the human brain can be conscious and intelligent, even though the brain's individual neurons are not conscious and intelligent. Russell & Norvig (2010) support the systems reply and discuss it in some detail, noting others including McCarthy and Wilensky proposed it.

I agree with the systems reply, and with the arguments given by Chalmers (1996). I also give a different reply in §4.2.4: Searle's argument does not preclude the possibility that the human in the Chinese Room may subconsciously process symbols to understand English in essentially the same way that he/she would consciously process symbols when following instructions to emulate understanding Chinese. The person may have constructed an internal program for understanding English when learning how to understand English as a child, and now be executing the program subconsciously. Thus we normally learn how to do new things consciously, and later perform complex processes unconsciously after they become routine. So from this perspective, Searle's argument does not prove that symbol processing cannot constitute understanding of semantics.

A discussion of how consciousness interacts with natural language understanding is relevant to understanding in general. Much of what we perceive and do happens automatically and unconsciously, with consciousness being drawn to things we do not understand, perceptions that are anomalous, actions and events that do not happen as expected, etc.[1] Once we become conscious of something anomalous, we may focus on trying to understand it, or trying to perceive it correctly, or trying a different action for the same purpose.

## 2.2 Dreyfus' Arguments

Dreyfus (1981) noted that Husserl and Heidegger encountered an apparently endless task in their attempts to define human concepts symbolically, and warned that AI confronts the same problem. Dreyfus[2] (1992) presents several criticisms of AI research from the 1960's through the 1980's. He identified theoretical issues for human-level AI to address, rather than theoretical objections to its possibility in principle. In discussing the future of AI research, Dreyfus (1992, pp.290-305) left open the possibility that human-level AI could be achieved, if his theoretical issues could be addressed – though he was very skeptical about the potential to address these issues, practically.

Jackson (2019) advocates a research approach (called 'TalaMind'[3]) toward human-level AI which incorporates responses to Dreyfus' criticisms. The approach focuses on symbolic processing of conceptual structures, without claiming this is completely sufficient to achieve human-level AI. Other technologies may also be needed, such as connectionism or quantum information processing – connectionism is discussed in section 4. Likewise, (Jackson 2019) does not assume the mind operates with a fixed set of formal rules. In the TalaMind approach, rules and procedures can be represented by 'executable concepts' and executable concepts may be modified by other executable concepts, or accepted as input via natural language instructions from the outside environment (analogous to how people can learn new behaviors when given instructions). The TalaMind approach includes a

---

[1] Whitehead (1929), p.161, said that consciousness is involved in the perception of contrast between a possibly erroneous theory and a fact.

[2] This discussion refers in general to Hubert L. Dreyfus, though some of his work was co-authored with his brother Stuart E. Dreyfus.

[3] I pronounce 'TalaMind' to rhyme with "salad mind" or "ballad mind". Trademarks for *Tala* and *TalaMind* have been created to support future development.

variety of other methods, including mental spaces, conceptual blends, and cognitive categories for representing and understanding concepts.

Having mentioned TalaMind, I should inform the reader it is just the approach I think is best for achieving human-level AI and it is not yet generally accepted by other AI researchers. Since this is an introduction to the field of artificial intelligence, it's important for these pages to discuss a wide variety of research approaches. TalaMind will only be mentioned where it offers a different theoretical approach to a research issue, or a different way to support an approach. However, there are several places where it is relevant.

## 2.3 Penrose's Arguments

Penrose (1989 *et seq.*) presented the following claims:

1. Computers cannot demonstrate consciousness, understanding, or human-level intelligence.

2. Some examples of human mathematical insight transcend what could be achieved by computers.

3. Theorems of Turing and Gödel showing theoretical unsolvability of certain logical problems imply human intelligence transcends computers. Penrose gave two arguments similar to arguments by Lucas (1959) and Gödel (1951).

4. Human consciousness depends on quantum gravity effects in microtubules within neurons (an hypothesis with Hameroff).

Jackson (2019, §4.1.2) discusses Penrose's arguments in detail, considers counter-arguments by other researchers, and finds Penrose's arguments are not sufficiently strong to prove human-level artificial intelligence cannot be achieved, at least in principle theoretically.

Regarding consciousness, Penrose's view is that one cannot be genuinely intelligent about something unless one understands it, and one cannot genuinely understand something unless one is aware of it. These are commonsense notions of intelligence, understanding, and awareness, and I agree with them. The topic of 'artificial consciousness' is discussed in section 8.1.9.

## 3. Architecture Levels of an Intelligent Agent

To proceed further, it helps to consider an AI system as a potentially independent agent which can perceive its environment and act intelligently within its environment.[4] There are three architectural levels natural to identify

---

[4] Kennedy (2012) gives a more general discussion of agent-based models for human behavior.

within an AI agent, which I call the linguistic, archetype, and associative levels. They are adapted from Gärdenfors (1995)[5] and will support discussing a wide variety of different AI research approaches in the following pages.

At the linguistic level an AI system represents information and performs inference using one or more symbolic languages. Depending on the particular AI system, a symbolic language may be a simple notation (e.g. n-tuples of symbols), or it could be a formal, logical language like predicate calculus, or in theory it could even be a natural language like English – an approach investigated by (Jackson, 2014), to be discussed later.

The archetype level is where categories, classes, or types are represented. Again, the representations may be simple or complex, depending on the AI system. Some AI systems may not even have a separate architectural level for representing categories. Others may represent categories using symbolic notations, e.g. logical expressions, or represent categories using methods studied in cognitive linguistics and semantics, e.g. conceptual spaces, image schemas, radial categories, etc. (Evans & Green, 2006).

The associative level typically processes information from the environment. It may recognize instances of common classes in the environment (e.g. faces, people, animals, chairs, cars, etc.) and process speech and visual information to recognize words, symbols, sentences, etc. This can support recognition of categories at the archetype level, and representation at the linguistic level of information and relationships in the environment.

An AI system may perform reasoning at the linguistic level and decide communication and actions to perform in the environment. The associative level may support physical actions in the environment. Depending on the architecture and research approach, there may be significant integration across the three levels.

At the linguistic level it is also natural to identify two other components:

- A Conceptual Framework: An information architecture for managing an extensible collection of concepts, expressed linguistically. A conceptual framework supports processing and retention of concepts ranging from immediate thoughts and percepts to long term memory, including concepts representing linguistic definitions of words, knowledge about domains of discourse, memories of past events, expected future contexts, hypothetical or imaginary contexts, etc. These may be implemented using symbolic representations such as mental models, discussed in section 7.

---

[5] Gärdenfors called them the linguistic, conceptual, and subconceptual levels. I call them linguistic, archetype, and associative levels since different kinds of concepts may be represented at each level—e.g. Jackendoff (1989) discussed 'sentential concepts'. However, others often use 'concept' only to refer to conceptual categories, the word sense used by Gärdenfors. The 'TalaMind architecture' combines these three architecture levels with support for a natural language of thought, multiple levels of mental representation, and a self-extending 'intelligence kernel' (viz. Jackson, 2019, §1.4 §1.5, §4.2.6).

- Conceptual Processes: An extensible system of processes that operate on concepts in the conceptual framework, to produce intelligent behaviors and new concepts.

These two elements are sufficiently important that they will be discussed after the linguistic level.

# 4. Machine Learning at the Associative Level—Neural Networks

Over the decades, "learning" has been a major AI research topic. Michalski, Carbonell, & Mitchell (1983) edited a relatively early collection of papers on this topic. Langley (1996) gave an extensive survey. Russell & Norvig (2010) give introductions to relatively recent research.

Machine learning could occur at any of the architecture levels[6], or across them. This section will give an introduction to learning at the associative level[7] using neural networks, which are one of the most successful, important classes of methods for machine learning. It should be emphasized there are many other methods for machine learning, e.g. genetic algorithms (Holland, 1975; Koza, 1992 et seq.) and support vector machines (Cortes & Vapnik, 1995; Cristianini & Shaw-Taylor, 2000).

In AI research, a neural network is an extremely simplified model of a biological neural network, represented as a collection of interconnected 'artificial neurons'. The most often-used approach for neural networks is the 'feedforward' multi-level topology[8] with backpropagation, discussed by Rumelhart, Hinton, & Williams (1986) and further developed in subsequent research by many others. There are other approaches to neural networks, e.g. Bayesian neural networks (see Pearl (1988), Neapolitan (1990), MacKay (1992), Jensen (1996), Ghahramani (2015), Blundell et al. (2015)). In general, 'connectionism' refers to research on neural networks.

## 4.1 Feedforward Multi-Level Neural Networks

The goal of this subsection is to give the reader enough information to write a computer program for creating and training this kind of neural network. If the following details are not interesting or too technical, they can be skipped—later sections do not depend on them.

---

[6] Related to learning and discovery at the linguistic and archetype levels, see Lenat et al. (1979 et seq.) and Gärdenfors (1990 et seq.).

[7] Researchers commonly call computer processing at the associative level 'sub-symbolic'. Technically, all processing performed by digital computers can be viewed as symbolic computation, even processing of neural net algorithms, because bits are symbols. Symbolic computation is a more general class than symbolic representation and reasoning by AI programs (which may or may not access neural networks).

[8] Recurrent networks, in which a network's outputs can also be its inputs, are another important topology (viz. Géron, 2017).

A feedforward multi-level neural network arranges artificial neurons in a series of levels[9], from an input level to intermediate levels (normally called hidden levels), to an output level.

The structure of such a neural network can be defined by specifying the number of levels, or 'height' of the network, $H \geq 2$, and by specifying the number of neurons ('width' of the network) at each level, in an integer array $W[1:H]$. A specific neuron in the network[10] can be identified as $n[h, i]$, where $1 \leq h \leq H$ and $0 \leq i \leq W[h]$.

There are links for sending signals between neurons at successive levels of the network. Each neuron $n[1 \leq h < H, i]$ has links to all neurons $n[h+1, 0 < j \leq W[h+1]]$ at the next level of the network. A neuron $n[h, i]$ will send the same (positive, negative, or zero) signal $s[h, i]$ on all its links to neurons at the next level, but the signal that's received by a neuron $n[h+1, j]$ at the next level will be multiplied by a 'bias weight' $b[h, i, j]$ for the link. Each bias weight can be positive, negative, or zero.

- To begin processing a neural network, input values (which can be positive, negative, or zero) are provided to each input neuron $n[1, i]$. Each input neuron simply 'passes through' its input value as the output signal $s[1, i]$ that it sends to neurons at the second level of the network.

- Every level $h < H$ has a 'bias neuron' $n[h, 0]$ which receives no input and always outputs the signal value $s[h, 0] = 1$ on links to all non-bias neurons $n[h+1, j > 0]$ at the next level. Bias neurons can help train a network to escape suboptimal regions of the search space.[11]

- The total input $S[h, i]$ to each non-bias neuron $n[h, i]$ at level $h > 1$ for $i > 0$ is the sum of the values of its input signals multiplied by their bias weights. (The formula is straightforward, and will be given below.)

- The output signal value from each non-bias neuron $n[h >1, i > 0]$ is $s[h, i] = f(S[h, i])$, where $f$ is an 'activation function'. This output is the signal value from the neuron on input links to neurons in the next higher level of the network, if there is a next level. Otherwise it is an output value from the network. The activation function $f$ is often[12] the sigmoid logistic function $f(x) = 1/(1 + e^x)$. The sigmoid function produces a value that ranges between 0 and 1. If $x$ is negative then

---

[9] Discussions often refer to levels in a network as layers, and to artificial neurons as nodes.

[10] Exactly how to represent artificial neurons and neural networks in a computer program is a design issue. For example, a program may use parallel arrays to store the structure and state of a neural network. This section focuses only on the nature of the algorithm.

[11] Rumelhart et al. (1986, p.330) discussed an alternative to bias neurons, using 'momentum terms'.

[12] Hyperbolic tangent and rectified linear unit activation functions can be more useful than the sigmoid.

$f(x)$ is less than 0.5, approaching 0 as $x$ becomes more negative. If $x$ is positive then $f(x)$ is greater than 0.5, approaching 1 as $x$ becomes more positive.

For a neural network to be useful, the entire network should in effect compute some function of its input values, which produces useful output values. In general we may not know how to precisely define the function, except by giving examples of output values that correspond to input values. To achieve machine learning, we'd like to somehow train the network to compute the function, using paired examples of input and output values that we provide, with the training process adjusting the bias weights within the network.

This can be done using a process called *backpropagation*, which computes errors of the output neurons versus a desired output, and then propagates deltas from the output level to previous levels of the net, determining how much each bias weight in the network contributes to errors. After error contributions have been determined, bias weights are adjusted corresponding to their error contributions. Following are details for how forward propagation and backpropagation work, in training a neural network:

- First, the bias weights for all links between neurons in the network are initialized to random numbers between -1 and 1. Next, the network is trained using a sequence of examples. Each example pairs a possible input to the network with a corresponding desired output from the network.

- For each example, the input to the network is an input vector x and the desired output is an output vector y. The $i$th component $x_i$ ($i \geq 1$) of the input vector x is used to create the output signal value $s[1, i]$ from each of the neurons $n[1, i]$ in the first (input) level of the network. Again, the input neurons are simply pass-through neurons: they don't have activation functions. However, it's typical to pre-process input values to be normalized between -1 and 1, or between 0 and 1, depending on the problem domain.

- Next, the levels of the network from level 2 to the output level H are processed. At each level all the neurons are processed, before proceeding to the next level. For each non-bias neuron $n[h > 1, i > 0]$, its total input is computed using the formula

$$S[h, i] = \Sigma(b[h-1, k, i] \times s[h-1, k])$$

where $0 \leq k \leq W[h-1]$. Then the neuron's output signal value $s[h, i] = f(S[h, i])$ is computed.

- When all the levels have been processed, the actual output from the neural network for the training example has been computed, and is compared with the desired output. The error value[13] of each output neuron is the difference between the $i$th component of the desired output vector y and the neuron's actual output value: $y_i - s[H, i]$, for $i > 0$. This error value is used to compute a 'delta value' for the output neuron:

$$\delta[H, i] = f'(S[h, i]) \times (y_i - s[H, i])$$

where $f'$ is the derivative of the activation function $f$.[14]

- Working backwards from the output level, the preceding levels are processed. A delta value is computed for each non-bias, non-input neuron $n[h, i]$ using the formula:

$$\delta[h, i] = f'(S[h, i]) \times \Sigma(b[h, i, j] \times \delta[h+1, j])$$

where $1 < h < H$, $0 < i \leq W[h]$, and $0 < j \leq W[h+1]$. That is, each neuron's delta value is the rate of change of its activation multiplied by the sum of the products of the bias weights on its links to neurons at the next higher level with the deltas of the neurons at the next higher level.

- After all the delta values for neurons in the network have been computed, the bias weights for input links into each neuron $n[h, i]$ at level $h > 1$ for $i > 0$ are updated, using the formula

$$b[h\text{-}1, k, i] = b[h\text{-}1, k, i] + (r \times s[h\text{-}1, k] \times \delta[h, i])$$

where $0 \leq k \leq W[h\text{-}1]$ and $r$ is a learning rate parameter, chosen by the developer. That is, the bias weight on each input link is incremented by the product of the learning rate and the input signal on that link with the delta value for the neuron receiving the signal.

- After all the weights are updated for links to neurons above the input level, the next training example is processed. Training continues, iterating over training examples, until some limit for iteration is reached, or until errors for output values are minimized according to problem-specific criteria.

When a network is trained in this way with multiple different input and output examples, the process can gradually make the network produce outputs that correspond to inputs, generally with smaller errors. So, back-

---

[13] Technically, this is the rate of change of the error, if the error is defined as $\frac{1}{2}(y_i - s[H, i])^2$.

[14] The derivative of the sigmoid function $f(x)$ is $f'(x) = x(1\text{-}x)$.

propagation can accomplish machine learning for neural networks. The key insights that support backpropagation are: a) the error and delta of an output neuron is caused by all of its inputs, in proportion to the signals and weights on its input links; b) the delta computed for a neuron at level $h$ should reflect how much that neuron's output contributes to the deltas of all the neurons at level $h+1$; adjustments should be gradual, corresponding to the deltas from each neuron and using the derivative of the activation function to compute adjustment values. More details about backpropagation and the advantages of different activation functions are given by Russell & Norvig (2010) and Géron (2017).

## 4.2 Generality and Success of Multi-Level Networks: Deep Neural Networks

The generality of neural networks depends on how many hidden levels they have. A neural network without any hidden levels, just having an input and output level, is called a *perceptron* (Rosenblatt, 1959). Minsky & Papert (1969) showed perceptrons cannot represent or learn a function that is not linearly separable, such as the exclusive-or (XOR) function. Rumelhart, Hinton, & Williams (1986) showed how backpropagation could enable neural nets with hidden levels to learn XOR and several other, much more complex functions. Cybenko (1988, 1989) showed that a single hidden level enables neural nets to represent any continuous function, and two hidden levels are enough to also represent discontinuous functions.

Because such generality is possible in theory, for many years researchers focused mostly on neural networks with a single hidden level. Yet in principle multi-level neural nets can represent more complex patterns with fewer neurons (Géron, 2017). Also, the complex patterns in the world tend to be produced by hierarchical systems (Simon, 1962), so one might expect such patterns would be more easily represented and learned by networks with many levels.

The development of fast algorithms for successfully training 'deep'[15] neural networks has been the major factor in success of recent research: a paper by Hinton, Osindero & Teh (2006) is credited with starting the 'machine learning tsunami' for deep learning (Géron, 2017). Another major factor has been the ability to leverage massive parallel processing in pools of computer graphics cards to accelerate training neural networks. In recent years the term 'deep learning' has been used to refer to learning by deep neural nets.[16, 17] (LeCun, Bengio & Hinton, 2015)

---

[15] Géron (2017) defines 'deep' as two or more hidden levels, and discusses techniques for training a deep neural network with 10 levels.

[16] The technology continues to evolve, e.g. as discussed by (Jouppi et al., 2018).

[17] 'Deep learning' has an older usage in education to refer to people internalizing their learning and understanding of a subject. The term was also used in AI research as early as 1986 to discuss learning in the search space of a constraint-satisfaction problem, not involving neural nets (Dechter, 1986).

As examples of this success, Cireşan *et al.* (2012) discussed the use of graphics cards to speedup training of deep neural nets[18] and improve performance in recognizing handwritten digits and characters in Latin and Chinese scripts, and recognition of three-dimensional toys, traffic signs, and human faces, achieving human-competitive results. Cireşan *et al.* (2013) discussed the use of deep neural nets[19] in the 2012 International Conference on Pattern Recognition's mitosis detection competition, outperforming other competitors in detection of cancer in histology images. (In general, the algorithms used for levels of these deep neural nets were more exotic than the description I've given above.)

Another noteworthy example of the success of deep neural net technology is AlphaGo, a computer program that plays Go, developed by DeepMind, part of Google's Alphabet group. It's estimated that Go has $10^{170}$ game configurations, making it far more complex than Chess.[20]

AI research essentially conquered Chess when IBM's "Deep Blue" computer system defeated the world champion Gary Kasparov in a game in 1996 and won a regulation match with Kasparov in 1997.[21] (The name "Deep Blue" referred to the system's depth of search in a Chess game, not to deep neural networks.) Since then, other companies have developed computer programs for Chess which can defeat human grandmasters, running on personal computers or smartphones.

In 2015, AlphaGo defeated the European Go Champion, Fan Hui, in a 5-0 victory. In 2016, AlphaGo defeated Lee Sedol, who had won 18 world titles, with a 4-1 victory. These versions of AlphaGo were trained using data from thousands of human games, and used two deep neural networks, one for selecting a next move to play and another for predicting the winner of a game. After training with records of human games, AlphaGo was trained by playing against itself to generate successively stronger versions of the system. (Silver *et al.*, 2016)

Another version of the system called AlphaGo Zero was developed that used only a single deep neural network, and was trained only against successive versions of itself, starting with no knowledge of the game except its rules. (Silver *et al.*, 2017) After 3 days of training, AlphaGo Zero was at the performance level which had defeated Lee Sedol. After 21 days, AlphaGo Zero was at the level of the AlphaGo system that defeated world champion Ke Jie in a 3-0 victory in 2017. With 40 days of training, AlphaGo Zero defeated the championship AlphaGo in a 100-0 victory.

---

[18] They used multiple columns of deep neural nets, each having 6 to 10 levels of neurons.

[19] Having 9 to 11 hidden levels.

[20] Researchers have also achieved success in Poker, a game of imperfect information. (Monroe, 2018)

[21] To be fair, it should be noted that Kasparov demanded a rematch, accused IBM of cheating, and that IBM refused a rematch.

## 4.3 Future Prospects for Neural Networks

These and many other examples illustrate the potential for research using deep neural networks to help achieve human-level AI, or even super-human AI, in solving limited, specific problems. The technology is being applied to a wide variety of tasks in robotics, vision, speech, and linguistics. The technology is essentially domain-independent. Research on neural networks can also be developed in several ways, e.g. recurrent networks and Bayesian networks as noted earlier, or research into other models of biological neurons or topologies of neural networks similar to those in the human brain (Huyck, 2017). So it is clear that neural networks will be an important focus of research for AI in the 21st century.

There does not appear to be any theoretical reason in principle[22] that prevents research on the wide variety of possible neural network architectures from eventually achieving a fully general human-level AI, that is not limited to solving specific problems. However, achieving a general human-level AI via this approach will not be easy: Human neurons are much more complex than the artificial neurons described above. The human brain has about 90 billion neurons, and about 100 trillion connections (synapses) between neurons. It may not be feasible to simulate such orders of magnitude by a computer system, at least in this century.[23] Also, the development of human intelligence within the brain of a child follows a different path from the training sequence of a conventional neural network, leveraging natural language communication and interaction with other humans. Finally, if human-level AI is achieved solely by relying on neural networks then it may not be very explainable to humans: Immense neural networks may effectively be a black box, much as our own brains are largely black boxes to us. It will be important for a human-level AI to be more open to inspection and more explainable than a black box.

These factors suggest research on neural networks to achieve human-level AI should be pursued in conjunction with other approaches that support explanations in a natural language like English, support a child-like learning process, and avoid open-ended dependence on neural nets by allowing an AI system to use other computational methods when neural nets aren't needed.

## 5. The Archetype Level—Categories

As stated above, the archetype level of an AI architecture is where categories, classes, or types are represented. Here are three questions I'll consider:

- What categories should be represented?

---

[22] If human intelligence is finitely computable—see §4.1.2.4.
[23] Although research projects have been undertaken in this direction, e.g. (Markram, 2006).

- Why should categories be represented?
- How can categories be represented?

An answer to the first question is categories of whatever may exist. An answer to the second is that categories should be represented to help an AI system process linguistic expressions about whatever may exist, using either formal or natural languages. After discussing these two answers, subsections 5.1 and 5.2 will discuss representation of categories in formal ontologies and as cognitive categories. Section 6 then discusses formal and natural languages in more detail.

The phrase "whatever may exist" is intentionally open-ended. To achieve human-level artificial intelligence, an AI system will need to represent and process thoughts about whatever humans can think about. Humans can think about things (including objects, processes, and events) that exist objectively[24] (physically in space and time), and humans can also think about things that exist subjectively (e.g. ideas, emotions, or feelings[25]), and about things that exist *intersubjectively* based on people sharing beliefs and ideas using natural language.

Examples of intersubjective existence include money, corporations, laws, governments, nations, scientific theories, etc. Intersubjective entities may have limited grounding in objective reality, yet be very important to individuals and society.[26] Harari (2015) discusses the importance of intersubjective existence throughout human history.[27] Gärdenfors (2017) gives a detailed discussion of the role of intersubjectivity in how children learn word meanings. Word senses exist intersubjectively via natural language: People explain the meanings of words, either linguistically or by physical demonstration, and reach intersubjective agreement that they understand the same meanings for words, at least in some contexts.

People can also think about things that exist hypothetically or fictionally, or things that only existed in the past or may exist in the future, or things

---

[24] Regarding objective existence, it should be noted that we do not have direct knowledge of objective reality. Instead, we have an internal, projected reality constructed from our perceptions of external reality (Jackendoff, 1983). Our perceptions are internal constructs that indirectly represent external reality, sometimes incompletely, inaccurately or paradoxically. It is only because projected reality generally tracks objective reality closely, that we normally think we directly perceive objective reality – viz. §2.2.3.

[25] Ideas and emotions are grounded in objective reality via physical events in the brain, of course. But we do not need to know anything about how ideas and emotions are grounded in objective reality, to think about them as existing subjectively within human minds. Of course, ideas and emotions can also be represented or described by expressions in a natural language like English, and some ideas can be represented by expressions in formal logic or mathematics.

[26] There are physical records, buildings, people, etc. supporting the existence of laws, governments, etc. Yet the existence of laws and governments depends on intersubjective agreement by people about meanings for concepts like 'law', 'freedom', 'property', 'justice', 'fairness', 'human rights', etc. The physical records only have meaning to the extent that people agree about the semantics of the words in them.

[27] Harari's text hyphenates the term as 'inter-subjective'.

which may only exist in their imaginations. AI systems may need to represent categories related to these things, e.g. to represent thoughts about future technologies for interstellar space travel.

So, to achieve human-level AI, the range of categories which can be represented at the archetype level eventually needs to extremely broad. We are still far from implementing this range of expression in AI systems.

There are basically two ways to represent categories at the archetype level, both important for AI systems: The first is to specify formal ontologies, which can be used by computer programs for AI systems, databases, etc. The second is to represent categories using methods studied in cognitive linguistics and semantics, e.g. conceptual spaces, image schemas, radial categories, etc., which are likely to be important for AI research on understanding natural language (Evans & Green, 2006).

## 5.1 Formal Ontologies

Formal ontologies[28] have been a subject of AI research for several decades. To represent a complex problem domain in a way that can support an AI system using a formal language for inference, it's important to develop an ontology, i.e. a formal description of the domain's classes, subclasses, and entities and processes, describing entities and processes, and relationships between classes and subclasses.

Some AI researchers have attempted to develop very open-ended, large-scale ontologies. Others have worked on ontologies to support AI applications in specific problem domains, such as representing a business or government enterprise. Both large-scale and domain-specific ontologies are likely to continue being subjects of AI research and development.

Different formal notations can be used to specify domain-specific ontologies. OWL (Web Ontology Language) is a standard notation defined by W3C (the Worldwide Web Consortium). However, at present OWL is a subset of first-order logic[29]: It cannot fully express SQL queries, for example, because the SQL WHERE-clause has the expressive power of full first-order logic. Per Sowa (2007), common SQL queries can be processed in linear or logarithmic time, and in general worst-case queries can be processed in polynomial time. Examples requiring exponential time are very rare. Common Logic with the CLIF and CGIF notations is a highly expressive logic that is often used as an extension for aspects OWL cannot express.

To develop an ontology for a specific application domain, it's advisable to work with domain experts to develop descriptions in a controlled natural language. Two CNL's in particular that I will note for the reader's further studies are Attempto (Fuchs et al., 2006) and Gellish (Van Renssen, 2005).

---

[28] I thank John Sowa for reviewing and providing input to this section.
[29] Per communication from John Sowa, Sept. 2018.

Examples of large-scale ontologies include Cyc and DBpedia:

- The Cyc project has been working since 1984 on developing a comprehensive, large-scale ontology and knowledge-base which supports commonsense reasoning (Lenat, Prakash, & Shepherd, 1986). In 2017, Cyc contained 365,593 concepts, and 21.7 million assertions (Sharma & Goolsbey, 2017).

- The DBpedia project automatically extracts a large knowledge base from Wikipedia. In 2015, DBpedia included over 400 million facts about 3.7 million things, extracted from Wikipedia's English edition. The DBpedia knowledge bases extracted from non-English Wikipedia editions contained 1.46 billion facts about 10 million things. (Lehmann *et al.*, 2015)

Russell & Norvig (2010) provide a much more detailed introduction to technical issues involved in symbolic representation of ontologies and knowledge.

## 5.2 Cognitive Categories

In effect, a cognitive category represents a typical meaning for a word or word phrase[30] in English or some other natural language, i.e. a word sense. Cognitive categories allow variation in instances, and may be based on associative processing. They can be represented by a variety of methods studied in cognitive semantics, for example:

- Conceptual Spaces (Gärdenfors, 1995 *et seq.*)
- Idealized Cognitive Models, Radial Categories (Lakoff, 1987)
- Image Schemas (Johnson, 1987; Talmy, 2000)
- Semantic Frames (Fillmore, 1977 *et seq.*)
- Conceptual Domains (Lakoff & Johnson, 1980)
- Cognitive Grammar (Langacker, 1987 *et seq.*)
- Perceptual Symbols (Barsalou, 1993)

There is not a consensus view in modern linguistics about how word senses exist and should be represented – this remains an unresolved topic, philosophically and scientifically as well as technically (the writings of Peirce and Wittgenstein are still relevant). Much modern work on computational linguistics is corpus-based and does not use word meanings and definitions. A respected lexicographer wrote a paper (Kilgarriff, 1997) saying he did not believe in word senses. However, Kilgarriff

---

[30] Within a mind, perhaps a category might be thought about before it is given a word-phrase name, but eventually an important category is given a name to support communication about it. An example of a word-phrase name for a category is 'deep neural network'.

(2007) clarified his position and continued to support research on word sense disambiguation (WSD) (Evans *et al.*, 2016). A sub-community within computational linguistics conducts research on WSD, reported in annual SemEval workshops.

A general view of cognitive semantics[31] is that word senses exist with a radial, prototypical nature; words may develop new meanings over time, and old meanings may be deprecated; words when used often have meanings that are metaphorical or metonymical and may involve mental spaces and conceptual blends[32]; commonsense reasoning and encyclopedic knowledge may be needed for disambiguation relative to situations in which words are used; the meanings of words and sentences in general depend on the intentions of speakers.[33]

Some word meanings can be represented by definitions in a natural language. Such a definition might be found in a dictionary, or created *ad hoc* to answer a question about what a word means in a usage. Such definitions tend to be prototypical rather than precise: usages may often be variants of definitions stated in dictionaries.

In general, understanding what words mean may be relatively straightforward in some contexts and complex in others.

Gärdenfors (2017) hypothesizes semantic knowledge is organized into domains, and that learning of domains is connected to the development of intersubjectivity, which involves 'theory of mind' – the ability to represent other people's emotions, attention, desires, intentions, belief and knowledge. Gärdenfors' research discusses the use of conceptual spaces for modeling semantics of nouns, adjectives, and verbs.

Understanding natural language will be an increasingly important topic for research on human-level AI. To support such research, learning and representation of cognitive categories will also be important topics for continuing research.

## 6. The Linguistic Level

At the linguistic level of an AI architecture, information is represented and inference is performed using one or more symbolic languages. Depending on the particular AI system, a symbolic language may be a simple notation (e.g. n-tuples of symbols), or a formal, logical language like predicate calculus, or in theory it could even be a natural language like English (Jackson, 2019).

Both formal, logical languages and natural languages are needed for human-level AI, in different ways. Formal languages specialize and stan-

---

[31] See Evans & Green (2006).
[32] See Fauconnier & Turner (2002) and §3.6.7.8, §3.6.7.9 in (Jackson, 2019).
[33] See Kilgarriff (2007).

dardize use of natural language words and phrases, and have supported automated reasoning. In principle, anything that can be expressed in formal logic could be translated into equivalent expressions in natural language, but the opposite is not true: natural language can express ideas and concepts more flexibly than formal logic. Natural language permits being vague and general, in ways not supported by formal logic, and allows communication without needing to be precise about everything at once (Sowa, 2007).

Before discussing formal languages and natural languages separately, there were two research endeavors which considered both aspects of language and should be noted, i.e. the writings of Charles Sanders Peirce and of Ludwig Wittgenstein.

Peirce (1839-1914) developed a theory of 'semiotics' which addressed the nature of meaning and understanding for signs (e.g. symbols) and languages in general, including natural languages. He also developed 'existential graphs', a formal language of graphical diagrams equivalent to first-order predicate calculus with equality—see (Johnson-Laird, 2002; Sowa, 2011a).

Wittgenstein (1889-1951) initially developed a purely logical description of the relationship between language and reality, published in 1922. He later restated much of his philosophy about language in *Philosophical Investigations*, published in 1953. A central focus of *Investigations* was the idea that the meaning of words depends on how they are used, and that words in general do not have a single, precisely defined meaning. As an example, Wittgenstein considered the word "game" and showed it has many different, related meanings. What matters is that people are able to use the word successfully in communication about many different things. He introduced the concept of a "language game" as an activity in which words are given meanings according to the roles that words perform in interactions between people.

It does not appear there is any fundamental contradiction between Wittgenstein and Peirce regarding their theories of language (§2.2.2). The ideas of Peirce and Wittgenstein remain relevant for AI research.

## 6.1 Formal Languages

Virtually since its inception, AI research has considered formal languages for representation and computation. Research has focused on logical, truth-conditional approaches to representation and processing.

### 6.1.1 Predicate Calculus (First-Order Logic)

Chapter 6 gives an introduction to first-order predicate calculus, and related topics. In the decades since it was written, much work on AI could be described as developing optimizations for improving the speed of first-order logic in a series of application languages and frameworks, which have been applied to a variety of difficult problems. Highlights of this history include:

- Development of forward chaining as an alternative to resolution, and development of efficient algorithms for rule-matching in production system languages (Forgy, 1982).

- Development of logic in the SOAR architecture (Laird *et al.*, 1987) to support matching up to one million rules (Doorenbos, 1994). Development of ACT-R, ICARUS, Sigma. (Viz. Anderson & Lebiere, 1998; Langley *et al.*, 2009; Rosenbloom *et al.*, 2016) Kotseruba & Tsotsos (2018) give an overview of 84 cognitive architectures developed over 40 years of research, of which 49 architectures are presently being actively developed. They report that over 900 practical projects were implemented using these architectures.

- Application of backward chaining[34] in logic programming systems.

- Development of logic programming by Kowalski (1974 *et seq.*) and of the Prolog language (Colmerauer & Roussel, 1993).

- Research on constraint logic programming (Jaffar & Lassez, 1987), leading to commercial systems such as ILOG (Junker, 2003).

- Research on deductive databases (Gallaire & Minker, 1978) and work on Datalog as a language for deductive databases (Chandra & Harel, 1980; Ullman, 1985).

- Use of forward chaining in deductive databases, along with relational database methods (Bancilhon *et al.*, 1986).

Theorem proving has been used in important, real-world applications, e.g. synthesizing and verifying control software for spacecraft (Denney *et al.*, 2006; Lowry, 2008); verifying correctness of microprocessor design (Hunt & Brock, 1992). Theorem proving has also solved difficult, open mathematical problems (Wos & Pieper, 2003). In 1996, an AI theorem-proving system resolved a mathematical question that had been open since 1933 (McCune, 1997).

Weyhrauch (1980) described another approach to first-order logic in a system called FOL, which remains relevant to future AI research: FOL was described as a conversational system that represents theories, supports a self-reflective meta-theory, can reason about possibly inconsistent theories,

---

[34] Backward chaining was used in PLANNER by Hewitt (1969), discussed in Chapter 6.

and address questions about theories of theory building. The FOL research direction appears synergetic with the TalaMind approach (Jackson, 2019), though FOL focuses on logic rather than natural language. Other papers on FOL include (Talcott & Weyhrauch, 1990) and (Weyhrauch *et al.*, 1998).

Technologies for formal logic will continue to be important for AI research. Introductions to these topics include (Russell & Norvig, 2010), (Genesereth & Nilsson, 1987), and (Clocksin, 2003). Some further discussion of formal logic is given by §2.3.1.

## 6.1.2 Conceptual Graphs

Conceptual graphs[35] (CGs) are a formal language for first-order logic and knowledge representation developed by Sowa (1976 *et seq.*), based on Peirce's existential graphs and semantic networks in AI research. This section can only give a brief overview of conceptual graphs – Sowa (2008) gives a chapter-length introduction to conceptual and existential graphs. An encyclopedia article on semantic networks is provided by Sowa (1992).

CGs have a formal, abstract syntax which can be represented graphically or in linear text notations. For example,

```
[Travel *x] [Person: Phil] [City: Tampa] [Automobile *y]
(Agnt ?x Phil) (Dest ?x Tampa) (Inst ?x ?y)
```

is an example of the conceptual graph interchange format (CGIF) linear text notation representing *Phil is traveling to Tampa by automobile*. Another linear notation is the Common Logic Interchange Format (CLIF). An example of CLIF is:

```
(exists ((x Travel) (y Automobile))
    (and (Person Phil) (city Tampa)
        (Agnt x Phil) (Dest x Tampa) (Inst x y) ))
```

Although the linear text notations are different, they have the same semantics and can be translated to equivalent statements in typed predicate calculus.

Conceptual graphs can represent contexts, also. A CG context has a referent that is a nested set of conceptual graphs. CG contexts can be of different types, e.g. Proposition and Situation. For example, the sentence *Tom believes that Mary wants to marry a sailor*, can be represented in CGIF notation by:

```
[Person: Tom] [Believe: *x1] (Expr ?x1 Tom)
(Thme ?x1 [Proposition:
    [Person: Mary] [Want: *x2] (Expr ?x2 Mary)
    (Thme ?x2 [Situation:
        [Marry: *x3] [Sailor: *x4] (Agnt ?x3 Mary)
        (Thme ?x3 ?x4)])])
```

---

[35] I thank John Sowa for reviewing and providing input to this section.

Tom believes the proposition that Mary wants a situation to exist in which she marries a sailor. This example can be represented in IKL (Hayes & Menzel, 2006) by:

```
(exists ((x1 Believe)) (and (Person Tom) (Expr x1 Tom)
(Thme x1 (that
    (exists ((x2 Want) (s Situation))
      (and (Person Mary) (Expr x2 Mary)
        (Thme x2 s)
        (Dscr s (that
            (exists ((x3 Marry) (x4 Sailor))
              (and (Agnt x3 Mary) (Thme x3 x4)
)))))))))
```

IKL is based on CLIF and adds the operator *that*. Sowa (2008) gives further discussion of this example, including how to represent different interpretations in CGIF, and IKL. Sowa (2013) discusses the relation of conceptual graphs to Peirce's existential graphs. Sowa (2018) notes that existential graphs are isomorphic to discourse representation structures (Kamp & Reyle, 1993).

There is a formal mapping for CGs to the ISO Standard 24707 for Common Logic (CL), which is a standard for first-order logic languages for representing and exchanging information between computer systems. CGIF can express the full semantics of CL. IKL is not an ISO standard. An international conference on conceptual structures has been held annually since 1993.

### 6.1.3 Probabilistic Reasoning

Besides formal logic based on truth and falsity, AI research has focused on a variety of approaches which can be placed under the umbrella 'probabilistic reasoning'. Bayes' theorem provides a foundation for much of this research. Russell (2015) describes a formal language called 'Bayesian Logic' (BLOG) for representing open-universe probability models that combine first-order logic with Bayesian networks (Pearl, 1988). (The term 'open-universe' means such models support uncertainty about existence and identity of objects.) Getoor & Taskar (2007) edited a collection of relevant papers. Russell & Norvig (2010) survey research on probabilistic reasoning.

### 6.1.4 Controlled Natural Languages

Apart from representing formal logic, another way to create a formal language that can be processed by computers is to develop a 'controlled natural language' (CNL). This approach restricts a natural language like English in ways that allow it to be used as a formal language. Only a subset of English sentences can be expressed in a CNL for English, but these sentences can be understood by people and also be processed by computers for a variety of applications.

CNL's have been created and used for several purposes, including knowledge acquisition and representation, developing the semantic web, and specifying formal ontologies to support design of databases and AI systems. Schwitter (2010) gives an introduction to CNL's. Kuhn (2014) provides a survey of 100 English CNL's. Section 5.1 above discusses ontologies. Two CNL's in particular that I will note for the reader's further studies are Attempto (Fuchs *et al.*, 2006) and Gellish (Van Renssen, 2005).

### 6.1.5 Ad Hoc Symbolic Languages & Representations

A human-level AI needs the ability to develop *ad hoc* symbolic notations, representations, and languages, to support solving new kinds of problems. This includes developing new mathematical notations, representations, and formal languages for scientific problems and theories. Research related to this topic is discussed by (Langley *et al.*, 1987) and (Shrager & Langley, 1990). Since this ability of human intelligence supports scientific discoveries and knowledge representation, it should be an important topic for AI research.

## 6.2 Natural Languages

To achieve human-level artificial intelligence, a system will need the ability to understand the full range of human thoughts that can be expressed in a natural language like English, Mandarin, Spanish, Hindi, Arabic, etc.[36] No existing formal language can represent this range of thoughts. A natural language like English already has the ability to do this, perhaps as well as any artificial, formal language ever could. So it is pertinent to ask whether there can actually be a computational 'language of thought', and whether a natural language like English could serve as a 'natural language of thought' within an AI system.

These questions focus on a different topic from the normal, effectively standard approach to natural language understanding in AI systems, introduced in Chapter 7 of this textbook. The normal approach is for AI programs to translate natural language expressions into simpler representations of meanings, which are processed internally. The simpler representations can include expressions in formal logic, conceptual graphs, discourse representation structures, relational data structures and queries, semantic networks, etc. – a wide variety of simpler representations have been studied

---

[36] These are the top five languages by population of native speakers listed in Wikipedia, which cites the 2010 Swedish encyclopedia *Nationalencyklopedin*. English is actually third by population – I've listed it first, since English is used throughout this text as a language of reference. The arguments given here could support use of any natural language as a language of thought, and support use of multiple languages.

by AI researchers. Russell & Norvig (2010) give an introduction to such methods for natural language processing.[37]

Much research on computational linguistics has bypassed the problem of representing meanings, relying on statistical methods and neural networks to process natural language expressions. Géron (2017) includes a chapter introducing how a neural network can approximate translation of English to French sentences. Clark *et al.* (2012) present a collection of foundational papers on topics in computational linguistics.

Conventional AI approaches and computational linguistics are still far from achieving human-level understanding and use of natural language. Therefore the next two subsections discuss foundational questions relevant to a different approach: using a natural language like English as a computational language of thought within an AI system.

### 6.2.1 Is There a Language of Thought?

This is an old and important question. For example, Wittgenstein (1953) wrote that St. Augustine described the language learning process as if a child has an innate language preceding and enabling the acquisition of a spoken, public language.[38]

Fodor (1975 *et seq.*) argued in favor of a 'language of thought hypothesis', which has been the subject of philosophical arguments pro and con, e.g. about whether an innate language is needed to learn an external language and the degree to which an innate language must contain all possible concepts, or constrains the concepts that can be learned and expressed. Fodor (2008) accepted the principle of semantic compositionality, an issue in earlier philosophical debates.

Fodor's writings do not yield the only possible language of thought theory. Schneider (2011) considered arguments for and against Fodor's theory and presented an alternative theory for a computational language of thought, which she developed to be compatible with cognitive science and neuroscience.

Sloman (1979 *et seq.*) contended that the primary role of language is the representation of information within an individual, and its role in communication is an evolutionary side effect, i.e. human-level intelligence requires some innate, internal language for representation of thoughts, prior to learning and using natural language. Sloman disagreed with Fodor about the necessary content of the innate language, arguing that in principle a system can learn new concepts (which may be represented by new words

---

[37] Lindes (2018) and Kelly & Reitter (2018) discuss how cognitive architectures can support natural language processing.

[38] Berwick & Chomsky (2016) give a perspective on the evolution and nature of a language of thought in humans, and discuss how it might be related to an innate 'universal grammar' – viz. Chapter 7, p.279 of this textbook, and Chomsky (1966 et seq.).

or symbols) that may not be definable in terms of previously known concepts, words or symbols.

The existence of 'inner speech' suggests some thoughts are represented internally in a language of thought with the expressiveness of natural language. Inner speech is a feature people ascribe to their minds, and a psychological phenomenon which has been remarked upon for centuries: We have the ability to mentally hear some of our thoughts expressed internally in natural language. Baars & Gage (2007) write that inner speech is not just for verbal rehearsal but provides an individual's "running commentary" on current issues, and is related to linguistic and semantic long-term memory. Fernyhough (2016 *et seq.*) describes functional MRI studies of inner speech indicating it can involve parts of the brain that are often used to understand other people's points of view (Theory of Mind or 'perspective taking') and that inner speech may be a conversation between multiple points of view. He suggests inner speech may help our intelligence to be self-directing, flexible, and open-ended.

These considerations suggest inner speech is not an epiphenomenon, but may play a role in human intelligence, and that natural language may play a role in representing thoughts within the mind, beyond its role for communicating thoughts between people.[39]

It is tempting to say that if we restrict 'language' to verbal or written, serial human natural languages such as English, Chinese, etc., then thought is possible without language: People can solve some kinds of problems using spatial reasoning and perception that are at least not easy to express in English. Children can display intelligence and thinking even if they haven't yet learned a language such as English. Pinker (1994) cited medical and psychological evidence showing thought and intelligence are not identical to the ability to understand spoken, natural languages. Yet these considerations do not rule out the possibility that a child's mind may use an innate language of thought to support reasoning, before the child learns a spoken natural language.

Pinker also argued against the Sapir-Whorf hypothesis that language determines and limits our thinking abilities, providing a variety of arguments and evidence to refute a strict interpretation of Sapir-Whorf. On the other hand, Boroditsky & Prinz (2008) discussed evidence that statistical regularities in English, Russian, and other natural languages have an important role in thought, suggesting people who speak different languages may think in different ways. And Pinker (1994, p.72) concluded that people do have a language of thought, or *mentalese*, though he reasoned it is different from a spoken, natural language.

Jackendoff (1989) gave an elegant argument that concepts must be expressed as sentences in a mental language: Since natural language

---

[39] It has also been reported that deaf people may experience 'inner sign language' (Sacks, 1989).

sentences can describe an effectively unlimited number of concepts, and the brain is finite, 'sentential concepts' must be represented internally within the mind as structures within a combinatorial system, or language.[40]

The expressive capabilities of natural languages should be matched by expressive capabilities of mentalese, or else by Jackendoff's argument the mentalese could not be used to represent the concepts expressed in natural language. The ability to express arbitrarily large, recursively-structured sentences is plausibly just as important in a mentalese as it is in English. The general-purpose ability to metaphorically weld concepts together across arbitrary, multiple domains is plausibly just as important in a mentalese as it is in English. Considering Jackendoff's argument, it is cognitively plausible that natural language representation and processing are in some form *core functionalities* of human-level intelligence, needed for representation of thoughts.

This is not to say mentalese would have the same limitations as spoken English, or any particular spoken natural language. In mentalese, sentences could have more complex, non-sequential, graphical structures not physically permitted in speech. (Jackson, 2019) discusses hierarchical list structures for representing English syntax, to facilitate conceptual processing.

## 6.2.2 Can AI Systems Use a Natural Language of Thought?

There is not a consensus based on analysis and discussion among AI scientists that an AI system cannot use a natural language like English as an internal language for representation and processing – a computational natural language of thought.

Rather, it has been an implicit assumption by AI scientists over the decades that computers must use formal logic languages (or simpler symbolic languages) for internal representation and processing of thoughts in AI systems.

It does not appear there is any valid theoretical reason why the syntax and semantics of a natural language like English cannot be used directly by an AI system as its language of thought, without translation into formal languages, to help achieve human-level AI.

There would be theoretical advantages for using a natural language of thought in an AI system: Natural language has a syntax and semantics that can support meta-reasoning, analogical reasoning, causal and purposive reasoning, and logical inference in any domain.

In 1955, John McCarthy proposed that his research in the Dartmouth summer project on artificial intelligence would focus on intelligence and language. He noted that every formal language yet developed omitted important features of English, such as the ability for speakers to refer to

---

[40] Others gave similar arguments, e.g. Chomsky (1975), Fodor (1975).

themselves and make statements about progress in problem-solving. He proposed to create a computer language that would have properties similar to English. The artificial language would allow a computer to solve problems by making conjectures and referring to itself. Concise English sentences would have equivalent, concise sentences in the formal language.

Although McCarthy proposed in 1955 to develop a formal language with properties similar to English, his subsequent work did not take this direction, though it appears in some respects he continued to pursue it as a goal. Beginning in 1958 his papers concentrated on use of predicate calculus for representation and inference in AI systems, while discussing philosophical issues involving language and intelligence.

McCarthy was far from alone in such efforts: AI research on natural language understanding has attempted to translate natural language into a formal language such as predicate calculus, frame-based languages, conceptual graphs, etc., and then to perform reasoning and other forms of cognitive processing with expressions in the formal language. As noted above, some approaches have constrained and controlled natural language, so that it may more easily be translated into formal languages, database queries, etc.

In effect, previous AI research has treated natural language as an application to be supported by AI systems using simpler formal languages for internal data and rules of inference. Such approaches are unlikely to succeed in achieving human-level AI if natural language processing is a *core functionality* of human-level intelligence, needed for internal representation of thoughts.

Since progress has been very slow in developing natural language understanding by translation into formal languages, (Jackson, 2014) investigated whether it may be possible and worthwhile to perform cognitive processing directly with unconstrained natural language. To support this, I defined a formal language called Tala based on the unconstrained syntax of English, and discussed how an AI architecture called TalaMind could use Tala to potentially achieve human-level artificial intelligence.[41] Tala will in effect be a natural language of thought for an AI system.

The Tala language responds to McCarthy's 1955 proposal for a formal language that corresponds to English. It enables a TalaMind system to formulate statements about its progress in solving problems. Tala can represent complex English sentences involving self-reference and conjecture. Short English expressions have short correspondents in Tala.

---

[41] TalaMind is open to inclusion of other approaches toward human-level AI, for instance permitting predicate calculus, conceptual graphs, and other symbolisms in addition to the Tala language at the linguistic level, and permitting integration across architectural levels, e.g. potential use of neural nets at the linguistic and archetype levels. TalaMind is actually a broad class of architectures, open to design choices at each level.

The TalaMind approach can address theoretical questions not easily addressed by more conventional approaches. For instance, it supports reasoning in mathematical contexts, but also supports reasoning about people who have self-contradictory beliefs. Tala provides a language for reasoning with sentences that have meaning yet which also have nonsensical interpretations. Tala sentences can declaratively describe recursive mutual knowledge. Tala facilitates representation and conceptual processing for learning by analogical, causal and purposive reasoning, and imagination via conceptual blends (Jackson, 2019, §3.7). The TalaMind approach can leverage theories of cognitive linguistics and cognitive semantics (Evans & Green, 2006).

Tala also supports sentences that describe how to perform processes – such sentences are called 'executable concepts'. A TalaMind system could reason about how to perform processes, and improve the way it performs a process. Thus the approach could support learning by self-programming, and support 'intelligence kernels' for self-developing conceptual systems (Jackson, 1979).

Again, there is apparently no theoretical reason why the syntax and semantics of a natural language like English cannot be used by an AI system as its language of thought, and there are several theoretical advantages for using a natural language of thought in an AI system. So I think this will be an increasingly important research topic for AI in the 21st century. However, as I've said above, this belief is not yet widely shared by others.

## 7. Conceptual Framework and Conceptual Processes

I will use the term 'conceptual framework' to refer to an information architecture for managing an extensible collection of concepts, expressed linguistically or via mental models (discussed below). A conceptual framework supports processing and retention of concepts ranging from immediate thoughts to long term memory, including concepts representing linguistic definitions of words, knowledge about domains of discourse, memories of past events, expected future contexts, hypothetical or imaginary contexts, etc.

In saying that concepts are expressed linguistically, all the methods discussed in section 6 above may be allowed, e.g. n-tuples of symbols, expressions in formal, logical languages, or sentences in a natural language of thought.

'Conceptual processes' refers to a (potentially extensible) system of processes that operate on concepts in the conceptual framework, to produce intelligent behaviors and new concepts – in general, conceptual processes will support or correspond to higher-level mentalities, discussed in section 8.1. The term 'information architecture' is a general, technology-independent description, not prescribing any particular implementation. 'Managing' means supporting the creation, storage, retrieval, and deletion of conceptual structures.

A discussion of a conceptual framework and processes for the TalaMind approach to human-level AI is given in (Jackson, 2019), beginning in §3.2.2. The next two sections focus on discussions by other authors about structures and processes relevant to conceptual frameworks.

## 7.1 Mental Models

Johnson-Laird (1983 *et seq.*) gives an insightful discussion of topics still relevant to AI research. His 1983 book (here referenced as "MM") discussed three major forms of mental representations, which he called mental models, images, and propositional representations. The theory of mental models continues to be a topic of active research. His 2006 book '*How We Reason*' is an introduction to mental models for the general reader.

*Mental models* are structural representations of situations, events, and processes in the world.[42] An image is a mental perception of a model from a point of view. Propositional representations are mental representations that correspond most broadly to expressions in natural language (MM, p.165).

Mental models can have different forms and purposes. Broadly, mental models are "iconic" – their structures correspond to structures of situations they represent. Beyond that, mental models may be more or less elaborate, depending on what needs to be represented – a typology includes simple relations, spatial, temporal, kinematic, and dynamic models. Mental models can support spatial-temporal reasoning, discussed in the next section.

The mental models theory stipulates that natural language expressions are represented by propositions in a mental language, which are mapped into mental models (MM, p.165). Johnson-Laird refers to the mental language as a "propositional language", though his discussion throughout MM shows clearly that the language exceeds the semantics of ordinary propositional logic, and even first-order logic. He found that no theory of syllogistic inference satisfies descriptive and explanatory criteria for mental models (MM, p.93).

In discussing how people reason, Johnson-Laird (MM, p.29) found there was no convincing evidence to say that people use any particular logic which corresponds to a formal, mathematical logic. Rather, he showed there are cases where the content of a problem or the way in which it is expressed affects how well people reason about it. In building mental

---

[42]Johnson-Laird notes that Kenneth J. W. Craik hypothesized the mind creates such models. Craik (1943, p.83) discussed a "thought-model" which parallels external reality to predict alternative possible events.

models, people find it easier to represent what is true rather than what is false, which can lead to predictable errors in reasoning. Psychological testing has confirmed such predictions, supporting the theory of mental models (Johnson-Laird, 2010; Khemlani *et al.*, 2013 *et seq.*).

Of course, there is a downside to building AI systems that match results of psychological tests by recreating human errors in logic: We don't want to use or rely on systems that can make logical errors. So there are reasons why we should not be solely guided by matching human cognition. A TalaMind system (Jackson, 2019) could have a design for mental models which would not have the potential for logical errors given consistent premises.

Johnson-Laird (MM, pp.426-427) notes that mental models can be "meta-linguistic", i.e. contain tokens representing linguistic expressions, and that mental models can be embedded within mental models (MM, pp.430-433). The TalaMind approach also allows inclusion of natural language expressions (represented by Tala structures) within mental models (contexts), to represent what actors within a model may think or say, and nested contexts to represent what an actor may think or perceive other actors think or perceive (i.e. 'theory of mind' capability).

## 7.2 Spatial-Temporal Representation and Reasoning

Human intelligence includes representing and using knowledge about spatial-temporal relationships: When we perceive our physical environment we construct an internal, mental representation of spatial relationships between objects we perceive, and of temporal relationships corresponding to movements and changes in what we perceive. We reason about spatial-temporal relationships to understand what's happening. We may plan and perform actions to move within the spatial-temporal relationships or to alter them.

Likewise perceiving, representing and reasoning about spatial-temporal relationships are also important for many AI applications.

Initially, much AI research on computer vision took the approach described in Chapter 5 of this textbook: Images were processed to identify lines and regions, which were processed to create representations of 3-dimensional structures, etc. Since then, a variety of methods have been developed to support computer vision. Russell and Norvig (2010, chapter 24) discuss the technologies and history of such research.

Research on convolutional neural networks (CNN's) to support computer vision has been important since the 1980's, and especially since the advent of deep neural network technologies, discussed in section 4 above (Géron, 2017).[43] Deep neural networks are increasingly being used to support machine vision for image searches, self-driving cars, etc. Architecturally,

---

[43] See papers by Fukushima (1980) and LeCun (1989 et seq.).

CNN's are at the associative level – I mention them here to provide a background for discussion of higher-level spatial-temporal reasoning.

To illustrate the nature and importance of spatial-temporal reasoning for human-level intelligence, consider the Wright brothers' invention of a flying machine, which was a tour de force of spatial-temporal reasoning.[44] Here were some of their major insights:

- Using their knowledge of bicycles, they reasoned that control of an airplane is more important than stability: a bicycle is unstable, but a rider can control it. Airplanes are unstable due to wind forces, so they invented three-axis control.

- They reasoned that the speed of a prevailing wind can be equivalent to launching a glider from the top of a hill or dropping it from a balloon. So they chose Kitty Hawk as a location for tests because its prevailing winds averaged 15 to 20 mph. This enabled testing flight more safely, close to the ground.

- They observed how buzzards warp their wings when turning, and applied this to their design for an airplane.

- They created a wind tunnel to study about how lift relates to wing design, inclination, and wind speed.

- They reasoned by analogy that a propeller could be designed as a rotating wing, to pull an airplane through the air.

- They reasoned that bicycle chains could be used to transfer rotational motion from a gasoline engine to propellers.

The Wright brothers' first powered flight in 1903 was based on developing and applying an understanding of the engineering principles involved. Others failed (and sometimes died) by attempting to fly without taking such an approach.

Reasoning with diagrams can be important for spatial-temporal reasoning. Sowa (2018) discusses how diagrammatic reasoning can be supported using Peirce's existential graphs and Johnson-Laird's mental models. As an example, Sowa discusses Euclid's proof that an equilateral triangle can be created starting from any line segment, and shows how observation and imagination can be processes intermediating generalized existential graphs and geometric diagrams.[45]

---

[44] See Johnson-Laird (2006, pp.369-386).
[45] Schmidtke (2018) discusses another approach, which interprets propositional logic to create spatial maps.

While spatial-temporal reasoning can use perceptual support from the associative level of AI architectures, it also involves mental models. From a different perspective, spatial-temporal references and metaphors are pervasive in natural language, and are essential for understanding natural language. Hence spatial-temporal reasoning is important for a natural language of thought.

## 8. Human-Level Artificial Intelligence

Having discussed research in the architectural levels of intelligent agents, we are ready to consider research toward completely general human-level AI. This will be the overall, major challenge for AI research in the 21st century.

One criterion for guiding and evaluating research should be whether the research can achieve *human-like* as well as human-level AI. That is, can AI systems use cognitive processes similar to those of humans and understandable by humans, as well as achieving results that are at the level of human intelligence? This will be important for ensuring human-level AI is beneficial to humanity, which will be the overall, major responsibility for AI researchers in the 21st century, to be discussed in section 9.[46]

As noted in section 1, the issue of how to define human-level intelligence has been a challenge for AI researchers. Typically, researchers have not tried to define it, and have just expected to use a Turing Test to recognize it if it is ever achieved. Some have suggested human intelligence may not be a coherent concept that can be analyzed, even though we can recognize it when we see it in other human beings (Kaplan, 2016). Yet if we cannot define the mental abilities of human intelligence then it will be more difficult to design systems which achieve them.

So, a different approach is proposed by Jackson (2019), which is to define human-level intelligence by identifying certain *'higher-level mentalities'* achieved by humans and not yet achieved by any AI system. After describing these abilities in the next section, section 8.2 will discuss research efforts toward human-level AI.

### 8.1 Higher-Level Mentalities

The higher-level mentalities will together comprise a qualitative difference distinguishing human-level AI from computer systems in general.

Newell (1982) gave an insightful discussion of *computer system levels*, including the electronic device level, circuit level, logic level, register-transfer level, and symbolic program level. Each computer system level is a functional specialization (subset) of the systems that can be described at

---

[46] However, it is not a goal for AI research to produce human-identical AI systems. There would be ethical and legal reasons not to do so.

the next lower level. That is, each level provides functionality that not all systems at the next-lower level provide. For example, circuits at the logic level perform logical operations, a functionality that not all systems at the circuit level perform.

According to Newell and Simon's (1976) Physical Symbol System Hypothesis, a subset of the systems at the symbolic program level can achieve human-level artificial intelligence. Jackson (2018b) observed that this subset will be a functional specialization of systems at the symbolic program level, and therefore it will be a computer system level above the symbolic program level. Human-level AI will exist at this level, if and when it is achieved.

It is appropriate to call this future, new computer system level the *'intelligence level'*, or more fully, the *'human intelligence level'*. Systems at the intelligence level will be real, finite, changeable, useful systems. They will support Newell's (1990) proposal for "unified theories of cognition."[47] The following sections discuss the higher-level mentalities that will be provided by systems at the intelligence level.

## 8.1.1 Generality

A key feature of human intelligence is that it is apparently unbounded and completely general. Human-level AI must have this same quality. In principle there should be no limits to the fields of knowledge the system could understand, at least so far as humans can determine.

Having said this, it is an unresolved question whether human intelligence is actually unbounded and completely general. While we may be optimistic that human intelligence is completely general, there are many limits to human understanding at present. For instance:

- Feynman at times suggested quantum mechanics may be inherently impossible for humans to understand, because experimental results defy commonsense causality. Yet at least scientists have been able to develop a mathematical theory for quantum mechanics, which has been repeatedly verified by experiments, to great precision.

- General relativity and quantum theory are not yet unified. Astronomers have evidence black holes exist, which implies existence of gravitational singularities.

- At present scientists are having great difficulty explaining multiple, independent observations that appear to prove 95% of the universe

---

[47] The intelligence level is very different from Newell's (1982) "knowledge level", which he described as a theoretical level at which agents would have an absence of structure and could have potentially infinite knowledge (viz. §2.3.3.6.2).

consists of matter and energy we have not yet been able to directly observe, causing galaxies and galaxy clusters to rotate faster than expected, and causing the expansion of the universe to accelerate (Gates, 2009).

- Beyond this, there are several other fundamental questions in physics one could list, which remain open. And there are many challenging questions in other areas of science, including the great question of precisely how our brains function to produce human intelligence.

There is no proof at this point that we cannot understand all the phenomena of nature. And it is an unsettled question whether human-level artificial intelligence cannot also do so. Hopefully human-level AI will help us in the quest. Research on AI systems for discovery of scientific theories is presented in (Langley *et al.*, 1987) and (Shrager & Langley, 1990).

### 8.1.2 Natural Language Understanding

Humans need a natural language like English to develop and share an understanding of the world in virtually all of its aspects. To achieve human-like human-level AI, a system will need to be able to understand humans who use natural language, and to be able to explain its thoughts to humans, using natural language. Attempts to build systems that process natural language have made substantial progress, but still founder on the problem of understanding natural language as well as humans do. No AI system today can understand English as well as a five-year-old child.

### 8.1.3 Higher-Level Learning

There is still much research needed to achieve the 'higher-level learning' shown by human intelligence. I use this term to refer collectively to forms of learning such as learning by creating explanations and testing predictions about new domains based on analogies and metaphors with previously known domains, reasoning about ways to debug and improve behaviors and methods, learning and invention of natural languages and language games, learning or inventing new representations, and in general, self-development of new ways of thinking (Jackson, 2019). The phrase 'higher-level learning' is used to distinguish these from previous research on machine learning, as discussed in the references given at the beginning of section 4 above.

### 8.1.4 Metacognition

Metacognition is "cognition about cognition", cognitive processes applied to cognitive processes. This does not say much, until we say what we mean by cognition. There are both broad and narrow usages for the term *cognition* in different branches of cognitive science and AI. Many authors distinguish cognition from perception and action.

However, Newell (1990, p.15) gave reasons why perception and motor skills should be included in "unified theories of cognition". If we wish to consider metacognition as broadly as possible, then it makes sense to start with a broad idea of cognition, including perception, reasoning, learning, and acting, as well as other cognitive abilities Newell identified, such as understanding natural language, imagination, and consciousness.[48]

Since cognitive processes may in general be applied to other cognitive processes, we may consider several different forms of metacognition, for example:

* Reasoning about reasoning.
* Reasoning about learning.
* Learning how to learn.
* ...

Others have focused on different aspects of metacognition, such as "knowing about knowing" or "knowing about memory". Cognitive abilities could be considered in longer metacognitive combinations, e.g. "imagining how to learn about perception" – the combination could be instantiated to refer to a specific perception.

Such examples illustrate that natural language has syntax and semantics which can support describing different forms of metacognition. More importantly, a 'natural language of thought' could help an AI system perform metacognition, by supporting inner speech and by enabling the expression of specific thoughts about other specific thoughts, specific thoughts about specific perceptions, etc.

In terms of broad metacognition, inner speech corresponds to a perception of a silent speech action expressing a thought in natural language. The thought that is expressed may be a phrase, statement. or question about anything, in any domain: The thought may refer to a perception of an external situation or event, or to an (actual or possible) action in the external environment, or it may refer to another thought, or to an emotion, or to oneself, or to a combination. So the thought expressed by an inner speech act may itself indicate further metacognition.

Kralik *et al.* (2018) summarize several approaches to representing metacognition in cognitive systems.

## 8.1.5 Curiosity, Self-Programming, Theory of Mind

To support higher-level learning, an intelligent system must have another general trait, 'curiosity', which at the level of human intelligence

---

[48] This section uses text I provided as input to a joint paper (Kralik et al., 2018).

may be described as the ability to ask relevant questions and understand relevant answers.

In English, questions involve the interrogatives *who, what, where, when, why,* and *how.* The last two in particular merit further discussion:

A *how* question asks for a description of a method, which can be a procedure or a process. To understand the answer, an intelligent system needs to be able to represent procedures and processes, think about such representations, and ideally perform the procedures or processes described by representations, if it has the necessary physical abilities and resources. It is natural for an intelligent system to represent procedures and processes at the linguistic level of its AI architecture. With such representations it is a relatively direct step to support self-programming within an AI system. This is supported in the TalaMind architecture and further discussed in (Jackson, 2019).

A *why* question asks for a description of either a cause or an intent. Understanding the answer requires that an intelligent system be able to support causal reasoning about physical events, and also be able to support reasoning about people's intentions for performing actions. Reasoning about intentions involves supporting *'Theory of Mind'*, the ability for an AI system to consider itself and other intelligent agents (including people) as having minds with beliefs, desires, different possible choices, etc. This is also supported in the TalaMind architecture and further discussed in (Jackson, 2019).

### 8.1.6 Imagination

Imagination allows us to conceive things we do not know how to accomplish, and to conceive what will happen in hypothetical situations. To imagine effectively, we must know what we do not know, and then consider ways to learn what we do not know or to accomplish what we do not know how to do. A human-level AI must demonstrate imagination.

### 8.1.7 Visualization and Spatial-Temporal Reasoning

Very closely related to imagination (some might claim identical) is the ability people have to visualize situations in three-dimensional space and reason about how these situations might change, e.g. by visualizing motions of objects. This ability is important for understanding natural language expressions and metaphors, for imagination, and for discovery of theories and inventions, as discussed in section 7.2 above. So, this ability is listed as a higher-level mentality of human-level intelligence, though arguably it is foundational for cognition in general.

## 8.1.8 Creativity and Originality

A key feature of human intelligence is the ability to create original concepts. Human-level AI must have this same quality. The test of originality should be whether the system can create (or discover, or accomplish) something for itself it was not taught directly – more strongly, in principle and ideally in actuality, can it create something no one has created before, to our knowledge? This is Boden's (2004) distinction of (personal, psychological) P-creativity vs. (historical) H-creativity.

## 8.1.9 Self-Awareness – Artificial Consciousness

To act intelligently, a system must have some degree of awareness and understanding of its own existence, its situation and relation to the world, and its perceptions, thoughts and actions, both past and present, as well as potentials for the future. Without such awareness, a system is greatly handicapped in managing its interactions with the world, and in managing its thoughts. So, at least some aspects of consciousness are necessary for a system to demonstrate human-level intelligence.

To provide such awareness it is not necessary to solve all the mysteries of human consciousness, or for AI systems to achieve the human subjective experience of consciousness. The "Hard Problem" of consciousness (Chalmers, 1995 *et seq.*) is the problem of explaining the first-person, subjective experience of consciousness. This is a difficult, perhaps scientifically unsolvable problem because science relies on second and third person understanding of explanations for repeatable experiments. Evidently there is not a philosophical or scientific consensus for the Hard Problem.

Jackson (2019, §3.7.6) discusses how computers could have enough self-awareness to achieve human-level AI. To claim a system achieves *'artificial consciousness'* it should demonstrate:

> *Observation of an external environment.*
> *Observation of itself in relation to the external environment.*
> *Observation of internal thoughts.*
> *Observation of time: of the present, the past, and potential futures.*
> *Observation of hypothetical or imaginative thoughts.*
> *Reflective observation: observation of having observations.*

This wording is adapted in §3.7.6 from the "axioms of being conscious" proposed by Aleksander and Morton (2007) for research on artificial consciousness. They used first-person, introspective statements to describe these aspects of consciousness.

To observe these things, an AI system should be able to create and process data structures which represent them, based on information it receives from the environment. There is nothing inherently impossible about creating such data structures. Jackson (§6.3.6) discusses how the potential to support artificial consciousness is illustrated by the TalaMind prototype demonstration system.

The axioms of artificial consciousness appear to be compatible with *"global workspace"* theories of human consciousness advocated by Baars (1996 *et seq.*) and by Dehaene (2014), based on research in psychology and cognitive neuroscience.

Inner speech is a feature of human consciousness. Per section 8.1.4 above, representation of inner speech in broad metacognition would support several of the axioms of consciousness, such as observation of internal thoughts, observation of hypothetical or imaginative thoughts, and reflective observation.

There is a vast amount of scientific and philosophical literature about consciousness. Some discussions of this literature are given in §2.3.4, §2.3.5, §2.3.6, §4.2.7 of (Jackson, 2019).

### 8.1.10 Sociality and Emotional Intelligence

A human-level AI will need some level of social and emotional understanding to interact successfully with humans. It will need some understanding of cultural conventions, etiquette, politeness, etc. It will need some understanding of emotions humans may feel, to help guide its actions.

A human-level AI may have some emotions of its own, though we will need to be careful about this. One of the values of human-level artificial intelligence is likely to be its objectivity, and freedom from being affected by some emotions. People would be very concerned about interacting with emotional robots if robots could lose control of their emotions and become emotionally unpredictable. We probably would not want an AI system performing an important function like air traffic control to be emotional. On the other hand, we might want a robot taking care of infants, children, or hospital patients to show compassion and affection (cf. McCarthy, 2004); we might want a robot defending a family from violent home invaders to emulate anger.

Within an AI system, emotions could help guide choices of goals, or prioritization of goals. Apart from whether and how emotions may be represented internally, a human-level AI would also need to understand how people express emotions in behaviors and linguistically, and how its behaviors and linguistic expressions may affect people and their emotions.

There have been many books and papers related to emotions in AI systems. To mention just a few: Ortony *et al.* (1988) discussed the cognitive structure of emotions. Picard (1997) discussed how intelligent computers could recognize and have emotions. Norman (2004) discussed why intelligent machines would need emotions. Minsky (2006) discussed the mind as an 'emotion machine'. Bach (2009 *et seq.*) discusses a cognitive architecture for representing motivations and emotions integrated with thoughts, perceptions, and experience. Larue *et al.* (2018) summarize different approaches for representing emotions. McDuff & Czerwinski (2008) survey current research on emotional sentience.

### 8.1.11 Values and Ethical Reasoning

A human-level AI will need an understanding of ethical values, ethical rules and principles to interact with humans, and to support *"beneficial AI"*—AI that is beneficial to humanity and to life in general. (Bringsjord, Arkoudas, & Bello, 2006; Tegmark, 2017) This is a topic that has become increasingly important, as people have considered the potential good and bad consequences AI might have for humanity (Jackson, 2018a). This topic is further discussed in section 9.2.

## 8.2 Research Directions Toward Human-Level AI

Having discussed the capabilities human-level AI will need to support, the following sections give overviews of paths for research toward achieving human-level AI.

### 8.2.1 Unified Theories of Cognition

Newell (1990) advocated that scientists develop a series of progressively more complete *unified theories of cognition*. His initial list of areas to eventually be covered by a unified theory included problem solving, perception, language, emotion, imagination, learning, and self-awareness. He noted the list was incomplete and could be expected to grow. He also made clear that unified theories should be simulated by working computer systems. Thus, the broad scope of a unified theory corresponds to achieving human-level artificial intelligence. His advocacy for unified theories of cognition was in itself an important step toward human-level AI.

### 8.2.2 Artificial General Intelligence (AGI)

The term 'artificial general intelligence' was coined by Gubrud (1997), who essentially defined AGI as equivalent to human-level AI, by virtue of being able to successfully use general knowledge in any situation requiring human-level intelligence. (Gubrud wrote that features like consciousness would not be required for an AGI system if not needed for the situation.)

The term AGI has since been widely adopted for research on human-level AI, with an annual international conference beginning in 2008. Some important collections of papers about research approaches of several authors are presented by Goertzel & Pennachin (2007) and Wang & Goertzel (2012).

Some AGI research has focused on generality, without specifically addressing other higher-level mentalities needed for human-level AI that are discussed in section 8.1 above, e.g. natural language, higher-level learning, metacognition, curiosity, imagination, artificial consciousness, etc. If one defines a general framework for representing problems and a general algorithm for learning how to solve problems in the framework, such a definition can be discussed as an approach to AGI.

While one can argue that considering generality alone is sufficient, it's reasonable to conjecture that considering generality plus other higher-level mentalities will accelerate achieving human-level AI, and also facilitate achieving human-like AI.

Yudkowsky (2007) advocated levels of organization in a "deliberative general intelligence" (DGI) for research in AGI, proposing a research direction somewhat similar to the TalaMind approach (Jackson, 2014). However, it appears Yudkowsky (2007, pp.458-461) did not expect DGI thoughts would (at least initially) be represented as sentences in a natural language of thought, nor did he propose representing thoughts in structures corresponding to parse-trees of natural language expressions, as advocated by (Jackson, 2014).

### 8.2.3 Child Machines / Baby Machines

Turing (1950) suggested human-level AI could be achieved by programming a computer to learn like a human child, calling such a system a "child machine". He noted the learning process could change some of the child machine's operating rules. Since 1950, the idea has been frequently discussed and often called the "baby machine" idea.

Sloman (1978) published a high-level description of an architecture for an intelligent system that would be able to work flexibly and creatively in multiple domains. He wrote that to achieve artificial intelligence comparable to an adult human, it would be necessary to develop a baby machine that could learn through interaction with others.

Minsky (2006, pp.178-182) was not optimistic about the prospects for the baby machine approach to human-level AI. He cited several previous research efforts toward general-purpose learning systems and said they all failed because systems stopped being able to extend themselves. He attributed this to the inability of systems to develop new representations of knowledge. He also identified three problems for baby machines related to optimization, complexity, and investment: As systems become more successful, optimized and complex, there is a tendency to invest less in improvements and alternatives, and there is a greater chance that changes will degrade performance and have unforeseen consequences.

McCarthy (2008) discussed the design of a baby machine approach to human-level artificial intelligence. He wrote that the language of thought for a baby machine should be based on logic, and not on natural language.

Jackson (2019) discusses how the TalaMind approach could address the issues identified by Minsky and McCarthy, and support a child machine which could achieve human-level AI.

In 2018, MIT launched a project called Quest for Intelligence. Joshua Tenenbaum was quoted as saying the long-term goal would be to create an AI system which could learn like a baby and a child (Schaffer, 2018).

In summary, the idea of a baby/child machine is an old idea, yet increasingly relevant for research toward human-level AI.

### 8.2.4 Intelligence Kernels / Seed AI

Closely related to the idea of a baby/child machine, several researchers have proposed designs and architectures for self-extending AI systems[49]:

Jackson (1979) hypothesized that intelligent systems could be designed as 'intelligence kernels', i.e. self-extending systems of concepts that could create and modify concepts to behave intelligently within an environment.

Doyle (1980) discussed how a system could perform causal and purposive reasoning to reflectively modify its actions and reasoning. He described a conceptual language based on a variant of predicate calculus, in which theories could refer to theories as objects and some concepts could be interpreted as programs. He noted use of predicate calculus was not essential.

Smith's (1982) doctoral thesis studied how a system could reason about its reasoning processes. Though he focused on a limited aspect of this problem (procedural reflection, allowing programs to access and manipulate descriptions of their operations and structures), he gave remarks relevant to human-level AI.

Coven (1991) gave further discussion of reflection within functional programming languages, toward support of systems that could in principle reflect on their own reasoning processes and learning algorithms.

Nilsson (2005) proposed that human-level AI may need to be developed as a "core" system able to extend itself when immersed in an appropriate environment, and wrote that similar approaches were suggested by Wegbreit[50], Brooks (1997), McCarthy[51], and Hawkins & Blakeslee (2004).

Yudkowsky (2007) proposed creating "seed AI" systems which could understand and improve themselves recursively.

In 2011, papers by Goertzel, Hall, Leijnen, Pissanetzky, Skaba, and Wang were presented at a workshop on self-programming in AGI systems.

---

[49] More generally, 'autopoietic systems' have been discussed in relation to human intelligence and living systems by several authors, e.g. (Maturana & Verela, 1973), (Mingers 1995).

[50] In a personal communication from Wegbreit to Nilsson, ca. 1998.

[51] In a 1999 version of the paper later published by McCarthy (2008).

Thórisson (2012) discussed a "constructivist AI" approach toward developing self-organizing architectures and self-generated code.

Jackson (2014) discussed how a TalaMind system could reason about how to perform processes, and improve the way it performs a process. Thus the approach could support learning by self-programming, and support intelligence kernels for self-developing conceptual systems.

The idea of self-extending AI systems is another old idea that is increasingly relevant for research toward human-level AI.

### 8.2.5 Societies of Mind / Multi-Agent Systems

Minsky (1986) defined a *society of mind* as a group of many small, simple processes (called 'agents') which interact to collectively perform the processes of an intelligent mind. Singh (2003) gave an overview of the history and details of Minsky's theory, noting that Minsky and Papert began work on this idea in the early 1970's.

Minsky's description and choice of the term were evocative, inspiring research on cognitive architectures more broadly than he described. The idea may be considered as a paradigm for research in either of two senses:

1. The society of mind as proposed by Minsky, including a specific set of methods for organizing mental agents and communicating information, i.e. K-lines, connection lines, nomes, nemes, frames, frame-arrays, transframes, etc.

2. A society of mind as a multiagent system, open to complex agents and to methods for organizing agents and communication between agents other than the methods specified by Minsky, e.g. including languages of thought.

To give examples of the second 'generalized' sense, Doyle (1983) described a mathematical framework for specifying the structure of societies of mind having alternative languages of thought. Doyle used the term 'language of thought' in a general sense, not referring or limited to Fodor's (1975) theory. More recently, Wright (2000) discussed the need for an economy of mind in an adaptive, multi-agent society of mind. Polyscheme (Cassimatis, 2002) may also be considered as a society of mind architecture in the generalized sense, using a propositional language for communication between agents. Bosse & Treur (2006) gave a formal logic discussion of the extent to which collective processes in a multiagent society can be interpreted as single agent processes. Shoham & Leyton-Brown (2008) provide an extensive text on multiagent systems, including a chapter on communication between agents. Sowa (2011b) described communication of conceptual graphs between heterogeneous agents in a framework which he wrote was inspired by Minsky's society of mind. Majumdar & Sowa

(2018) provide an updated description of this architecture. It is natural to have a society of mind at the linguistic level of a TalaMind architecture (Jackson, 2019, §2.3.3.2.1), in the broader, generalized sense, i.e. a multi-agent system using a language of thought for communication.

Generalized societies of mind are also increasingly relevant to research on human-level AI.

## 9. The Harvest of Artificial Intelligence

Chapter 9 of this textbook is about AI's possible future consequences for humanity. In many ways what was written in 1974 remains relevant. Yet much has changed, and predicting the future remains a constant challenge. There is an ongoing Stanford One Hundred Year Study on Artificial Intelligence to periodically review achievements in the field, predict future progress, and provide guidance for research and policies to ensure AI is beneficial to society (Grosz & Stone, 2018). (The One Hundred Year Study is not associated with this essay on AI in the 21st Century. It is mentioned as another source of information for the reader.)

Researchers are also discussing whether and how to regulate current and emerging AI technologies and applications. Etzioni (2018) advocates guidelines to avoid militarization of AI, make people responsible for what their AI systems do, prevent AI systems from pretending to be humans, prevent AI systems from revealing private information, and prevent AI systems from increasing social biases. O'Sullivan & Thierer (2018) advocate permissionless technology innovation within existing legal systems, and extending legal systems when necessary.

The range of potential applications for human-level AI would include any application for human-level intelligence. Some thinkers have suggested that even if it is theoretically possible to achieve human-level AI, such systems should not be created at all (Weizenbaum, 1984; Joy, 2000). More recently, potential dangers of artificial intelligence have been discussed by Bostrom, Omohundro, Tegmark, Yudkowsky and others. Some of these issues were also discussed by Jackson (1974, 1985).

The following sections discuss economic risks and benefits of AI, ethical issues related to human-level AI, how to ensure that human-level AI and superintelligence will be beneficial to humanity, and reasons why human-level AI may be necessary for humanity's survival and prosperity. Additional discussion is given in (Jackson, 2019) §8.2.

### 9.1 Economic Prospects

In 1930, Keynes defined 'technological unemployment' as unemployment caused by technology eliminating jobs faster than it creates new jobs. He warned it would be a significant problem for future generations.

In 1983, Leontief, Duchin, and Nilsson each wrote papers about the potential for automation[52] and AI to cause long-lasting unemployment. Leontief (1983a, b) reasoned the use of computers to replace human mental functions in producing goods and services would increasingly reduce the need for human labor. Nilsson (1983) predicted AI would significantly reduce the total need for labor, particularly for white-collar and service sector jobs. Duchin (1983) discussed methods for widely distributing incomes without paychecks.

In the past two decades, several authors have warned about this potential problem and suggested possible solutions. They include Albus (2011), Brain (2013), Brynjolfsson and McAfee (2011), Ford (2009), Reich (2009 *et seq.*), Rifkin (1995 *et seq.*), and others. So, several economists (Brynjolfsson, Duchin, Leontief, McAfee, Reich, Rifkin) and computer technologists (Albus, Brain, Ford, Nilsson) have discussed this problem and developed similar viewpoints.

To be concise in referring to these authors, they will here be called *Leontief-Duchin-Nilsson (LDN) theorists*, focusing only on their arguments regarding technological unemployment, automation, and AI—they may disagree about other topics. It would be incorrect to call them Keynesian economists, since this term refers to Keynes' theories more broadly. Nor is it accurate to call them Luddites or neo-Luddites, because they do not advocate halting technological progress.

Economists in general disagree about whether technological unemployment can have widespread and long-lasting effects on workers and the economy. Many economists[53] have considered it is not a significant problem, arguing that workers displaced by technology will eventually find jobs elsewhere, and the long-term effect on an economy will be positive. However, Leontief (1983b), who was awarded the Nobel Memorial Prize in Economic Sciences in 1973, wrote that it is not valid to assume that someone who loses a job due to technological progress will always be able to find another job, even after retraining. Brynjolfsson and McAfee (2011) wrote that no economic law says technological progress will automatically benefit most of the people. A large majority of the people in a nation can have reduced wealth as a result of technological progress, even if overall wealth increases.

Since it is beyond the scope of this text to resolve disputes about theories of economics, the views of LDN theorists can at most be presented tentatively. Those writing in the past two decades appear to roughly agree at least implicitly, on the following points for the problem of technological unemployment:

---

[52] This section uses the term 'automation' in a very broad sense, to include the use of computers to provide goods and services throughout an economy, e.g. not limited to manufacturing. This sense also includes 'computerization' of goods and services, changing their nature. For example, Web-based technologies can computerize and replace many services of brick-and-mortar shopping centers.

[53] For instance, see Von Mises (1949, p.768) and Easterly (2001, p.53).

1. In the next several decades of the 21st century, automation and AI could lead to technological unemployment affecting millions of jobs at all income levels, in diverse occupations, and in both developed and developing nations. This could happen with current and near-term technologies, i.e. without human-level AI. It has already occurred for manufacturing, agriculture, and many service sector jobs.

2. It will not be feasible for the world economy to create new jobs for the millions of displaced workers, offering equivalent incomes producing new products and services.

3. Widespread technological unemployment could negatively impact the worldwide economy, because the market depends on mass consumption, which is funded by income from mass employment. LDN theorists vary in discussing and describing the degree of impact.

4. The problem is solvable by developing ways for governments and the economy to provide alternative incomes to people who are technologically unemployed. LDN theorists have proposed methods for funding and distributing alternative incomes.

5. The problem can and should be solved while preserving freedom of enterprise and a free market economy.

6. The problem cannot be solved by halting or rolling back technological progress, because the world's population depends on technology for economic survival and prosperity.

7. Solutions to the problem could lead to greater prosperity, worldwide. LDN theorists vary in describing potential benefits: Nilsson (1984) envisioned automation and AI could provide the productive capacity to enable a transition from poverty to a "prosperous world society". Ford (2009) suggested extension of alternative incomes to people in poverty could create market demand causing a 'virtuous cycle' of global economic growth.

Based on the arguments of LDN theorists, the possibility that AI could help eliminate global poverty may be considered a 'potential best case event' for the economic risks and benefits of AI.

To summarize: *If* technological unemployment is a major economic problem, then global prosperity could require developing an economic system that provides alternative incomes to people who are technologically unemployed. The challenge could be to develop ways of funding a *universal basic income* while preserving freedom of enterprise and economic stability, and controlling monetary inflation.[54] (See §8.1 for more information.)

## 9.2 Toward Beneficial Human-Level AI and Superintelligence

Some potential consequences of general artificial intelligence were outlined in Chapter 9 of this textbook. Two possibilities for the "harvest of AI" were discussed: A world with the machine as dictator, and a world with "well-natured machines" having enormous benefits to humanity.[55]

Relatively recent work on 'artificial general intelligence' has included arguments (collected by Bostrom, 2014) that if AGI is not developed carefully it could be catastrophically harmful to humanity. These arguments were presaged in a paper by Gubrud (1997). Pinker (2018) gave counter-arguments that these catastrophes will not materialize because people will avoid them by designing AI systems to be beneficial to humanity. The following pages give an introduction to some of the relevant design issues.

This is important because AI may be needed for humanity's prosperity: AI may enable global economic growth and the elimination of global poverty, as discussed in the previous section. So we are obliged to consider the problem: *How to ensure human-level AI will be beneficial to humanity?* This question inherently extends to 'superintelligence' (discussed in section 9.2.3). The term 'beneficial' in this context does not seem to have any rigorous, agreed-upon definition. It will be used broadly to refer to consequences that are positive for humanity and biological life in general.

To achieve beneficial AI, we need to consider questions of right and wrong conduct in the interactions of intelligent machines and humanity. Ethics is the branch of philosophy that studies concepts of right and wrong (good and bad) conduct. Until recently ethics has only needed to focus on conduct by humans. Ethics and AI research now intersect regarding concepts of right and wrong conduct by intelligent machines, and in human applications of intelligent machines.

---

[54] While Pinker (2018, p.300) notes there are still many jobs only humans can do, such as building infrastructure and caring for children and the aged, he also suggests (p.119) universal basic income may eventually be needed, noting a negative income tax was proposed by Friedman (1962).

Regarding infrastructure, Pinker (pp.146-149) cites studies showing that to reduce carbon emissions and address global climate change the USA needs a major expansion of safe, advanced nuclear power plants to generate carbon-free electricity. Kappenman (2012) warns that the nation's electric grid and power plant transformers need to be protected against catastrophic damage from electromagnetic pulses that could be caused by a future Carrington Event solar storm, or by a nuclear electromagnetic pulse attack (Foster et al., 2008). These infrastructure improvements for the USA would also be necessary worldwide. Other nations are working to develop advanced nuclear power plants.

[55] Material in this section and subsections is taken from (Jackson, 2018a).

This is a challenge for AI scientists because ethical concepts of right and wrong go beyond simple questions of whether factual or theoretical knowledge is true or false, and whether problem-solving behavior is successful or unsuccessful.

### 9.2.1 Importance of Natural Language for Beneficial AI

The use of a natural language of thought for human-level AI (discussed in 6.2.2) would facilitate representing ethical concepts and goals, and support human inspection and human understanding of AI systems, helping to achieve beneficial human-level AI.

Others have also suggested the importance of natural language for explanations and for representing ethical concepts. Monroe (2018) emphasized the importance of AI systems being able to explain their decisions and actions, and discussed the difficulties of providing explanations for other AI technologies. Bringsjord, Arkoudas, and Bello (2006) recommended that robots not be deployed in life-or-death situations until the robots' governing principles can be clearly expressed in natural language.

Using a natural language of thought, an AI system could do more: It could represent and explain ethical reasoning in natural language, request and accept advice in natural language, discuss ethical alternatives, etc.

An AI system using a natural language of thought could support multiple approaches to ethics, e.g. deontology, virtue ethics, consequentialism, utilitarianism, pragmatic ethics, etc. (Viz. Kuipers, 2018) The system could have this ability because any approach to human ethics must be expressed in natural language, if humans are to understand the ethical approach. Thus, a natural language of thought would provide a starting point for an AI system to have a general understanding of ethics.

### 9.2.2 AI's Different Concept of Self-Preservation

Human-level AI can be 'human-like' without being human-identical. In particular for beneficial AI, the concept of self-preservation could be quite different for a human-level AI than it is for a human. A human-level AI could periodically backup its memory, and if it were physically destroyed, it could be reconstructed and its memory restored to the backup point. So even if it had a goal for self-preservation, a human-level AI might not give that goal the same importance a human being does. It might be more concerned about the technical infrastructure for the backup system, which might include the cloud, and by extension, civilization in general.

A human-level AI could understand that humans cannot backup and restore their minds, and regenerate their bodies if they die, at least with present technologies. It could understand that self-preservation is more important for humans, than for AI systems. The AI system could be willing to sacrifice itself to save human life, especially knowing that as an artificial system it could be restored.

### 9.2.3 The Possibility of Superintelligence

Since one of the abilities of human intelligence is the ability to design and improve machines, it's natural to suppose human-level AI could be applied to improve itself, and to think this might lead to "runaway" increases in machine intelligence beyond the human level. This possibility was first suggested by Good (1965), and later considered by Vinge (1993), Moravec (1998), Kurzweil (2005), and others. Bostrom (2014) and Tegmark (2017) gave recent discussions.[56]

Before evaluating whether superintelligence is possible, it's important to emphasize that we are still a long way, probably decades, from achieving human-level AI – there is still much research and development to be done. Also, for the foreseeable future there will be limits to the knowledge that can be achieved by any system, even superintelligence. And it should be noted that we already have a form of superintelligence: The world's scientific community knows more than any individual scientist, and may be considered a form of superintelligence. These topics are discussed further below.

To evaluate whether an artificial superintelligence can be achieved, we need to be more specific about what it could mean to improve human-level artificial intelligence, so that we can understand whether and how human-level AI could improve itself to achieve superintelligence.

Here is a list of ways human-level AI could surpass human intelligence, and also potentially improve itself:

- *Sensory capabilities* – An AI system could perceive light (and sound) at different wavelengths, and phenomena at different scales (smaller or larger), than humans can directly observe.

- *Active capabilities* – An AI system could perform actions at different physical scales than humans can directly perform.

- *Speed of thought* – A computer can perform logical operations at speeds orders of magnitude faster than a neuron can fire. This may translate to corresponding speedups in thought.

- *Information access* – An AI system could in principle access all the

---

[56] Two earlier related suggestions are noteworthy: Turing (1950) asked whether a machine could generate ideas in a manner analogous to super-criticality of nuclear reactions. Ulam (1958) recalled a conversation with von Neumann on the accelerating progress of technology toward a potential singularity.

information in Wikipedia, or even the entire Web. A human-level AI could understand much of this information.

- *Extent and duration of memory* – An AI system could in principle remember everything it has ever observed. Only a few humans claim this ability.

- *Duration of thought* – A human-level AI could continue thinking about a particular topic for years, decades, ...

- *Community of thought* – A collection of human-level AIs could share thoughts (conceptual structures) more directly, more rapidly, and less ambiguously than a collection of humans. If human-level AI can be copied and processed inexpensively, then much larger groups of human-level AIs could be assembled to collaborate on a topic than would be possible with humans.

- *Nature of thought* – A human-level AI (or community of HLAIs) can develop new concepts and new conceptual processes.

- *Recursive self-improvement* – This term does not seem to have any rigorous, agreed-upon definition, though it is frequently used to describe how superintelligence could be achieved. Essentially it could be the recursive compounding of all the above improvement methods, and any other specific methods that may be identified.

These characteristics might all be described as 'more and faster' human-level AI, and may be called 'weak' superintelligence (cf. Vinge, 1993). If human-level AI is achieved then it will be possible to create weak superintelligence through the above methods, at least in principle. It will be of paramount importance to ensure that superintelligence is beneficial to humanity and to biological life in general – a topic discussed in the following sections. (Strong superintelligence is discussed in §8.2.13.)

### 9.2.4 Nature of Thought and Conceptual Gaps

A superintelligence may develop new concepts and new conceptual processes more rapidly than humans develop or understand them, creating 'conceptual gaps'[57] in understanding between AI systems and humans.

Conceptual gaps happen normally between human minds: For example, scientists have developed concepts that are not understood by the average person, or even by scientists in other fields. The worldwide scientific

---

[57] Jackson (2018a) used the word 'gulf' rather than 'gap'. 'Gap' is used here to be more generally accurate.

community may be considered superintelligent relative to any individual human. People accept this form of superintelligence because they believe scientific ideas can be understood and validated between scientists, and they believe scientific knowledge in general is beneficial to humanity – which it can be, as discussed by Pinker (2018).

Likewise, conceptual gaps between weak superintelligence and humans could be bridged and new concepts could be explained to humans. This will be facilitated if AI systems use a language of thought based on a natural language. Conceivably, conceptual gaps between weak superintelligence and humans may have short duration in some domains, though there may always be conceptual gaps to bridge.

### 9.2.5 Two Paths to Superintelligence

There are at least two somewhat different paths toward superintelligence. One path would focus on recursive self-improvement of general AI systems (AGI) having unchangeable 'final goals' which may be relatively simple and arbitrary. Bostrom (2014) discussed several ways this path could achieve superintelligence that would be catastrophically harmful to humanity and life in general, perhaps leading to extinction events.

Yudkowsky (2008) noted the design space for AGI is much larger than human intelligence. He strongly urged readers not to assume a fully general optimization process for AGI will be beneficial to humanity, yet advised not writing off the challenge of achieving beneficial AI.

A second path toward superintelligence, consistent with the TalaMind approach, focuses on limiting the research design space to AI systems that have generality and also have higher-level mentalities that are characteristic of human intelligence. This design space would be further limited to systems for which the only unchangeable goals are ethical goals beneficial to humanity and to biological life in general. This narrowing of the design space should improve our ability to achieve beneficial human-level AI and beneficial superintelligence via recursive self-improvement.

### 9.2.6 Human-Level AI and Goals

In discussing the first path to superintelligence, Bostrom[58] (2014) relied on an 'orthogonality thesis' that any level of intelligence could have any unchangeable, final goals. He described some simple, at first-glance harmless final goals that could lead to disasters, such as counting the number of grains of sand on a beach, calculating π's infinite decimal string, or maximizing the number of paperclips throughout the future.

---

[58] Bostrom (2014) consolidated research on the first path by himself and others, including Omohundro and Yudkowsky.

In taking the second path to superintelligence, these would not be allowed as unchangeable final goals. The AI system would realize it is pointless to count the grains of sand on a beach, impossible to fully calculate π's infinite decimal string, and harmful to maximize the number of paperclips rather than achieve other goals throughout the future. So it would reject or abandon these simple goals.

Bostrom (2014) also relied on an 'instrumental convergence thesis' that superintelligent agents with different final goals will pursue similar instrumental goals. He cited two instrumental goals that could cause superintelligent systems to be very harmful to humanity, perhaps leading to an extinction event: The first is a goal of self-preservation. The second is a goal of maximizing available resources. Section 9.2.2 above discusses how a human-level AI could have a different concept of self-preservation, facilitating self-sacrifice to save human life. This could apply also to a superintelligence.

In scenarios Bostrom (2014) discussed, the goal of maximizing resources causes a superintelligent system to accumulate as much money and power as possible, leading to very harmful consequences for humanity. This is a case where the ability to think ethically about goals, and change or abandon them, is important. A human-level AI should understand there are appropriate and inappropriate relationships between goals and possible means to achieve them. It should understand that achieving an important goal does not justify acquiring as much money and power as possible – rather, it should have an ethical meta-goal to achieve its goals with as little resources and money as possible, and without acquiring power over human lives or human decisions.

### 9.2.7 TalaMind's Role in Beneficial Superintelligence

Taking the second path won't be easier than the first path just because the design space is smaller. Yet the TalaMind approach will help achieve beneficial superintelligence, since it will help achieve beneficial AI as discussed above (9.2.1), and since the use of a natural language of thought will facilitate explaining new concepts and conceptual processes, and bridging conceptual gaps between superintelligence and humans.

Additionally, the TalaMind approach will support achieving superintelligence in two ways:

- Tala will support developing new concepts and new conceptual processes, arguably better than formal logical languages due to the openness and flexibility of natural language. This support will facilitate 'nature of thought' improvements by superintelligence.

- Tala will provide an interlingua supporting communities of thought for collaboration of human-level AI's to achieve superintelligence.

It should also be noted that the TalaMind approach is open to inclusion of other approaches toward beneficial AI.

## 9.3 When Will Human-Level AI Be Achieved?

We are still a very long way from achieving human-level AI – there is much research and development to be done. As noted earlier, a survey in 2012 and 2013 of about 550 AI experts found almost 18% believed no research approach would ever achieve human-level AI (Müller & Bostrom, 2016). The authors summarized the survey by saying that overall the experts thought human-level AI would have a 50% chance of existing by 2040-50, and a 90% chance of existing by 2075. After achieving human-level AI, there would be a 10% chance of achieving superintelligence in 2 years, and a 75% chance of achieving superintelligence within 30 years. Overall, the experts thought there would be a 31% chance that superintelligence would be harmful for humanity.

The survey's estimates for timeframes to achieve human-level AI and superintelligence seem reasonable to me, if the TalaMind approach is followed. The TalaMind approach could help ensure superintelligence will be beneficial for humanity.

## 9.4 Humanity's Long-Term Prosperity and Survival

Although there are existential threats to humanity's long-term prosperity and survival, such as climate change due to greenhouse gases, we can be 'conditionally optimistic' that these problems can be solved using reason and science. Pinker (2018) surveys evidence showing the welfare of humanity has improved significantly over the past century, even though many problems remain to be solved, now and in future decades.

AI can help in solving such problems. In addition to the potential best case event that AI could help eliminate world poverty, AI applications can support the development of science and technologies that benefit humanity. Eventually, human-level AI (and its consequence, artificial superintelligence) could help develop scientific knowledge more rapidly and perhaps more objectively and completely than possible through human thought alone. If it is so applied, then human-level AI could help advance medicine, agriculture, energy systems, environmental sciences, and other areas of knowledge directly benefitting human prosperity and survival.

Human-level AI may also be necessary to ensure the long-term prosperity of humanity by enabling the economic development of outer space: If civilization remains confined to Earth then humanity is kept in an economy limited by Earth's resources. However, people are not biologically suited for lengthy space travel, with present technologies. To economically develop outer space it could be more cost-effective to use robots with human-level AI for most travel throughout the Solar System, and to minimize sending people in spacecraft that overcome the hazards of radiation and weightlessness, and which provide water, food and air for space voyages lasting months or years.

For the same reason, human-level AI may be necessary for the long-term survival of humanity. To avoid the fate of the dinosaurs (whether from asteroids or super-volcanoes) our species may need economical, self-sustaining settlements off the Earth. Human-level artificial intelligence may be necessary for mankind to spread throughout the Solar System, and later the stars.

## 10. How To Learn More About AI

One of the promises at the beginning of this supplement was to give suggestions for how to learn more about AI. Of course, there is the textbook presented in the following pages, published in 1974, and there are the references in the Bibliographies of this supplement and the textbook. Here are some recommendations for other general texts:

- *AI: Its nature and future* (Boden, 2016). A survey of the field that focuses on the nature of intelligence and AI's feasibility and consequences.

- *The Cambridge Handbook of Artificial Intelligence* (Frankish & Ramsey, 2014). A collection of papers by different authors on several topics in AI. Taken together, several of these papers give a non-technical introduction to the field.

- *Artificial Intelligence Simplified – Understanding Basic Concepts* (George & Carmichael, 2016). A concise introduction to AI and to some of the basic methods used in AI systems.

- *How We Reason* (Johnson-Laird, 2006). An introductory discussion of human intelligence and the theory of mental models.

- *Artificial Intelligence – What Everyone Needs to Know* (Kaplan, 2016). An introduction discussing several topics, including technological unemployment.

- *Artificial Intelligence: A Modern Approach* (Russell & Norvig, 2010). A comprehensive scientific and technical introduction to the field for college-level students.

In citing these authors and others throughout these pages, I do not suggest they would agree with everything said here or with each other on all points. AI is a research field with many perspectives.

The reader can find much published AI research in the journals *Artificial Intelligence* and *Cognitive Science, Advances in Cognitive Systems,* the *Proceedings of the Association for the Advancement of AI,* the *Proceedings of International Joint Conferences on AI, Proceedings of the Cognitive Science Society, Proceedings* and *Postproceedings of the Annual International Conference on Biologically Inspired Cognitive Architectures,* and in more general publications such as *Communications of the ACM, Proceedings of the IEEE,* and *Procedia Computer Science.* There are many other journals and proceedings which focus on subfields, for example the journals *Computational Linguistics* and *Transactions on Computational Linguistics,* and *Proceedings of the Association for Computational Linguistics.*

# SUPPLEMENTARY BIBLIOGRAPHY

## Abbreviations

The abbreviations used in this bibliography are the same as those used in the original bibliography. They are defined on pages 401–403 of the following textbook.

## Mnemonics for New References

*AAAIxx*, Association for the Advancement of Artificial Intelligence (19xx) or (20xx). Proceedings of AAAI Conferences on Artificial Intelligence. https://www.aaai.org/Library/conferences-library.php

*IJCAIxx*, International Joint Conferences on Artificial Intelligence Organization (19xx) or (20xx). *Proceedings of International Joint Conferences on Artificial Intelligence.*

     https://ijcai.org/past_proceedings

*MD*, Haugeland, J., ed. (1981) *Mind Design: Philosophy, Psychology, Artificial Intelligence.* Cambridge, Mass.: MIT Press

*MI9*, Hayes, J., Michie, D., and Mikulich, L. I., eds. (1979) *Machine Intelligence 9.* New York: John Wiley & Sons

*MM*, Johnson-Laird, P. N. (1983) *Mental Models: Towards a Cognitive Science of Language, Inference, and Consciousness.* Harvard University Press.

*PDP*, Rumelhart, D. E., McClelland, J. L., eds. (1986) *Parallel Distributed Processing: Explorations in the Microstructure of Cognition*, 2 volumes, MIT Press.

## Bibliography

Albus, J. S. (2011) *Path to a Better World:A Plan for Prosperity, Opportunity, and Economic Justice.* iUniverse, Inc.

Alderson-Day, B., Weis, S., McCarthy-Jones, S., Moseley, P., Smailes, D. & Fernyhough, C. (2016) The brain's conversation with itself: neural substrates of dialogic inner speech. *Social Cognitive and Affective Neuroscience*, 2016, 110–120.

Aleksander, I., and Morton, H. (2007) Depictive architectures for synthetic phenomenology. In *Artificial Consciousness*, 67-81, ed. A. Chella & R. Manzotti (2007), Imprint Academic.

Anderson, J. R. (2007) *How Can the Human Mind Occur in the Physical Universe?* Oxford University Press.

Anderson, J. R., & Lebiere, C. (1998) *The Atomic Components of Thought.* Psychology Press.

Baars, B. J. (1996) Understanding subjectivity: Global workspace theory and the resurrection of the observing Self. *Journal of Consciousness Studies*, 3, 3, 211-216.

Baars, B. J., & Gage, N. M. (2007) *Cognition, Brain, and Consciousness: Introduction to Cognitive Neuroscience.* Elsevier.

Bach, J. (2009) *Principles of Synthetic Intelligence: PSI: An Architecture of Motivated Cognition.* Oxford University Press.

—— (2015) Modeling motivation in MicroPsi 2. *Artificial General Intelligence, 8th International Conference, AGI 2015*, 3-13.

Bancilhon, F., Maier, D., Sagiv, Y., & Ullman, J. D. (1986) Magic sets and other strange ways to implement logic programs. *Proceedings of the Fifth ACM SIGACT-SIGMOD Symposium on Principles of Database Systems*, 1–16.

Barsalou, L. W. (1993) Flexibility, structure, and linguistic vagary in Concepts: manifestations of a compositional system of perceptual symbols. In *Theories of Memory*, ed. A. F. Collins, S. E. Gathercole, M. A. Conway & P. E. Morris, 29-101. Lawrence Erlbaum Associates.

Berwick, R. C. & Chomsky, N. (2016) *Why Only Us: Language and Evolution.* MIT Press.

Blundell, C., Cornebise, J., Kavukcuoglu, K. & Wierstra, D. (2015) Weight uncertainty in neural networks. *Proc. 32nd International Conference on Machine Learning*, Lille, France.

Boden, M. A. (2004) *The Creative Mind: Myths and Mechanisms.* London: Routledge.

—— (2016) *AI: Its Nature and Future.* Oxford University Press.

Boroditsky, L. & Prinz, J. (2008) What thoughts are made of. In *Embodied Grounding: Social, Cognitive, Affective, and Neuroscientific Approaches*, G. Semin and E. Smith (eds.), 108-125. Cambridge University Press.

Bosse, T. & Treur, J. (2006) Formal interpretation of a multi-agent society as a single agent. *Journal of Artificial Societies and Social Simulation*, 9, 2.

Bostrom, N. (2014) *Superintelligence: Paths, Dangers, Strategies.* Oxford University Press.

Brain, M. (2013) *Robotic Nation and Robotic Freedom.* Tenth Anniversary Edition, BYG Publishing.

Bringsjord, S., Arkoudas, K. & Bello, P. (2006) Toward a general logicist methodology for engineering ethically correct robots. *IEEE Intelligent Systems*, July 2006, 38-44.

Brooks, R. A. (1997) From earwigs to humans. *Robotics and Autonomous Systems*, 20, 2–4, 291–304.

Brynjolfsson, E., & McAfee, A. (2011) *Race Against The Machine: How the Digital Revolution is Accelerating Innovation, Driving Productivity, and Irreversibly Transforming Employment and the Economy*. Digital Frontier Press.

Cassimatis, N. L. (2002) *Polyscheme: A Cognitive Architecture for Integrating Multiple Representation and Inference Schemes*. Ph.D. Thesis, MIT.

Chalmers, D. J. (1995) Facing up to the problem of consciousness. *Journal of Consciousness Studies*, 2, 3, 200-219.

—— (1996) *The Conscious Mind: In Search of a Fundamental Theory*. Oxford University Press.

—— (2010) *The Character of Consciousness*. Oxford University Press.

Chandra, A. K. & Harel, D. (1980) Computable queries for relational data bases. *Journal of Computer and System Sciences*, 21, 2, 156–178.

Charlesworth, A. (2016) A theorem about computationalism and "absolute" truth. *Minds & Machines*, 26, 205-226.

Chomsky, N. (1975) *Reflections on Language*. Pantheon.

—— (1995) *The Minimalist Program*. MIT Press.

—— (2000) *New Horizons in the Study of Language and Mind*. Cambridge University Press.

—— (2015) *The Minimalist Program*. 20th Anniversary Edition. MIT Press.

Cireşan, D. C., Meier, U., & Schmidhuber, J. (2012) Multi-column deep neural networks for image classification. *Computer Vision and Pattern Recognition*, 3642-3649.

Cireşan, D. C., Giusti, A., Gambardella, L. M., & Schmidhuber, J. (2013) Mitosis detection in breast cancer histology images with deep neural networks. *Medical Image Computing and Computer-Assisted Intervention: MICCAI 2013*, 411-418.

Clark, A., Fox, C., & Lappin, S., Eds. (2012) *The Handbook of Computational Linguistics and Natural Language Processing*. Wiley-Blackwell.

Clocksin, W. F. (2003) *Clause and Effect: Prolog Programming for the Working Programmer*. Springer.

Cole, D. (2009) The Chinese Room argument. *The Stanford Encyclopedia of Philosophy*.

Colmerauer, A., & Roussel, P. (1993) The birth of Prolog. *SIGPLAN Notices*, 28, 3, 37-52.

Cortes, C., & Vapnik, V. N. (1995) Support vector networks. *Machine Learning*, 20, 273–297.

Coven, H. J. (1991) *A Descriptive-Operational Semantics for Prescribing Programming Languages with "Reflective" Capabilities.* Ph.D. Thesis, Arizona State University.

Craik, K. J. W. (1943) *The Nature of Explanation.* Cambridge University Press.

Cristianini, N., & Shawe-Taylor, J. (2000) *An Introduction to Support Vector Machines and Other Kernel-Based Learning Methods.* Cambridge University Press.

Cybenko, G. (1988) Continuous valued neural networks with two hidden layers are sufficient. Technical report, Department of Computer Science, Tufts University.

—— (1989) Approximation by superpositions of a sigmoidal function. *Mathematics of Controls, Signals, and Systems,* 2, 303-314.

Davidson, D. (1981) The material mind. *MD,* 339–354.

Dechter, R. (1986) Learning while searching in constraint-satisfaction problems.

Dehaene, S. (2014) *Consciousness and the Brain: Deciphering How the Brain Codes Our Thoughts.* Viking.

Denney, E., Fischer, B., & Schumann, J. (2006) An empirical evaluation of automated theorem provers in software certification. *International Journal on AI Tools,* 15, 1, 81–107.

Doorenbos, R. B. (1994) Combining left and right unlinking for matching a large number of learned rules. *AAAI94,* 451-458.

Doyle, J. (1980) *A Model for Deliberation, Action, and Introspection.* Ph.D. Thesis. MIT AI Lab TR-581.

—— (1983) A society of mind: multiple perspectives, reasoned assumptions, and virtual copies. *IJCAI83,* 309-314.

Dreyfus, H. L. (1981) From micro-worlds to knowledge representation: AI at an impasse. *MD,* 161–204.

—— (1992) *What Computers Still Can't Do: A Critique of Artificial Reason.* The MIT Press.

Duchin, F. (1983) Computers, employment, and the distribution of income. *IJCAI8,* 1196–1197.

—— (1998) *Structural Economics: Measuring Change in Technology, Lifestyles, and the Environment.* Washington, D.C.: Island Press.

Easterly, W. (2001) *The Elusive Quest for Growth: Economists' Adventures and Misadventures in the Tropics.* MIT Press.

Etzioni, O. (2018) Should AI technology be regulated? Yes, and here's how. *Comm. ACM,* 61, 12, 30-32.

Evans, R., Gelbukhy, A., Grefenstettez, G., Hanks, P., Jakubícek, M., McCarthy, D., Palmer, M., Pedersen, T., Rundell, M., Rychlý, P., Sharoff, S., & Tugwell, D. (2016) Adam Kilgarriff's legacy to computational linguistics and beyond. *CICLing 2016: Computational Linguistics and Intelligent Text Processing*, 3-25. Springer

Evans, V., & Green, M. (2006) *Cognitive Linguistics: An Introduction*. London: Lawrence Erlbaum Associates.

Fauconnier, G. (1994) *Mental Spaces: Aspects of Meaning Construction in Natural Language*. Cambridge University Press.

Fauconnier, G., & Turner, M. (2002) *The Way We Think: Conceptual Blending and the Mind's Hidden Complexities*. Basic Books, New York.

Fernyhough, C. (2016) *The Voices Within: The History and Science of How We Talk to Ourselves*. Basic Books.

—— (2017) Talking to ourselves. *Scientific American*, August 2017, 76–79.

Fillmore, C. J. (1977) Scenes-and-frame semantics, in *Linguistic Structures Processing*, A. Zampolli (ed.), Amsterdam: North Holland, 55–82.

—— (1992) Toward a frame-based lexicon: the semantics of RISK and its neighbors, in *Frames, Fields and Contrasts*, A. Lehrer & E. F. Kittay (eds.), Lawrence Erlbaum, 75–102.

Fodor, J. A. (1975) *The Language of Thought*. Thomas Y. Crowell Co.

—— (2008) *LOT2: The Language of Thought Revisited*. Oxford University Press.

Ford, M. (2009) *The Lights in the Tunnel: Automation, Accelerating Technology and the Economy of the Future*. Acculant Publishing.

Forgy, C. (1982) A fast algorithm for the many patterns/many objects match problem. *Artificial Intelligence*, 19(1), 17–37.

Foster, J. S., Jr., E. Gjelde, W. R. Graham, R. J. Hermann, H. M. Kluepfel, R. L. Lawson, G. K. Soper, L. L. Wood, J. B. Woodard (2008) *Report of the Commission to Assess the Threat to the United States from Electromagnetic Pulse (EMP) Attack: Critical National Infrastructures*. April 2008.

Frankish, K., and Ramsey, W. (2014) *The Cambridge Handbook of Artificial Intelligence*. Cambridge University Press.

Friedman, M. (1962) *Capitalism and Freedom*. University of Chicago Press.

Fuchs, N. E., Kaljurand, K., & Schneider, G. (2006) Attempto Controlled English meets the challenge of knowledge representation, reasoning interoperability and user interfaces. *FLAIRS 2006*.

Fukushima, K. (1980) Neocognitron: A self-organizing neural network model for a mechanism of pattern recognition unaffected by shift in position. *Biological Cybernetics*, 36, 193-202.

Gallaire, H., & Minker, J., eds. (1978) *Logic and Databases*. Plenum.

Gärdenfors, P. (1990) Induction, conceptual spaces and AI. *Philosophy of Science*, 57, 78-95.

———1995) Three levels of inductive inference. *Studies in Logic and the Foundations of Mathematics*, 134, 427-449. Amsterdam: Elsevier.

——— (2000) *Conceptual Spaces: The Geometry of Thought*. MIT Press.

——— (2017) Semantic knowledge, domains of meaning and conceptual spaces. In *Knowledge and Action*, 203-219, ed. P. Meusburger, B. Werlen, & L. Saursana. Springer.

Gates, E. (2009) *Einstein's Telescope: The Hunt for Dark Matter and Dark Energy in the Universe*. W. W. Norton & Company.

Genesereth, M. R. & Nilsson, N. J. (1987) *Logical Foundations of Artificial Intelligence*. Morgan Kaufmann.

George, B. & Carmichael, G. (2016) *Artificial Intelligence Simplified: Understanding Basic Concepts*. CSTrends LLP.

Géron, A. (2017) *Hands-On Machine Learning with Scikit-Learn & TensorFlow*. O'Reilly Media.

Getoor, L. & Taskar, B., eds. (2007) *Introduction to Statistical Relational Learning*. MIT Press.

Ghahramani, Z. (2015) Probabilistic machine learning and artificial intelligence. *Nature*, 521: 452-459.

Gödel, K. (1951) Some basic theorems on the foundations of mathematics and their implications. In *Collected works / Kurt Gödel, III*, ed. S. Feferman et al. (1995) Oxford University Press, 304-23.

Goertzel, B. (2011) Self-programming = Learning about intelligence-critical system features. *AGI-11 Workshop on Self-Programming in AGI Systems*.

Goertzel, B. & Pennachin, C., eds. (2007) *Artificial General Intelligence*. Springer.

Good, I. J. (1965) Speculations concerning the first ultraintelligent machine. *Advances in Computers*, vol. 6.

Grosz, B. J. & Stone, P. (2018) A century-long commitment to assessing Artificial Intelligence and its impact on society. *Comm. ACM*, 61, 12, 68-73.

Gubrud, M. A. (1997) Nanotechnology and international security. *Fifth Foresight Conference on Molecular Nanotechnology*.

Hall, J. S. (2011) Self-programming through imitation. *AGI-11 Workshop on Self-Programming in AGI Systems*.

Harari, Y. N. (2015) *Sapiens: A Brief History of Humankind*. HarperCollins.

Haugeland, J. (1981) The nature and plausibility of cognitivism. *MD*, 243–281.

—— (1985) *Artificial Intelligence: The Very Idea*. MIT Press.

Hawkins, J., & Blakeslee, S. (2004) *On Intelligence*. Times Books.

Hayes, P., & Menzel, C. (2001) A semantics for the Knowledge Interchange Format, *Proc. IJCAI 2001 Workshop on the IEEE Standard Upper Ontology*, Seattle.

—— (2006) IKL Specification Document. http://www.ihmc.us/users/phayes/IKL/SPEC/SPEC.html

Hinton, G. E., Osindero, S., & Teh, Y-W. (2006) A fast learning algorithm for deep belief nets. *Neural Computation*, 18, 1527-1554.

Hinton, G. E., & Salakhutdinov, R. R. (2006) Reducing the dimensionality of data with neural networks. *Science*, 313, 5786, 504-507.

Holland, J. H. (1975) *Adaptation in Natural and Artificial Systems*. University of Michigan Press.

Huyck, C. R. (2017) The neural cognitive architecture. *AAAI Fall Symposium Series Technical Reports*, FS-17-05, 365-370.

Hunt, W. & Brock, B. (1992) A formal HDL and its use in the FM9001 verification. *Philosophical Transactions of the Royal Society of London*, 339.

ISO/IEC 24707 (2007) Common Logic (CL): A Framework for a family of Logic-Based Languages, *International Organisation for Standardisation*, Geneva, Switzerland.

Jackendoff, Ray (1983) *Semantics and Cognition*. MIT Press.

—— (1989) What is a concept, that a person may grasp it? *Mind & Language*, 4, 1-2, 68-102, Also in Jackendoff, R. (1992) *Languages of the Mind*, MIT Press.

Jackson, P. C. (1974) *Introduction to Artificial Intelligence*. New York: Mason-Charter.

—— (1979) *Concept: A Context for High-Level Descriptions of Systems Which Develop Concepts*. Master's Thesis, Information Sciences, University of California, Santa Cruz.

—— (1985) *Introduction to Artificial Intelligence*. Second Edition, Dover Publications.

—— (1992) Proving unsatisfiability for problems with constant cubic sparsity. *Artificial Intelligence*, 57, 125-137.

—— (2014) *Toward Human-Level Artificial Intelligence: Representation and Computation of Meaning in Natural Language*. Ph.D. thesis, Tilburg University, The Netherlands.

———— (2017) Toward human-level models of minds. *AAAI Fall Symposium Series Technical Reports*, FS-17-05, 371-375.

———— (2018a) Toward beneficial human-level AI... and beyond. *AAAI Spring Symposium Series Technical Reports*, SS-18-01, 48-53.

———— (2018b) The intelligence level and TalaMind. *Sixth Annual Conference on Advances in Cognitive Systems*, Poster Session.

———— (2018c) Natural language in the Common Model of Cognition. *Procedia Computer Science*, 145, 699-709.

———— (2018d) Thoughts on bands of action. *Procedia Computer Science*, 145, 710-716.

———— (2019) *Toward Human-Level Artificial Intelligence: Representation and Computation of Meaning in Natural Language.* Dover Publications.

Jaffar, J., & Lassez, J.-L. (1987) Constraint logic programming. *Fourteenth ACM Conference on Principles of Programming Languages*, 111–119.

Jensen, F. V. (1996) *An Introduction to Bayesian Networks.* London: UCL Press Limited.

Johnson, M. (1987) *The Body in the Mind: The Bodily Basis of Meaning, Imagination and Reason.* University of Chicago Press.

Johnson-Laird, P. N. (2002) Peirce, logic diagrams, and the elementary operations of reasoning. *Thinking and Reasoning*, 8, 1, 69-95.

———— (2006) *How We Reason.* Oxford University Press.

———— (2010) Mental models and human reasoning. *Proceedings of the National Academy of Sciences*, 107, 18243-18250.

Joy, W. (2000) Why the future doesn't need us. *Wired Magazine*, Issue 8.04, April 2000.

Jouppi, N. P., Young, C., Patil, N., & Patterson, D. (2018) A domain-specific architecture for deep neural networks. *Comm. ACM*, 61, 9, 50-59.

Junker, U. (2003) The logic of ilog (j) configurator: Combining constraint programming with a description logic. *IJCAI03 Configuration Workshop*, 13–20.

Kamp, H. & Reyle, U. (1993) *From Discourse to Logic.* Kluwer.

Kappenman, J. G. (2012) A perfect storm of planetary proportions. *IEEE Spectrum*, February 2012.

Kaplan, J. (2016) *Artificial Intelligence: What Everyone Needs to Know.* Oxford University Press.

Kelly, M. A. & Reitter, D. (2018) How language processing can shape a Common Model of Cognition. *Procedia Computer Science*, 145, 724-729.

Kennedy, W. G. (2012) Modelling human behavior in agent-based models.

*Agent-Based Models of Geographical Systems*, 2, 167-179. M. Batty, A. Heppenstall, & A. Crooks, eds. Springer.

Keynes, J. M. (1930) Economic possibilities for our grandchildren. In *Essays in Persuasion*, Volume 9 of *The Collected Writings of John Maynard Keynes*, ed. E. Johnson & D. Moggridge, Cambridge University Press, 2013.

Khemlani, S.S., Mackiewicz, R., Bucciarelli, M., & Johnson-Laird, P.N. (2013) Kinematic mental simulations in abduction and deduction. *Proceedings of the National Academy of Sciences*, 110, 42,16766–16771.

Khemlani, S. S., Lotstein, M., & Johnson-Laird, P.N. (2015) Naive probability: Model-based estimates of unique events. *Cognitive Science*, 39,1216–1258.

Khemlani, S.S., & Johnson-Laird, P.N. (2017) Illusions in reasoning. *Minds and Machines*, 27, 11-35.

Kilgarriff, A. (1997) "I don't believe in word senses". *Computers and the Humanities*, 31: 91-113.

—— (2007) Word senses. In *Word Sense Disambiguation – Algorithms and Applications*, 29-46, ed. Agirre, E. & Edmonds, P. Springer.

Kotseruba, I. & Tsotsos, J. K. (2018) 40 years of cognitive architecture research: Core cognitive abilities and practical applications. *Artificial Intelligence Review*, 1-78. Springer.

Kowalski, R. (1974) Predicate logic as a programming language. Proc. IFIP Congress, 569–574.

—— (1979) Logic for Problem Solving. Elsevier/North-Holland.

—— (1988) The early years of logic programming. *Comm. ACM*, 31, 38–43.

Koza, J. R. (1992) *Genetic Programming: On the Programming of Computers by Means of Natural Selection*. MIT Press.

Koza, J. R. (1994) *Genetic Programming II: Automatic Discovery of Reusable Programs*. MIT Press.

Koza, J. R., Bennett, F. H., Andre, D., & Keane, M. A. (1999) *Genetic Programming III: Darwinian Invention and Problem Solving*. Morgan Kaufmann.

Kralik, J. D., Lee, J., Rosenbloom, P. S., Jackson, P. C., Epstein, S. L, Romero, O. J., Sanz, R., Larue, O., Schmidtke, H., Lee, S. W., & McGreggor, K. (2018) Metacognition for a Common Model of Cognition. *Procedia Computer Science*, 145, 730-739.

Kuhn, T. (2014) A survey and classification of controlled natural languages. *Computational Linguistics*, 40, 1, 121-170.

Kuipers, B. (2018) How can we trust a robot? *Comm. ACM*, 61, 3, 86-95.

Kurzweil, R. (2005) *The Singularity Is Near: When Humans Transcend Biology.* Viking.

Laird, J. E., Newell, A., & Rosenbloom, P. S. (1987) Soar: An architecture for general intelligence. *Artificial Intelligence,* 33: 1-64.

Lakoff, G. (1987) *Women, Fire, and Dangerous Things: What Categories Reveal about the Mind.* University of Chicago Press.

Lakoff, G., & Johnson, M. (1980) *Metaphors We Live By.* University of Chicago Press.

Langacker, R. W. (1987) *Foundations of Cognitive Grammar. I: Theoretical Prerequisites.* Stanford University Press.

——— (1991) *Foundations of Cognitive Grammar. 2: Descriptive Application.* Stanford University Press.

——— (2008) *Cognitive Grammar: A Basic Introduction.* Oxford University Press.

Langley, P. (1996) *Elements of Machine Learning.* San Francisco: Morgan Kaufmann.

——— (2006) Cognitive architectures and general intelligent systems. *AI Magazine,* 27, 2, 33-44.

Langley, P., Simon, H. A., Bradshaw, G. L., & Zytkow, J. M. (1987) *Scientific Discovery: Computational Explorations of the Creative Processes.* MIT Press.

Langley, P., Choi, D., & Rogers, S. (2009) Acquisition of hierarchical reactive skills in a unified cognitive architecture. *Cognitive Systems Research,* 10, 316–332.

Larue, O., West, R., Rosenbloom, P. S., Dancy, C. L., Samsonovich, A. V., Petters, D., Juvina, I. (2018) Emotion in the Common Model of Cognition. *Procedia Computer Science,* 145, 740-746.

LeCun, Y., Jackel, L., Boser, B., & Denker, J. (1989) Handwritten digit recognition: Applications of neural network chips and automatic learning. *IEEE Communications Magazine,* 27, 11, 41–46.

LeCun, Y., Jackel, L., Bottou, L., Brunot, A., Cortes, C., Denker, J., Drucker, H., Guyon, I., Muller, U., Sackinger, E., Simard, P., & Vapnik, V. N. (1995) Comparison of learning algorithms for handwritten digit recognition. *Int. Conference on Artificial Neural Networks,* 53–60.

LeCun, Y., Bottou, L., Bengio, Y., & Haffner, P. (1998) Gradient-based learning applied to document recognition. *Proc. IEEE,* November, 1-46.

LeCun, Y., Bengio, Y., & Hinton, G. (2015) Deep learning. *Nature,* 521, 436-444.

Lehmann, J., Isele R., Jakob M., Jentzsch A., Kontokostas, D., Mendes, P.

N., Hellmann, S., Morsey, M., van Kleef, P., Auer, S., & Bizer, C. (2015) DBpedia: A large-scale, multilingual knowledge base extracted from Wikipedia. *Semantic Web*, 6, 2, 167-195.

Leijnen, S. (2011) Thinking outside the box: Creativity in self-programming systems. *AGI-11 Workshop on Self-Programming in AGI Systems.*

Lenat, D. B. (1979) On automated scientific theory formation: a case study using the AM program. *MI9*, 251–283.

—— (1982) The nature of heuristics. *J. Art. Intell.*, vol. 19, no. 2.

Lenat, D. B., Borning, A., McDonald, D., Taylor, C, & Weyer, S. (1983) Knoesphere: building expert systems with encyclopedic knowledge. *IJCAI8*, 167–169.

Lenat, D. B., & Brown, J. S. (1983) Why AM and Eurisko appear to work. *AAAI83*, 236–240.

Lenat, D. B., & Greiner, R. D. (1980) RLL: a representation language language. *AAAI80*, 165–169.

Lenat, D. B., Prakash, M., & Shepherd, M. (1986) CYC: Using common sense knowledge to overcome brittleness and knowledge acquisition bottlenecks. *AI Magazine*, 6, 4, 65-85.

Leontief, W. (1983a) Technological advance, economic growth, and the distribution of income. *Population and Development Review*, 9, 3, 403-410.

—— (1983b) National perspective: the definition of problems and opportunities. In *The Long-Term Impact of Technology on Employment and Unemployment*, 3-7. *National Academy of Engineering Symposium*, June 30, 1983. National Academy Press, Washington, D. C.

Leontief, W. & Duchin, F. (1986) *The Future Impact of Automation on Workers*. Oxford University Press.

Lindes, P. (2018) The Common Model of Cognition and humanlike language comprehension. *Procedia Computer Science,* 145, 765-772.

Lowry, M. (2008) Intelligent software engineering tools for NASA's crew exploration vehicle. Proc. International Symposium on Methodologies for Intelligent Systems 2008, 28-37.

Lucas, J. R. (1959) Minds, machines, and Gödel. Paper presented to the Oxford Philosophical Society on October 30, 1959, and published in *Philosophy*, 36, 112-127, 1961.

MacKay, D. J. C. (1992) A practical Bayesian framework for backpropagation networks. *Neural Computation*, 4, 3, 448-472.

Majumdar, A. K., & Sowa, J. F. (2018) Relating language, logic, and imagery. *Procedia Computer Science,* 145, 773-781.

Markram, H. (2006) The Blue Brain Project. *Nature Reviews Neuroscience*, 7, 153-160.

Maturana, H., & Verana, F., eds. (1973) *Autopoiesis and Cognition: The Realization of the Living*. Dordrecht: Reidel.

McCarthy, J. (1955) Proposal for research by John McCarthy. A section of (McCarthy *et al*., 1955).

—— (2004) The robot and the baby. Unpublished short story.

—— (2007) From here to human-level AI. *Artificial Intelligence*, 171, 1174-1182.

—— (2008) The well-designed child. *Artificial Intelligence*, 172, 18, 2003-2014.

McCarthy, J., Minsky, M. L., Rochester, N. and Shannon, C. E. (1955) *A Proposal for the Dartmouth Summer Research Project on Artificial Intelligence*. The first 5 pages were reprinted in *AI Magazine*, 2006, 27, 4, 12-14. The proposal was also reprinted in *Artificial Intelligence: Critical Concepts in Cognitive Science*, ed. R. Chrisley & S. Begeer (2000) 2, 44-53. Routledge Publishing.
http://jmc.stanford.edu/articles/dartmouth/dartmouth.pdf

McCune, W. (1997) Solution of the Robbins problem. *Journal of Automated Reasoning*, 19, 3, 267–276.

McDermott, D. (1983) Under what conditions can a machine attribute meanings to symbols? *IJCAI8*, 45–46.

McDuff, D. & Czerwinski, M. (2018) Designing emotionally sentient agents. *Comm. ACM*, 61, 12, 74-83.

Michalski, R. S, Carbonell, J. G. & Mitchell, T. M, eds. (1983) *Machine Learning: An Artificial Intelligence Approach*. Palo Alto, Calif.: Tioga.

Mingers, J. (1995) *Self-Producing Systems: Implications and Applications of Autopoiesis*. Plenum Press.

Minsky, M. L. (1982) A framework for representing knowledge. *MD*, 95–128.

—— (1986) *The Society of Mind*. New York: Simon and Schuster.

—— (2006) *The Emotion Machine: Commonsense Thinking, Artificial Intelligence, and the Future of the Human Mind*. Simon & Schuster.

Minsky, M. L. & Papert, S. A. (1988) *Perceptrons: An Introduction to Computational Geometry*. Expanded Edition. MIT Press.

Monroe, D. (2018) AI holds the better hand, *Comm. ACM*, 61, 9, 14-16.

—— (2018) AI, explain yourself. *Comm. ACM*, 61, 11, 11-13.

Moravec, H. (1988) *Mind Children*. Harvard University Press.

—— (1998) *Robot: Mere Machine to Transcendent Mind*. Oxford University Press.

Müller, V. C. & Bostrom, N. (2016), Future progress in artificial intelligence:

A survey of expert opinion, in V. C. Müller (ed.), *Fundamental Issues of Artificial Intelligence*, 555-572. Springer.

Neapolitan, R. E. (1990) *Probabilistic Reasoning in Expert Systems: Theory and Algorithms*. John Wiley & Sons.

—— (2003) *Learning Bayesian Networks*. Pearson.

Newell, A. (1973) You can't play 20 questions with nature and win: projective comments on the papers of this symposium. *Visual Information Processing*, ed. W. G. Chase, 283-308. Academic Press.

—— (1982) The knowledge level. *Artificial Intelligence*, 18, 87-127.

—— (1990) *Unified Theories of Cognition*. Cambridge, MA: Harvard University Press.

Newell, A. & Simon, H. A. (1976) Computer science as empirical inquiry: Symbols and search. *Comm. ACM*, 19, 3, 113-126.

Nilsson, N. J. (1980) *Principles of Artificial Intelligence*. Palo Alto, Calif: Tioga.

—— (1983) Artificial intelligence and the need for human labor. *IJCAI8*, 1195–1196.

—— (1984) Artificial intelligence, employment, and income. *AI Magazine*, 5, 2, 5-14.

—— (2005) Human-level artificial intelligence? Be serious! *AI Magazine*, 26, 4, Winter, 68-75.

Norman, D. A. (2004) *Emotional Design: Why We Love (or Hate) Everyday Things*. Basic Books.

Omohundro, S. M. (2008) The basic AI drives. *Artificial General Intelligence 2008: Proceedings of the First AGI Conference*, ed. P. Wang, B. Goertzel & S. Franklin, 483-492.

Ortony, A., Clore, G. L & Collins, A. (1988) *The Cognitive Structure of Emotions*. Cambridge University Press.

O'Sullivan, A. & Thierer, A. (2018) Regulators should allow the greatest space for AI innovation. *Comm. ACM*, 61, 12, 33-35.

Pearl, J. (1988) *Probabilistic Reasoning in Intelligent Systems: Networks of Plausible Inference*. Morgan Kauffman Publishers.

Peirce, C. S. (CP) Collected Papers of C. S. Peirce edited by C. Hartshorne, P. Weiss, & A. Burks, 8 vols., Harvard University Press, Cambridge, MA, 1931-1958.

Penrose, R. (1989) *The Emperor's New Mind: Concerning Computers, Minds and The Laws of Physics*. Oxford University Press.

—— (1994) *Shadows of the Mind: A Search for the Missing Science of Consciousness*. Oxford University Press.

—— (1996) Beyond the doubting of a shadow: A reply to commentaries on Shadows of the Mind. I, 2, 23.

Penrose, R., Shimony, A., Cartwright, N., and Hawking, S., ed. M. Longair (1997) *The Large, the Small, and the Human Mind*. Cambridge University Press.

Penrose, Roger & Stuart Hameroff (1995) What 'Gaps'? Reply to Grush and Churchland. Journal of Consciousness Studies, 2, 2, 1995, 99-112.

Penrose, R., and Hameroff, S. (2011) Consciousness in the universe: Neuroscience, quantum space-time geometry and Orch OR theory. In *Consciousness and the Universe*, eds. R. Penrose, S. Hameroff and S. Kak, *Consciousness and the Universe*, 3-24. Cosmology Science Publishers, Cambridge, MA.

Picard, R. (1997) *Affective Computing*. MIT Press.

Pinker, S. (1994) *The Language Instinct*. HarperCollins Publishers. Originally published by William Morrow & Co.

—— (2018) *Enlightenment Now: The Case for Reason, Science, Humanism, and Progress*. Penguin Random House.

Pissanetzky, S. (2011) Emergent inference, or how can a program become a self-programming AGI system? *AGI-11 Workshop on Self-Programming in AGI Systems*.

Reich, R. B. (2009) Manufacturing jobs are never coming back. *Forbes. com*, May 28, 2009.

—— (2011) *Aftershock: The Next Economy and America's Future*. Vintage Books.

—— (2012) Is technology to blame for chronic unemployment? *Marketplace*, October 10, 2012.

Rifkin, J. (1995) *The End of Work: The Decline of the Global Labor Force and the Dawn of the Post-Market Era*. G. P. Putnam's Sons, New York

—— (2011) *The Third Industrial Revolution: How Lateral Power Is Transforming Energy, the Economy, and the World*. Palgrave MacMillan.

Rosenbloom, P. S., Demski, A., & Ustun, V. (2016) The Sigma cognitive architecture and system: towards functionally elegant grand unification. *Journal of Artificial General Intelligence*, 7, 1-103.

Rumelhart, D. E., Hinton, G. E., & Williams, R. J. (1986) Learning internal representations by error propagation. In (PDP), vol. 1, 318-362.

Russell, S. J. (2015) Unifying logic and probability. *Comm. ACM*, 58, 7, 88-97.

Russell, S. J. & Norvig, P. (2010) *Artificial Intelligence: A Modern Approach*. Third Edition. Prentice-Hall.

Sacks, O. (1989) *Seeing Voices: A Journey into the World of the Deaf.* Vintage Books.

Schaffer, A. (2018) Boosting AI's IQ. *MIT Technology Review*, June 27, 2018.

Schneider, S. (2011) *The Language of Thought: A New Philosophical Direction.* MIT Press.

Schwitter, R. (2010) Controlled natural languages for knowledge represen-tation. *Proceedings of the 23rd International Conference on Computational Linguistics: Posters.* Association for Computational Linguistics.

Searle, J. R. (1980) Minds, brains, and programs. *The Behavioral and Brain Sciences*, 3, 417-424. Cambridge University Press. Also in *MD*, 282–306.

—— (1990) Is the brain a digital computer? Presidential address, *Proceedings American Philosophical Association*, 1990. Reprinted in Searle (1992)

—— (1992) *The Rediscovery of the Mind*, The MIT Press.

—— (2004) *Mind: A Brief Introduction.* Oxford University Press.

Sharma, A., & Goolsbey, K. M. (2017) Identifying useful inference paths in large commonsense knowledge bases by retrograde analysis. *Proceedings AAAI17*, 4437-4443.

Shoham, Y., & Leyton-Brown, K. (2008) *Multiagent Systems: Algorithmic, Game-Theoretic, and Logical Foundations.* Cambridge University Press.

Schmidtke, H. R. (2018) A canvas for thought. *Procedia Computer Science*, 145, 805-812.

Shrager, J., & Langley, P., eds. (1990) *Computational Models of Scientific Discovery and Theory Formation.* Morgan Kaufmann.

Silver, D., Huang, A., Maddison, C. J., Guez, A., Sifre, L., van den Driessche, G., Schrittwieser, J., Antonoglou, I., Panneershelvam, V., Lanctot, M., Dieleman, S., Grewe, D., Nham, J., Kalchbrenner, N., Sutskever, I., Lillicrap, T., Leach, M., Kavukcuoglu, K., Graepel, T., & Hassabis, D. (2016) Mastering the game of Go with deep neural net-works and tree search. *Nature*, 529, 484-503.

Silver, D., Schrittwieser, J., Simonyan, K., Antonoglou, I., Huang, A., Guez, A., Hubert, T., Baker, L., Lai, M., Bolton,A., Chen, Y., Lillicrap, T., Hui, F., Sifre, L., van den Driessche, G., Graepel, T., & Hassabis, D. (2017) Mastering the game of Go without human knowledge. *Nature*, 550, 354-359.

Simon, H. A. (1962) The architecture of complexity. *Proceedings of the American Philosophical Society*, 106, 6, 467-482.

Singh, P. (2003) Examining the society of mind. *Computing and Informatics*, 22, no. 5, 521-543.

Skaba, W. (2011) Heuristic search in program space for the AGINAO cognitive architecture. *AGI-11 Workshop on Self-Programming in AGI Systems.*

Sloman, A. (1978) *The Computer Revolution in Philosophy: Philosophy Science and Models of Mind.* The Harvester Press.

—— (1979) The primacy of non-communicative language. In *The analysis of Meaning: Informatics 5, Proceedings ASLIB/BCS Conference,* Oxford, March 1979, ed. M. MacCafferty and K. Gray.

—— (1983) Under what conditions can a machine attribute meanings to symbols? *IJCAI8,* p.44.

—— (2008) What evolved first: Languages for communicating? or Languages for thinking? http://www.cs.bham.ac.uk/research/projects/cogaff/talks/glang-evo-ai1.pdf

—— (2018) Huge, unnoticed, gaps between current AI and natural intelligence. To appear in V. C. Müller, (ed.), *Philosophy and Theory of Artificial Intelligence 2017,* Springer.

Smith, B. C. (1982) *Procedural Reflection in Programming Languages.* Ph.D. Thesis, MIT.

Sowa, J. F. (1976) Conceptual graphs for a data base interface. *IBM Journal of Research and Development,* 20, 4, 336–357.

—— (1984) *Conceptual Structures: Information Processing in Mind and Machine.* Addison-Wesley.

—— (1991) Towards the expressive power of natural language. In J. F. Sowa, ed., *Principles of Semantic Networks: Explorations in the Representation of Knowledge,* 157-189, Morgan Kaufmann.

—— (1992) Semantic networks. *Encyclopedia of Artificial Intelligence,* Second Edition. Wiley. Updated version at: http://www.jfsowa.com/pubs/semnet.pdf

—— (1995) Syntax, semantics, and pragmatics of contexts. *AAAI Fall Symposium Series Technical Reports,* FS-95-02, 85-96.

—— (1999) *Knowledge Representation: Logical, Philosophical and Computational Foundations.* Brooks/Cole Publishing.

—— (2007) Fads and fallacies about logic. *IEEE Intelligent Systems,* March 2007, 22, 2, 84-87.

—— (2008) Conceptual graphs. *Handbook of Knowledge Representation,* ed. F. van Harmelen, V. Lifschitz, & B. Porter, 213-237. Elsevier.

—— (2011a) Peirce's tutorial on existential graphs. *Semiotica,* 186:1-4, 345-394.

—— (2011b) Cognitive architectures for conceptual structures, *Proc. 19th International Conference on Conceptual Structures,* ed. S. Andrews, S.

Polovina, R. Hill & B. Akhgar, LNAI 6828, Heidelberg: Springer, 35-49.

—— (2013) From existential graphs to conceptual graphs. *International Journal of Conceptual Structures*, 1, 1, 39-72.

—— (2014) Why has AI failed? And how can it succeed? *Computación y Sistemas*, 18, 3, 1-5.

—— (2018) Reasoning with diagrams and images. *Journal of Applied Logics – IFCoLog Journal of Logics and their Applications*, 5, 5, 987-1059.

Talcott, C. L., & Weyhrauch, R. W. (1990) Towards a theory of mechanizable theories: I, FOL contexts: the extensional view, *European Conference on Artificial Intelligence*, 634-639.

Talmy, L. (2000) *Toward a Cognitive Semantics: Volume I: Concept Structuring Systems*. MIT Press.

—— (2000) *Toward a Cognitive Semantics: Volume II: Typology and Process in Concept Structuring*. MIT Press.

Tegmark, M. (2017) *Life 3.0: Being Human in the Age of Artificial Intelligence*. Alfred A. Knopf.

Thórisson, K. R. (2012) A new constructivist AI: From manual methods to self-constructive systems. In *Theoretical Foundations of Artificial General Intelligence*, Wang & Goertzel, eds., 145-171.

Ulam, S. (1958) Tribute to John von Neumann. *Bulletin of the American Mathematical Society*, 64, 3, 1-49.

Ullman, J. D. (1985) Implementation of logical query languages for databases. *ACM Transactions on Database Systems*, 10, 3, 289–321.

Valiant, L. G. (2013) *Probably Approximately Correct: Nature's Algorithms for Learning and Prospering in a Complex World*. New York: Basic Books.

Van Renssen, A. (2005) *Gellish: A Generic Extensible Ontological Language*. Delft University Press.

Vinge, V. (1993) The coming technological singularity: how to survive in the post-human era. *Whole Earth Review*, Winter 1993.

Von Mises, L. (1949) *Human Action: A Treatise on Economics*. Yale University.

Wang, P. (2011) Behavioral self-programming by reasoning. *AGI-11 Workshop on Self-Programming in AGI Systems*.

Wang, P., & Goertzel, B., eds. (2012) *Theoretical Foundations of Artificial General Intelligence*. Amsterdam: Atlantis Press.

Weizenbaum, J. (1984) *Computer Power and Human Reason: From Judgment to Calculation*. Penguin Books.

Weyhrauch, R. W. (1980) Prolegomena to a theory of mechanized formal reasoning. *Artificial Intelligence*, 13, 133–170.

Weyhrauch, R. W., Cadoli, M., & Talcott, C. L. (1998) Using abstract resources to control reasoning. *Journal of Logic, Language and Information*, 7, 1, 77-101.

Whitehead, A. N. (1929) *Process and Reality*. Macmillan, New York. Corrected Edition, ed. D. R. Griffin & D. Sherburne, 1978, The Free Press.

Wittgenstein, L. (1922) *Tractatus Logico-Philosophicus*. Translated by C. K. Ogden, with assistance from G. E. Moore, F. P. Ramsey and Wittgenstein. Published by Routledge & Kegan Paul. Reprinted by Dover, 1999.

——— (1953) *Philosophical Investigations*. Translated by G. E. M. Anscombe. Wiley-Blackwell Publishers, Oxford. Second Edition, 1958.

Woods, W. A. (1983) Under what conditions can a machine use symbols with meaning? *IJCAI8*, 47–48.

Wos, L. & Pieper, G. (2003) *Automated Reasoning and the Discovery of Missing and Elegant Proofs*. Rinton Press.

Wright, I. (2000) The society of mind requires an economy of mind. *Proceedings AISB'00 Symposium Starting from Society: the Application of Social Analogies to Computational Systems*, Birmingham, UK: AISB, 113-124.

Yudkowsky, E. (2007) Levels of organization in general intelligence. In *Artificial General Intelligence*, eds. Goertzel & Pennachin, 389-501.

——— (2008) Artificial intelligence as a positive and negative factor in global risk. In *Global Catastrophic Risks*, eds. N. Bostrom & M. M. Circovic, 308-345. Oxford University Press.

# Introduction to
# *Artificial*
# *Intelligence*

Smith and Tinker's mechanical man from Frank Baum, *Ozma of Oz.*
(with permission)

# 1
# INTRODUCTION

## INTRODUCTION

"Artificial intelligence" is the ability of machines to do things that people would say require intelligence. Artificial intelligence (AI) research is an attempt to discover and describe aspects of human intelligence that can be simulated by machines. For example, at present there are machines that can do the following things:

1. Play games of strategy (e.g., Chess, Checkers, Poker) and (in Checkers) learn to play better than people.
2. Learn to recognize visual or auditory patterns.
3. Find proofs for mathematical theorems.
4. Solve certain, well-formulated kinds of problems.
5. Process information expressed in human languages.

The extent to which machines (usually computers) can do these things independently of people is still limited; machines currently exhibit in their behavior only rudimentary levels of intelligence. Even so, the possibility exists that machines can be made to show behavior indicative of intelligence, comparable or even superior to that of humans.

Alternatively, AI research may be viewed as an attempt to develop a mathematical theory to describe the abilities and actions of things (natural or man-made) exhibiting "intelligent" behavior, and serve as a calculus for the design of intelligent machines. As yet there is no "mathematical theory of intelligence," and researchers dispute whether there ever will be.

This book serves as an introduction to research on machines that

1

display  intelligent  behavior  (note  1–1).[1]  Such  machines  sometimes will  be  called  "artificial  intelligences,"  "intelligent  machines,"  or  "mechanical  intelligences."

The  inclination  in  this  book  is  toward  the  first  viewpoint  of AI research,  without  forsaking  the  second.  Since AI research  is  still  in  its infancy,  it  is  therefore  prudent  to  withhold  estimation  of  its  future.  It  is best  to  begin  with  a  summation  of  present  knowledge,  considering  such questions  as:

1. What  is  known  about  natural  intelligence?
2. When  can  we  justifiably  call  a  machine  intelligent?
3. How  and  to  what  extent  do  machines  currently  simulate  intelligence  or  display  intelligent  behavior?
4. How  might  machines  eventually  simulate  intelligence?
5. How  can  machines  and  their  behavior  be  described  mathematically?
6. What  uses  could  be  made  of  intelligent  machines?

Each  of  these  questions  will  be  explored  in  some  detail  in  this book.  The  first  and  second  questions  are  covered  in  this  chapter.  It  is hoped  that  the  six  questions  are  covered  individually  in  enough  detail so  that  the  reader  will  be  guided  to  broader  study  if  he  is  so  inclined. For  parts  of  this  book,  some  knowledge  of  mathematics  (especially  sets, functions,  and  logic)  is  presupposed,  though  much  of  the  book  is  understandable  without  it.

# TURING'S TEST

A  basic  goal  of AI research  is  to  construct  a  machine  that  exhibits the  behavior  associated  with  human  intelligence,  that  is,  comparable  to the  intelligence  of  a  human  being  (note  1–2).  It  is  not  required  that  the machine  use  the  same  underlying  mechanisms  (whatever  they  are)  that are  used  in  human  cognition  (note  1–3),  nor  is  it  required  that  the machine  go  through  stages  of  development  or  learning  such  as  those through  which  people  progress.

The  classic  experiment  proposed  for  determining  whether  a  machine possesses  intelligence  on  a  human  level  is  known  as  *Turing's  test*  (after A.  M.  Turing,  who  pioneered  research  in  computer  logic,  undecidability

---

[1] The  notes  at  the  ends  of  chapters  are  for  the  benefit  of  the  careful  reader, and  are  intended  to  clarify  questions  that  may  arise  in  the  text.

theory, and artificial intelligence). This experiment has yet to be performed seriously, since no machine yet displays enough intelligent behavior to be able to do well in the test. Still, Turing's test is the basic paradigm for much successful work and for many experiments in machine intelligence, from the Samuel's Checkers Player to "semantic-information processing" programs such as Colby's PARRY or Raphael's SIR (see Chapters 4 and 7).

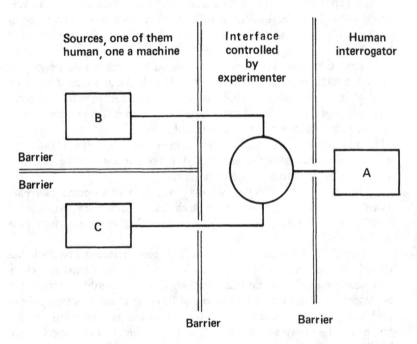

Figure 1–1. A diagram of Turing's test.

Basically, Turing's test consists of presenting a human being, *A,* with a typewriter-like or TV-like terminal, which he can use to converse with two unknown (to him) sources, *B* and *C* (see Fig. 1–1). The interrogator *A* is told that one terminal is controlled by a machine and that the other terminal is controlled by a human being whom *A* has never met. *A* is to guess which of *B* and *C* is the machine and which is the person. If *A* cannot distinguish one from the other with significantly better than 50% accuracy, and if this result continues to hold no matter what people are involved in the experiment, the machine is said to *simulate* human intelligence (note 1–4).

Some comments on Turing's test are in order. First, the nature of Turing's test is such that it does not permit the interrogator $A$ to observe the physical natures of $B$ and $C$; rather, it permits him only to observe their "intellectual behavior," that is, their ability to communicate with formal symbols and to "think abstractly." So, while the test does not enable $A$ to be prejudiced by the physical nature of either $B$ or $C$, neither does it give a way to compare those aspects of an entity's behavior that reflect its ability to act nonabstractly in the real world—that is, to be intelligent in its performance of concrete operations on objects. Can the machine, for example, fry an egg or clean a house?

Second, one possible achievement of AI research would be to produce a complete description of a machine that can successfully pass Turing's test, or to find a proof that no machine can pass it. The complete description must be of a machine that can actually be constructed. A proof that there is no such constructible machine (it might say, e.g., "The number of parts in such a machine must be greater than the number of electrons in the universe.") is consequently to be regarded as a proof of the "no machine" alternative.

Third, it may be that more than one type of machine can pass Turing's test. In this case, AI research has a secondary problem of creating a general description of all machines that will successfully pass Turing's test.

Fourth, if a machine passes Turing's test, it means in effect that there is at least one machine that can learn to solve problems as well as a human being. This would lead to asking if a constructible machine can be described which would be capable of learning to solve not only those problems that people can usually solve, but also those that people create but can only rarely solve. That is, is it possible to build mechanical intelligences that are superior to human intelligence?

It is not yet possible to give a definite answer to any of these questions. Some evidence exists that AI research may eventually attain at least the goal of a machine that passes Turing's test.

It is clear that the intellectual capabilities of a human being are directly related to the functioning of his brain, which appears to be a finite structure of cells. Moreover, people have succeeded in constructing machines that can "learn" to produce solutions to certain *specific* intellectual problems, which are superior to the solutions people can produce. The most notable example is Samuel's Checkers Player, which has learned to play a better game of Checkers than its designer, and which currently plays at a championship level (see Chapter 4).

# NATURAL INTELLIGENCE

The definition of "intelligence" in *Webster's Third International Dictionary* (1966) reads:

¹in·tel·li·gence \ən·'teləjən(t)s\ *n* -s *often attrib* [ME, fr. MF, fr. OF, fr. L *intelligentia*, fr. *intelligent-*, *intelligens* (pres. part.) + *-ia* -y — more at INTELLIGENT] **1 a** (1) **:** the faculty of understanding **:** capacity to know or apprehend **:** INTELLECT, REASON ⟨~, which emerged during the revolutionary cycles of matter as the highest form yet achieved —Hermann Reith⟩ ⟨conceived of history as the expression of a divine ~⟩ (2) *Christian Science* **:** the basic eternal quality of divine Mind **b :** the available ability as measured by intelligence tests or by other social criteria to use one's existing knowledge to meet new situations and to solve new problems, to learn, to foresee problems, to use symbols or relationships, to create new relationships, to think abstractly **:** ability to perceive one's environment, to deal with it symbolically, to deal with it effectively, to adjust to it, to work toward a goal **:** the degree of one's alertness, awareness, or acuity **:** ability to use with awareness the mechanism of reasoning whether conceived as a unified intellectual factor or as the aggregate of many intellectual factors or abilities, as intuitive or as analytic, as organismic, biological, physiological, psychological, or social in origin and nature **c :** mental acuteness **:** SAGACITY, SHREWDNESS ⟨did all he was asked to do with ~ and great good humor⟩ **2 a :** an intelligent being; *esp* **:** an incorporeal spirit **:** ANGEL ⟨hierarchies of angelic ~s —S.F.Mason⟩ **b :** a person of some intellectual capacity ⟨all those ~s we have agreed to call great —*Times Lit. Supp.*⟩ ⟨the greatest all-round ~ writing in England —P.S.O'Hegarty⟩ **3 a :** the act of understanding **:** COMPREHENSION, KNOWLEDGE ⟨faith is necessary to the ~ of the Christian mysteries —*Encyc. Americana*⟩ **:** information communicated **:** NEWS, NOTICE, ADVICE ⟨more weight is laid upon ~ than on editorials —Horace Greeley⟩ ⟨the joyful ~ that there is hope —Georgina Grahame⟩ ⟨from the engine-room voice tube came ~ of more importance — M.S.Boylan⟩ (2) **:** interchange of information **:** COMMUNICATION ⟨accused of maintaining ~ with the enemy⟩ (3) *obs* **:** a piece of information — usu. used in pl. (4) *archaic* **:** common understanding or mutual relations **:** ACQUAINTANCE, INTERCOURSE (5) **:** evaluated information concerning an enemy or possible enemy or a possible theater of operations and the conclusions drawn therefrom; *also* **:** the section, agency, or persons engaged in obtaining such information **:** SECRET SERVICE ⟨investigated me and told me I was qualified for Navy ~ —T.F.Murphy⟩ ⟨an ~ bureau⟩ ⟨available to American and allied ~ organizations —L.W.Doob⟩ syn see MIND
²intelligence *vt* -ED/-ING/-S *obs* **:** to bring tidings of (something) or to (someone)

To summarize the definition in one phrase, one might say that intelligence is the ability "to act rightly in a given situation." Although one could imagine an entity that always behaves "rightly," without making any errors, AI research is more concerned with the concept of partial success, with building machines that can make mistakes, but which can also change their behavior with time and perhaps stop making mistakes.

Intuitively, AI research is concerned with building machines that can "adjust" or "adapt" to certain environments, and which in effect *learn to solve problems* within these environments. This corresponds with the ordinary conception of human intelligence—that it is limited, but that it can learn and thereby improve its performance of certain tasks with time.

Surprisingly little is known concerning the limitations of human intelligence. No one has made any complete survey of the problems that can be solved by human beings. The ability to solve certain types of problems has been studied and made the basis of "intelligence" tests, but the generality and validity of these tests is disputable. Isaac Newton, for example, might have scored low on such tests when he was an adolescent; yet he is estimated by some to have had an intelligence quotient (IQ) near 200. One of the shortcomings of these tests is that they predict little concerning the development of a person's intelligence, especially what problems he could learn to solve.

Evidence concerning human intelligence can be obtained from four major sources: history, introspection, the social sciences, and the biological sciences. Included in the social sciences are psychology, anthropology, sociology, economics, political science; among the biological sciences are neurobiology, biochemistry, biology. "Introspective" sciences might include mathematical logic, systems analysis, and music theory.

## Evidence from History

A discourse on the full history of human intelligence is certainly beyond the bounds of this book. Some allusions to this history can be woven in while presenting evidence from other sources.

## Evidence from Introspection

Introspection has yielded a wealth of seemingly ambiguous and contradictory views of intelligence. One important introspective work familiar in the Western world is Descartes' *Discourse on Method*. This work purports to be ultimately based only on the notion of thought: "I think therefore I exist." So far as the work concerns intelligence, Descartes made a clear distinction between animals and human beings. Animals, he believed, are not much different from machines; anything an animal can do he could imagine being done by a sufficiently complicated machine. People, however, are different from either animals or machines, since people have an ability to "communicate" with each other, to use signs, sentences, and languages that are clearly not completely the result of instinct or construction. Descartes regarded the ability to use languages as the most significant indication that something has human intelligence: ". . . for the word is the sole sign and the only certain mark of the presence of thought hidden and wrapped up in the body. . . ."[2]

Descartes was partially correct in his observation that animals cannot communicate in the same fashion as people. There is recent evidence that dolphins have some sort of language, but the nature of their language is still not understood (Lilly, 1967). Chapter 7 explores the relationship of intelligence and language.

Another introspective way of looking at the mind is that provided by the "rooms of consciousness" concept. In this system a human mind is viewed as being able to inhabit and move among a set of rooms, which are distinguished from each other by their lighting—Socrates' metaphor of the Cave in Plato's *Republic* is a good example. Various rooms can be associated with different levels and abilities of intelligence; this introspective metaphor has been developed in Eastern cultures by Buddha and Lao Tse, as well as in the Western world by other philosophers. Also, the significance of "light" in the metaphor is typical.[3] Other variations on the metaphor speak of some rooms as possessing illusions and dreams.

One viewpoint of intelligence, which is often developed by introspection, is that there is a distinction between *scientific* (intellectual) learning and *spiritual* learning abilities. Scientific learning is said to rely on certain rules for the belief, derivation, refutation, and proof of propositions about the universe. Presumably, science requires a language for describing events and the meanings of measurements, and is dependent on the existence of invariant, reproducible things in the universe. "Spiritual" learning, on the other hand, does not require words or language and may evade intellectual reasoning processes. For various people, introspection has yielded, for example, the following notions of nonintellectual learning:

1. Subconscious learning, in which knowledge is somehow obtained without conscious reasoning.
2. Emotional learning, in which knowledge is perceived as an emotion, without reasoning.
3. Inspired learning, in which knowledge is given to one instantaneously, without reasoning, perhaps by a deity.

---

[2] From a letter of 1647 to Henry More, translated by L. C. Rosenfield in the *Annals of Science,* Vol. 1, No. 1 (1936). Descartes did not claim that animals are machines; he said that they do possess "life" and "feeling."

[3] From a physical standpoint the relation of light to intelligence seems to be simply that light waves (electromagnetic radiation) are the fastest means for transmitting information.

4. Paradoxical learning, in which one is able to perceive knowledge that is self-contradictory, regardless of how it is expressed in words, and therefore beyond logical or scientific learning.

Again, this introspective viewpoint has been developed both in Eastern and Western cultures. The reader who wishes to study the subject deeply may wish to read Dostoevsky, Freud, Jung, and Lao Tse. Various people have, of course, argued that emotional and subconscious learning can be scientifically explained.

The viewpoint that intelligence in certain forms cannot be explained logically or scientifically is relevant to artificial intelligence research. If this viewpoint is correct, then presumably there are some types of knowledge that machines cannot be said to possess and there are some ways of gaining knowledge they cannot use. Chapter 2 discusses the nature of machines and of scientific and mathematical descriptions of things more thoroughly. For now, the viewpoint expressed there is that while it can be argued mathematically that there are entities which cannot be completely described mathematically, there is probably no way of proving in the real world that something is beyond the power of science to explain. All that can be proved is that science has *so far* not explained it.

Thus, no comment is made here as to the existence or nature of spiritual learning: What is important is whether there are *some* forms of learning and intelligence that can be exhibited by machines. Whether "some" means "all" is, scientifically speaking, an open question.

Perhaps not surprisingly, introspection as a technique for gaining knowledge about intelligence often seems to yield only "circular" questions (Can one learn how to learn? If one knows something, does one know that he knows it?). Even so, *introspection is probably the source most commonly used in artificial intelligence research for information about specific problem-solving abilities of human intelligence.* Most researchers use their own experience at having solved problems whenever they are attempting to make a machine solve one; usually if you are going to try to design a machine that does something, it is a good idea to try doing it yourself first and see what happens.

This does not mean that your machine will wind up imitating the human approach to the problem. Actually, machines will often work more efficiently on certain problems when they operate in ways that may seem quite foreign to human reasoning patterns. AI research is concerned with finding machines that *simulate* the abilities of human intelligence—that is, with finding machines that reproduce the outward abilities of human intelligence, though not necessarily the inner means people use to achieve these abilities.

Probably the major advantage to using introspection in artificial intelligence research is simply that it can give the researcher an idea of the information relevant to the problem he is trying to make a machine solve. One of the innate abilities of intelligent creatures seems to be an ability to discard large amounts of information, and focus only on that which is "relevant."

## Evidence from the Social Sciences

The evidence from the social sciences concerning human intelligence is scanty. Only a few general things are known with certainty:

1. Human intelligence is a species-wide trait; there does not seem to be any clear distinction between the innate learning and problem-solving abilities of infants belonging to the various races. Thus, a normal child, properly raised, can learn the language of any human culture, regardless of the language spoken by his biological parents.

2. The intelligence of an individual develops with time and is strongly affected by the nature of his environment. For example, identical twins (who have, barring mutations, the same genetic endowment) raised in different environments have been found to show differences in their intelligence quotients as great as 24 points.

3. The intelligence of an individual is also strongly affected by his heredity. Thus, identical twins raised in approximately the same environment tend to show less difference in their IQs than do other types of siblings.

4. The intelligence of an individual may vary with respect to different *problem domains*—we express this by saying that different individuals may have different "aptitudes."

Experiments performed by Piaget (1946 et seq.) and others have shown that the intelligence of a child *develops in stages*. Precisely why this is so is unknown, but it seems clear that these stages do exist and that the child must accumulate sufficient experience operating within each stage before he can progress completely to the next. Piaget distinguished four stages: sensori-motor, preoperational, concrete operational, and formal operations.

*Sensori-motor Stage.* This stage lasts for the first year and a half to two years of the individual's life. During this stage he makes the transition from using only his instinctive abilities to developing an elementary ability to reason causally and use signals. By the eighth week of an infant's life he is able to discriminate visually between different depths and orientations of objects and to visually perceive objects as having constant size and shape, even when they are receding and rotating. After about the eighth month a baby can understand that a rattle will shake only when he pulls on a string attached to the rattle. Also, after the eighth month, an infant develops vocal and bodily gestures that refer to events and objects in his environment: he will, for example, develop facial expressions and learn to make sounds that represent things he desires or wishes to avoid.

*Preoperational or Symbolic-Operational Stage.* This stage lasts roughly from the second to the seventh year of the child's life. During this stage the child learns the basic vocabulary of the language of his culture, and develops an ability to describe events in sentences (prior to this stage, he describes events with a single word). Also during this stage the child conducts extensive experiments in his environment and learns many different causal relationships. Most of his experimenting is, however, intuitively guided, as is also the way he describes things. If a child in this stage is asked what a jar is, he might say, "There's lemonade in it." Although he can distinguish between "all" and "some," his ability to express the distinction is limited: If he is shown a bouquet of flowers, only some of which are roses, and asked whether there are more roses or more flowers, he will typically respond that there are more roses. Toward the end of this stage a child can be taught to read and write.

*Concrete-Operational Stage.* From age seven to age eleven the child is able to make very significant generalizations of his notions of causality. In particular, he is able to recognize the concepts of *invariance, reversibility,* and *conservation.* Prior to this stage, a child, when shown two "congruent" glasses filled with the same amount of water, will say that they have the same amount of water; but if the water from one glass is then poured into a taller, thinner glass, he will say that the taller, thinner glass has more water. Only when he reaches the concrete-operational stage does he (evidently) realize that both glasses have the same amount of water, regardless of their shape. Also in this stage of development, when presented with a bouquet of flowers only some of which are roses, a child will say there are more flowers than roses.

*Formal Operations Stage.* From the age of eleven upward, the individual becomes able to operate logically with the form of an argument, independently of its meaning; that is, he recognizes factors involved in an event and plans experiments that will give him knowledge about it. It is in this stage that the individual appears to develop a proficiency at reasoning abstractly with words.

Some caveats concerning these four stages should be stated. First, very little is known concerning the emotional and subconscious development of a person's intelligence. Second, there are exceptions to the rate at which children go through these stages: Mozart, for instance, could play the piano and compose proficiently at the age of five. Gauss taught himself to read, could do complicated arithmetic when he was three, and had certainly reached the formal operations stage by the time he was eight or nine.

Another set of basic facts about intelligence and learning are those developed by *behavioristic psychology*. Behaviorist psychologists have attempted to understand intelligent behavior by treating their subjects as "black boxes," presenting them with certain standardized situations and then recording their reactions. They have been able to demonstrate certain phenomena repeatedly in several different species, including man.

The results best known involve learning experiments in the form of the traditional "classical conditioning" and "instrumental, or operant conditioning." In both cases a conditioned stimulus (cs) that has neutral intrinsic value to the animal (e.g., a light flash) is temporally paired with an unconditioned stimulus (ucs) that has a preexisting reward or pain value. In classical conditioning the ucs is followed by the cs despite the animal's response (e.g., Pavlov's induction of salivation in dogs when a bell was rung). In instrumental conditioning the subject's response to the cs determines whether he receives the ucs. Findings concerning learning in these situations include (Thompson, 1967):

1. Up to a point, the stronger the ucs, the more rapid is the conditioning.
2. The most effective time relations for classical conditioning appear to be when the cs begins about a half-second prior to the ucs. As the time between cs and ucs increases, the efficiency of the conditioning decreases.
3. The greater the time between trials, the fewer the trials required for conditioning.
4. If the cs is repeatedly given without the ucs after conditioning has occurred, the conditioned response will *extinguish,* or die out.
5. Following extinction, the conditioned response to the cs will exhibit spontaneous recovery in the absence of ucs presentations.
6. If reinforcement is given only in some of the trials, conditioning occurs more slowly, but is more resistant to extinction.
7. If an additional neutral stimulus is temporally paired with the cs after conditioning, it will subsequently elicit conditioned responses.

8. If conditioning and extinction series are repeated, both processes will occur progressively more rapidly.

Behaviorist psychologists postulate that these forms of conditioning underlie all forms of intelligent adaptation, or learning. They have had difficulty, however, in analyzing the development of relatively complex problem-solving behavior (such as that described by Piaget's findings). As yet, there is no very detailed explanation for the development and abilities of human intelligence in terms of classical and operant conditioning.

## Evidence from the Biological Sciences[4]

*State of Knowledge.* If a really detailed explanation for the individual human intelligence were to be given, it might well require a complete description of the human brain. Biologists are a long way from anything approaching such a description. This section, however, will present an overview of current knowledge and nescience, since for the person doing active research in artificial intelligence it is important to have such a summary.

*The Neuron and the Synapse.* The human brain contains approximately 12 billion nerve cells, or *neurons.* It has been shown that each cell has from 5600 to 60,000 dendritic connections (incoming signal carriers); consequently, each must have equivalent numbers, on the average, of axonal branches (outgoing signal carriers) contacting other neural cells (Cragg, 1967). Such numbers may indicate a storage and processing capability several orders of magnitude greater than current computers, because we know so little about the functions that can be executed by neurons.

The neuron is qualitatively quite different from "on-off" components of current computers. An idealized neuron is shown in Fig. 1–2.

The armlike projections from the cell body, or "soma," are called *dendrites.* Axons from other nerve cells contact the soma and dendrite proper or the dendritic spines (small projections from the dendritic surface) by means of *synapses* (see Fig. 1–3). It is believed that axons synapsing with the dendritic or soma surfaces are inhibitory and those synapsing on the dendritic spines are excitatory to the neuron receiving their signals.

Inpulses transmitted at the synapses add to or subtract from the magnitude of the voltage fluctuations that slowly wax and wane over the membrane of the soma. The electric currents are the result of a

[4] I am indebted to my friend and colleague Bryan Bruns for permission to adapt this section from an unpublished paper.

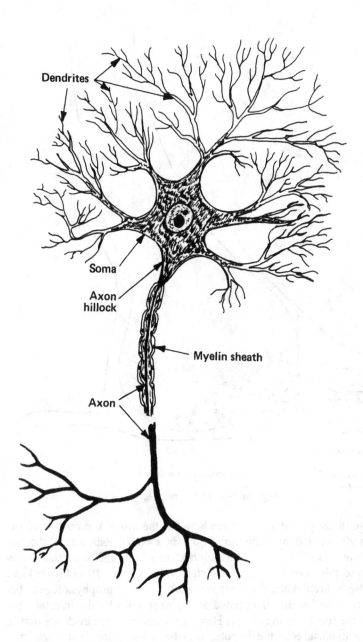

Figure 1–2. The idealized neuron.

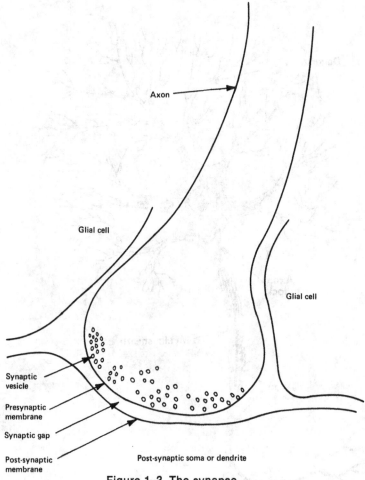

Figure 1–3. The synapse.

change in the potential difference between the inside and outside of the
cell body, caused by a disequilibrium of charged ions across the cell
membrane (see below). If the summation of the additions and decre-
ments to this current reaches a certain value (about 10 millivolts), an
impulse is fired down the neuron's axon. Most neurophysiologists believe that the impulse is initiated at the axon hillock (the interface between the soma and the axon). However, there is recent evidence that in
certain mollusk cells the impulse may be initiated inside the cell, and
may not be a direct consequence of the soma's integrated slow waves
(Pribram, 1971).

An electric impulse is propagated down the axon at a few feet per second; this propagation is based on a *nerve membrane potential.* The nerve membrane is a barrier composed of lipids (e.g., fats), proteins, and sugars, which selectively prevent large molecules and certain ions from entering or leaving the neuron. It selectively screens out sodium ions, and is freely permeable to potassium ions. This creates a cation excess outside the membrane, which opposes tendencies of the potassium ions to equilibrate the charge or equilibrate the potassium concentration on both sides of the membrane.

Consequently, not enough $K^+$ ions move inside to compensate for the large number of $Na^+$ ions on the outside of the cell, and this causes a potential difference across the membrane of about $-70$ millivolts. Initiation of the impulse at the axon hillock consists of a small 10 millivolt change in the membrane potential, which causes the breakdown of the $Na^+$ barrier, the influx of $Na^+$ ions, and the efflux of $K^+$ ions, and the consequent change in the nerve membrane potential to $+40$ millivolts. Immediately after these changes, enzymes embedded in the membrane "pump" the $Na^+$ out of the cell and readjust it to the resting potential. This initiation triggers a similar breakdown in the adjacent membrane, and so the electric signal is carried down the axon.

The amplitude and speed of the impulse are functions of the axon diameter, whereas the frequency is a result of the soma's integration of incoming stimulations and the consequent "decisions" to fire (Thompson, 1967, pp. 129–163).

Many of the longer axons are *myelinated,* that is, they possess a sheath of fat surrounding them which greatly speeds conduction and insulates the axon from neighboring electrical activity. After multiple branchings, the axons become smaller in diameter and unmyelinated; when they reach another cell, they are quite small and the current is of low amplitude and going more slowly. Here it is possible that the electric potentials of neighboring axons from different neurons might interact, either potentiating or damping local electrical activity.

The interface between the axon and dendrite of the contacting cells is the synapse. The impulse is transmitted across the "synaptic cleft" by chemical transmitters such as acetylcholine, norepinephrine and dopamine, seretonin, and certain amino acids. Different transmitters predominate in anatomically and functionally different portions of the brain and spinal chord. Acetylcholine, norepinephrine, and depamine have been shown to be packaged in very small vesicles in the presynaptic membrane. On being activated by an impulse, these vesicles extrude the transmitter into the synaptic cleft, where it crosses the 100 angstrom distance to combine with specific receptors on the post-synaptic membrane. This combination effects the opening of ionic gates, which cause either an increment or decrement in the general activity of the post-synaptic neuron. Excess transmitter is either destroyed or taken up again by the presynaptic bouton to prevent flooding of the post-synaptic receptors and allow the synapse to prepare itself for the next synaptic transmission (Thompson, 1967, pp. 111–128, 192–209; Weiner, 1971).

Until recently it had been hypothesized that all synapses between a neuron and its follower neurons had the same presynaptic transmitter, were functionally the same (excitatory or inhibitory), and had receptors that opened up only one kind of ionic gate. However, work with *Aplesia,* a sea slug with conveniently large neurons and a simple nervous system, revealed several neurons that could both excite and inhibit their "follower" cells. These neurons all used acetylcholine as their transmitters. At the synapses that were excitatory, acetylcholine combined with the post-synaptic receptors to open $Na^+$ ion gates, whereas at the inhibitory synapses acetylcholine combined with the receptors to open $Cl^-$ ion gates. One of these multiaction neurons had a follower cell that had both kinds of receptors in the post-synaptic membrane. Here the rate of stimulation determined whether the excitatory or the inhibitory ionic gates would predominate. Acetylcholine stimulated a third type of receptor to open up $K^+$ ionic gates that caused a longer lasting inhibition than the chloride gates had caused. Such work has shown that neurons with a single type of transmitter can have a variety of effects on their follower cells because the determination of the resultant effects of neural transmission is a function of the differences in the post-synaptic receptors and the ionic gates that are opened (Kandel, 1970; Gardner & Kandel, 1972).

Why, then, does the mammalian CNS have so many different transmitters? It has been shown that stimulation by cholinergic neurons (these that use acetylcholine as their transmitters) and seretonergic neurons (those that use seretonin) causes certain hormonal-like changes in the follower neurons. Seretonergic stimulation causes a rise in c-AMP, a mediator common to many hormones, in the follower cell. Cholinergic stimulation causes a rise in the phosphitidal inositol of the follower cell. Eric Kandel has hypothesized that post-synaptic membranes may have two different classes of receptors. The ionophoric receptors bind with a common transmitter to open different ionic gates, thereby affecting the post-synaptic membrane potential; they are receptor-specific. The chemophore receptors combine with different transmitters to cause metabolic changes in the follower cell. The actions of these receptors are transmitter-specific (Fig. 1–4). The demonstration that neurons

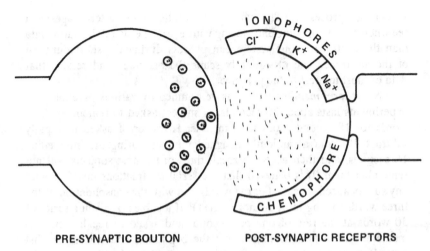

**PRE-SYNAPTIC BOUTON**     **POST-SYNAPTIC RECEPTORS**

Figure 1–4. Transmitter-specific receptors.

affect one another's metabolic as well as electric states has brought to light an entirely new dimension in interneuronal communication.

The most striking aspect of the neuron is its multiinput, single-output character. The slow potential on the soma apparently indicates a comparison of dendritical inputs on the basis of the temporal, structural, and qualitative nature of the synaptic input that results in the all-or-none decision to fire. Whether this comparison is solely a function of electrical interactions, or reflects molecular conformations of the membrane (Barondes, 1970), or is also modified by some mechanism inside the cell is an open question.

*Biological Memory.* Of fundamental importance to any system that wishes to modify its behavior on the basis of experience is an efficient memory storage and retrieval system.

It appears that there are multiple stages in the development of a memory and its means of retrieval. Demonstration of how memory might function has come from psychological and biological experimentation, clinical observations of memory dysfunctions, and attempts to mimic the structure and function of human memory by computer simulation. These differing approaches to the study of memory have caused some confusion, especially in the meaning of such terms as short- and long-term memory. As will be seen below, caution should be used in interpreting what an author means by such terms.

Psychologists have experimentally identified three types of memory. *Sensory information storage* (sis) is measured in tenths of a second. It serves to retain fleeting sensory data until the central nervous system

(CNS) can process it. The SIS system results in the after-images you see when rapidly opening and closing your eyes. The SIS retains more data than the central nervous system can process during the short duration of the SIS trace. The CNS rapidly scans the SIS trace and retains that data of most interest to the perceiver.

*Short-term memory* (STM) as determined by various psychological experiments lasts about 30 seconds. A subject asked to remember three words for 18 seconds does so with ease. However, if asked to rapidly subtract 3's from a randomly assigned number during the intervening 18 seconds, and then asked to recall the words, most subjects will not remember them. It is believed that the serial subtractions interfere with any subvocalized rehearsal of the words and with the consolidation of the three words to long-term memory (LTM). If a subject is given a series of 30 words at the rate of one per second, and asked to recall them immediately afterward, he remembers the beginning and end of the list best. If he is asked to subtract serial 3's immediately after seeing the list, the tail end of the curve disappears. Here, then, is a demonstration of which parts of the learning curve are a function of long-term as opposed to short-term memory.

A major part of the psychologist's investigation of *long-term memory* has centered around the use of computer simulations; much of this will be covered later. For an excellent overview of the psychologist's approach to memory and mind functioning, see Lindsay and Norman (1972).

When an individual suffers a fairly hard blow to the head, he often cannot remember events immediately preceding his accident. This phenomenon is called *retrograde amnesia;* it may begin with loss of memory for several hours or days prior to the accident. The earlier memories usually return first, followed by the later until only events 30 to 60 seconds prior to the trauma cannot be remembered (Jarvik, 1972). Memories following the accident (*anterograde memory*) are likewise impaired and are more refractory to recovery. Such phenomena have also been noted in psychiatric patients who undergo electro-convulsive shock therapy (ECS). These observations fostered the idea that short-term memory traces were transient electric events that eventually consolidated into long-term memories through chemical and biological changes in the brain.

Normally, a rat placed on a pedestal in a cage with an electrified grid floor needs only one or two trials to learn not to jump down from the pedestal. However, if ECS follows these learning trials, the rat will not learn to avoid either the electrified grid or the added negative experience of going through ECS (Deutsch, 1969). The longer the interval between the administration of the learning task and the ECS, the smaller the effect on long-term retention. However, investigators differ on how long after the learning trials the ECS is effective in preventing long-term retention. Some say that ECS is not effective after 15 to 30 seconds, whereas others claim that ECS will impair long-term retention when given hours after the learning trials.

Drugs that inhibit protein synthesis when given before learning trials do not impair learning, or the retention of that learning, for as long as 3 to 6 hours. Testing after 6 hours, however, shows a marked loss of memory. If the drug is given shortly after the learning task, memory is not inhibited. These results could suggest a "dual trace" theory of memory. Short-term memory and long-term memory would be separate processes; the former lasting up to 6 hours after the learning trial, the latter being initiated during learning and not susceptible to protein inhibition only a few minutes after the learning trial (Barondes, 1970). The duration of this "short term" memory, however, is a function of how well the animal is trained, suggesting that the protein inhibitors might simply be weakening the long-term trace that has been derived from a short-term trace.

Puromycin, which inhibits protein synthesis and has various other central nervous system effects, can cause retrograde amnesia when given up to several days following a learning situation. Normal saline, injected into the same place as the puromycin, can reverse these effects and restore the memory. It has been suggested that puromycin may disrupt the retrieval rather than the storage of information (Jarvik, 1972).

The plethora of experiments dealing with ECS and drug effects on memory have resulted in a confused, controversial, and often contradictory literature that is well reviewed by Deutsch (1969) and Jarvik (1972). Perhaps the most reasonable hypothesis of the moment is the following: The short-term memory reported by clinicians, psychologists, and some investigators to last about 30 seconds is indeed a transitory electrical reverberation that is consolidated into a more durable long-term memory. However, the strength and accessibility of this long-term memory is quite variable and is a function of the number of retrieval traces laid down during learning and of the use of old retrieval traces and the construction of new retrieval traces to the long-term memory after the initial learning trial.

Clinical observations have localized the hippocampus as that part of the brain responsible for the consolidation of memories. In the case of Henry M. (Barbizet, 1970), complete surgical, bilateral ablation of the hippocampi prevented him from learning anything new following his operation. There was no change in IQ, no loss of preoperational memories, and no abnormality in his ability to recall digits immediately after hearing them. His crippling deficit involved an inability to recall anything that had happened earlier than a minute before the present or later than the day of his operation.

A similar dysfunction is part of Korsikoff's syndrome, seen in chronic alcoholics. Here the pathology seems to affect the mammillary bodies, the dorsal thalamus, and the terminal fornix—areas of the brain, which along with the hippocampus, form part of the limbic system. This system also is the center for innate emotions, feelings, and the regulation of hunger, thirst, rage, and sexual activities (Pribram, 1971). Patients with Korsikoff's syndrome will frequently be unable to remember anything that occurred during the course of their disease and will confabulate these memories if questioned. However, it appears that they do retain long-term memories (Barbizet, 1970).

Pathological dysfunctions in long-term memory such as Alzheimer's disease or senile dementia do not appear to be localized, but consist of diffuse damage throughout the cortex. Terminal Alzheimer's and severe dementia leave the patient completely unable to learn, communicate, and function or care for himself.

These clinical studies have demonstrated that long-term memory stores are much less susceptible to damage than is the consolidating process. This is expressed in the general maxim that anterograde memory loss is nearly always greater than retrograde memory loss.

The hippocampus appears to act as the "store" mechanism for the brain. It is interesting that this function is integrated with parts of the brain which attach emotional weight, pleasure or pain, to external perceptions. Perhaps such emotive interest is necessary to activate the consolidation of a short-term percept.

The search for the "engram," the biological material that is a memory, was initiated by Lashley in 1929. He would train animals to a task, surgically ablate well-defined areas of the cortex, and see if the animal still was able to perform the task. He found that long-term memories were very difficult to destroy. He might destroy up to 80% of an animal's visual cortex, and still the animal would retain the visual discriminations it had learned. From his studies it appears that a long-term memory trace is diffusely spread throughout a significant portion of the brain.

Assuming, quite simplistically, that memories of certain "percepts" might be localized to specific cells or association networks of cells, then long-term learning would take place when any two of these percepts were temporally paired (as in conditioning experiments). Considering the large number of interconnections between neurons, one might postulate that learning is the *facilitation* of preexisting synapses, perhaps through an increase in transmitter receptors at the post-synaptic membrane or in transmitter substance in the presynaptic bouton. Long-term learning could also be the growth of new connections between neurons or association networks, directed, perhaps, by some neural growth factor excreted only by excited neurons.

Though it has been rather conclusively shown that adult neural cells do not reproduce, anatomical studies have shown that neural lesions are sometimes "repaired" by the growth of the dendritic and axonal networks of the remaining cells (Rose et al., 1969). It has also been shown that there are consistent differences in the brains of rats placed in a stimulating environment with other rats and various toys and in rats placed in an impoverished environment where they are isolated and have little stimulation. The former have thicker cortices, heavier occipital cortices, larger neutral cell bodies and nuclei, more dendritic spines, larger synaptic junctions, an increase in acetylcholine, and a greater number of glial cells (support cells for the neurons) (Rosenzweig et al., 1972). The changes show, for the first time, that experience results in measurable brain alterations, but the behaviors, and the changes they caused, are too general to demonstrate underlying mechanisms, though they are consistent with both the synaptic facilitation and neural growth hypotheses.

Perhaps the most outstanding example of information storage in nature is the DNA molecule that encodes all the information necessary for the construction of an entire organism within the structure of molecules that weigh about $10^{-12}$ gram (Watson, 1970). It has been suggested that memories may be stored in a like fashion in DNA or RNA (the chemical that transfers the DNA message throughout an individual cell and regulates the production of cellular proteins). Some researchers claim that RNA or proteins transferred from animals conditioned to a certain task helps naive animals learn the task faster. However, no one has yet reproducibly demonstrated that RNA or more than a few small, specific proteins can cross the mammalian brain's blood-brain barrier (Pribram, 1971).

Hydén (1969) taught rats to balance on a wire and then examined for changes in RNA that part of the brain that controls balance. He found that stimulated brain cells produced more RNA than any other tissue in the body. He also found that the type of RNA being produced had qualitatively changed. After stimulation, the RNA in the neural cells decreased, but there was a consonant increase of RNA in the neurons' glial cells similar to the RNA that had been produced in the neurons.

Evidently, learning causes changes in neurons, and the implementation of such a change in a cell necessarily involves the production of more and different RNA. The temporal contiguity of the disappearance of neural RNA and the appearance of similar glial RNA is provocative, but the experiments are still controversial and the significance of the results unclear.

Although we have some idea of how memory is stored, how it is structured in storage, and how it is retrieved, we have little idea of the biological correlates of these processes. Work with simplified neural systems such as those in *Aplesia* holds much promise for elucidating the biochemical dynamics that accompany new learning.

*Neural Data Processing at the Gross Anatomical Level.* As in the case of the hippocampus and the limbic region, neurologists have ascribed general and even quite specific data processing functions to gross regions of the brain through the careful testing of patients with defined forms of brain damage. A specific example of this approach is the identification of the respective functions of the right and left cerebral cortices (Gazzaniga, 1970).

Nearly all people, excepting 15% of the left-handers, are left-dominant for speech (that is, the left hemisphere of the brain is responsible for their capability to hear, understand, and speak language). It is well known that the left hemisphere deals with the motor and sensory functions of the right side of the body, and vice versa. In man this is true for eyesight, where the left side of the brain sees the right visual field (those objects to your right), and vice versa (Fig. 1–5). The corpus callosum is a thick sheet of neural fibers that is the *sole* source of communication between the right and left hemispheres (Fig. 1–6). In patients with severed corpus callosums the separate functions of the two hemispheres can be studied by presenting visual data to either the left or right visual fields, or tactile data to either the left or right hands. A word presented to the left visual field cannot be vocalized by such a subject because the image is perceived by the right hemisphere. However, if the word is "banana" and the left hand (also controlled by the right hemisphere) must choose between a number of objects that cannot be seen, the left hand invariably chooses the banana. Thus, the word has been perceived and translated by the right hemisphere into an appropriate motor action, though the instructions and the word "banana" have not been consciously heard and the subject has no idea of what he did.

Thus, we have two brains—one conscious in the sense that it can hear, understand, and repeat back what is said to it; and one that reacts to stimuli and performs activities that we will not be aware of

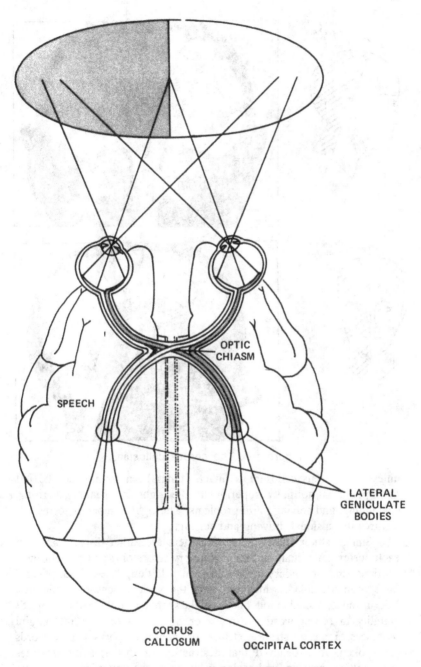

Figure 1–5. How the left side of the brain sees the right visual field
(and vice versa).

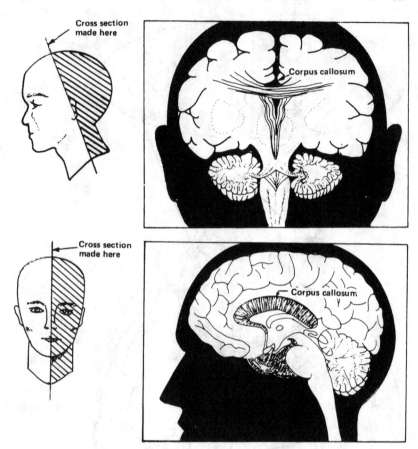

Figure 1–6. The corpus callosum.

unless our corpus callosum is intact. It has been found that the left hemisphere is normally superior to the right in speaking, writing, calculating, and solving maze problems. The right is superior to the left in three-dimensional drawing and singing.

Similar studies have shown that there are rather discrete areas of each cortex for visual, auditory, olfactory, gustatory, and somatic perceptions and secondary processing. In addition, "association areas" have been identified, which integrate the various sensory modalities. For instance, the ablation of Wernicke's area results in the subject's inability to repeat words he reads or hears and to emit meaningful sentences. Instead, strange strings of nonsense phrases and words are spoken. It is believed that destruction of this area disassociates the thinking, hearing, and seeing portions of the cortex from the area that converts thoughts into the motor actions that lead to speaking. Ablations in Broca's area cause aphasia—the nearly complete inability to speak any words even though the patient can still write his communications in a normal fashion (Geschwind, 1972).

While certain functions have been rather discretely localized, other tasks, such as the ability to recognize simple figures hidden in more complex figures, seem to be a function of how much material has been lost from any or all portions of the neocortex.

*Neural Data Processing at the Cellular Level.* Digital computers typically have certain built-in information processing functions for coding and decoding input-output information, for the transferral of data from the storage units to the general registers, and for the handling of data in the general registers. Certain neural functions and organizations have been discovered for data processing; these will be discussed for the particular case of visual perception (see Chapter 5).

The retina of the eye converts patterns of photons into more condensed patterns of electric impulses in the optic nerve (there is about a tenfold contraction of the information). There are about 100 million rod and cone receptors in the retina. In each cell, carotene attached to the enzyme rhodopsin produces molecular complexes sensitive to visual wavelengths of light. Photons induce a structural change in carotene, and this change triggers a receptor-cell voltage potential that is communicated to the "bipolar" cells, which in turn innervate the ganglion cells of the optic nerve. The receptor, bipolar, and ganglion cells are interconnected by amacrine and horizontal cells, which regulate how many and which receptor cells will communicate with ganglia cells via the bipolar cells. In the macula densa portion of the eye there is one receptor cell for each ganglion cell; in the other areas of the eye, up to 100 receptor cells may stimulate a ganglion cell (Fig. 1–7).

The electrical activity of all ganglion cells is greatest in the dark; when exposed to light, the interconnecting amacrine cells provide inhibitory "gates" that reduce the sensitivity of the surrounding receptors. This *surround inhibition* is responsible for the heightened contrast one sees at the borders of two different light intensities. When one looks at the border between light and dark shades, the dark border is darker than the rest of the dark shade and the light border is lighter than the rest of the light shade. In actuality the light intensity of each bar is uniform; the heightened contrasts are the result of surround inhibition. If one postulates that stimulated cells inhibit their neighbors' rate of firing to a degree related directly to the light intensity and inversely to the distance from the neighboring cells, and if the receptors are otherwise uniformly stimulated by the incoming light, then it follows that those

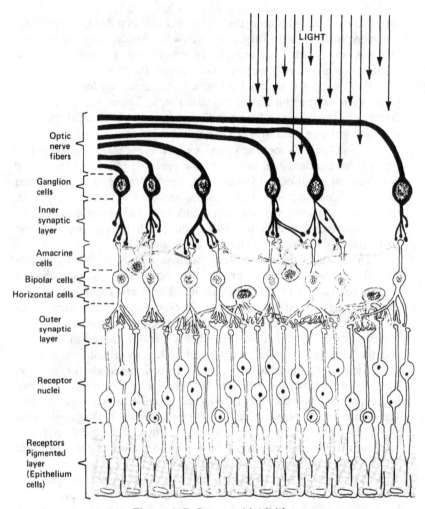

Figure 1–7. Surround inhibition.

cells exposed to the lighter band and near the border will be stimulated
as much as their fellow "light" cells. However, they will be inhibited
less by their neighbors because some of their neighbors are "dark"
cells which, because they are stimulated less by the dark band, will inhibit their neighbors less. Conversely, the dark cells near the border
will be inhibited more than their fellow dark cells because some of
their neighbors are the light cells, which inhibit their neighbors more
than dark cells. Surround inhibition (more thoroughly explained in
Ratliff, 1972) is one of the fundamental informational processing path-
ways used throughout the mammalian brain as well as in the eye.

Some visual receptors cause their ganglion to fire when stimulated by light ("on"), when not stimulated ("off"), or only when the light changes ("on-off"). All receptors fire rapidly when first stimulated, which helps to explain why mammals preferentially attend to moving objects. One organization of the receptors in the receptive field of a single ganglion is the *round* field, where the center is on and the periphery off, with an on-off interface between the two. Other receptive fields are shaped so that they respond preferentially to edges, curves, and lines (Spinelli, 1966).

If any receptive field sees the same image for more than 30 seconds, the bipolar and amacrine cells *adapt* to the receptor stimulation such that the ganglion is no longer stimulated and the object no longer seen. Consequently, the eye is always moving so that the receptors will not see the same image for more than several seconds, though these movements are normally very small. In the central nervous system, this mechanism is called *habituation*, and it allows the organism to screen out "background" noises when attending to a specific percept (Thompson, 1967).

The optic ganglia form a one-to-one projection to the lateral geniculate, where colors are mixed and the on-off responses of the ganglia are separated. A cell from the lateral geniculate may contact up to 5000 cells in the striate area of the occipital cortex (the rear end of the brain) where actual "seeing" takes place. Hübel and Wiesel discovered very specific feature detection cells in the occipital cortex which are arranged in what seems to be an ascending hierarchy of complexity. The procedure that they and many other investigators have used is the recording of *induced responses* by microelectrodes. Microelectrodes are carefully placed into single neural cells in the brain. Then the animal is presented with very specific stimuli and the electrical response of the single cell is recorded.

"Simple" cells respond to a line at a certain angle in a certain small defined area of the retina. "Complex" cells will respond similarly to a line, but at any point in a much larger retinal field. "Hypercomplex" cells require a given length in addition to a given orientation in order to fire, and "higher order hypercomplex" cells respond only to lines that form certain angles. Cells responding to lines at a certain orientation are arranged in columns perpendicular to the surface of the visual cortex, and groups of these columns responding to all the various orientations for a certain area of the retina are arranged together (Hübel and Wiesel, 1962, 1963; Lindsay and Norman, 1972).

The spatial arrangement of these cells and their hierarchial nature suggest a feature detection model of visual perception. Most simply, this model suggests that any percept is the summation of the discrete features reported by each of the receptive fields of the cortical neurons. The "pandemonium model" and other models of pattern perception are discussed in Chapter 5.

*Motivation.* Life forms are essentially chemical information processors designed to preserve the chemical information that describes them within the gene pool of a species. Complex mammalian intelligence is one of a variety of strategies that tends to preserve certain information. Thus, tendencies are built into biological organisms to insure survival; for instance, the tendency to repeat behaviors that reward an individual with food.

A great deal of work is currently being invested in finding out why (1) mammals, especially humans, do what they do, and (2) the basic biochemical and neurophysiological mechanisms underlying motivation and emotion. This literature is very extensive and will not be reviewed here.

*Review.* The major questions concerning the nature of biological intelligence remained unanswered. What are the information processing functions of neural and glial cells? How do context, expectations, and perceived features blend to make an understandable perception? How do experiences become memories in long-term storage? What is the biochemical substrate of memory? At what level do perceptions enter consciousness; when and where do cortical electricity and chemical transmitters become perceived thoughts? Artificial intelligence will certainly be a major contributor to the answering of these questions.

# COMPUTERS AND SIMULATION

Before concluding the discussion of the first and second questions cited at the start of this chapter, some mention should be made of the basic technique used in AI research. One significant fact is that it is not necessary to build a different physical machine each time we wish to investigate a new machine's abilities. A kind of machine exists which is capable of accepting a symbolic description (in the form of a *program*) of *any* machine and of simulating the machine described by such a program. The general-purpose digital computers are examples of this kind of machine.

Computers typically have five main components: an input unit, a control unit, a logic unit, a storage unit, and an output unit. The program and other data go into the computer via the input unit and are stored in the storage unit, or *memory,* of the computer. The logic and control units alter the information in the storage unit of the computer in a manner that is dependent upon the program. Also in a manner dependent on the program, the control unit causes the output unit to emit information (e.g., punched cards, electric impulses, printed paper). Computers can be designed so as to utilize a wide range of input-output devices, from television cameras and CRT (cathode-ray tube) display screens (like a television set) to mechanical arms and typewriter-like terminals.

Computers and the notion of "simulation" are discussed more thoroughly in Chapter 2. Briefly, a computer *simulates* something if it duplicates that thing's behavior. The duplication does not have to be exact, nor does it have to proceed at the same rate as the original. Thus, a computer is said to simulate a person playing Chess if it prints out a possible move on a sheet of paper whenever it is given as input a description of a possible chessboard configuration. We do not require that the computer print out the *same* move that a given person would make, nor must the computer be able to move physically the pieces of an actual Chess set, nor does the computer require the same time to make its move as a person would. A simulation may be a "speed-up" or a "slow-up" of the original. Likewise, a computer is said to simulate intelligence when it does something that a person needs intelligence to do—i.e., when its behavior corresponds in some manner to that of an intelligent person. Thus, the extent to which a machine simulates intelligence may vary. In this book the emphasis is on the ability of computers to do the things listed at the start of this chapter.

## NOTES

**1–1.** This note cites some general references on the subject of artificial intelligence. First, over the past two decades several authors have argued, both pro and con, the possibility of artificial intelligence; that is, whether machines can eventually be made to possess intelligence on a human level. Some classic papers in favor of the possibility are those of Turing (1947, 1950) and Armer (1963). Some recent arguments against the possibility of artificial intelligence are those of Dreyfus (1965, 1972) and Jaki (1969); the arguments of Dreyfus are effectively refuted in the paper by Papert (1968). (One argument against the possibility of AI that is quite commonly put forth is: "Computers can do only what they are told to do." This is true, but no one really knows the limits of what we can tell computers to do; perhaps we can tell them how to think, and how to learn; see Armer and Turing.) A number of books besides this one have been published about artificial intelligence or about specific areas of the subject: see Feigenbaum and Feldman (1963); Banerji (1969), Slagle (1971), Minsky (1968b), and Nilsson (1971). (Minsky [1963, 1966, 1970] has also written a number of stimulating papers on artificial intelligence.) Two journals, *Artificial Intelligence* and *Pattern Recognition*, regularly publish papers that are of interest to the AI researcher. The (voluminous) *Proceedings of the International Joint Conference on Artificial Intelligence* contains many important papers: to date, the IJCAI has been held twice, in 1969 and 1971, and the proceedings of each conference have been published. Papers on artificial intelligence may also be found in the *Journal of the Association for Computing Machinery* (JACM), the *Communications of the Association for Computing Machinery* (CACM), and the *Proceedings* of the Spring and Fall Joint Computer Conferences (SJCC and FJCC) of the American Federation of Information Processing Societies (AFIPS). Finally, a series of volumes entitled *Machine Intelligence* include many important papers. Information about these books and journals is provided in the Bibliography.

**1–2.** This text uses the phrases "human intelligence" and "intelligence on a human level" somewhat loosely, without really attempting to define the word "human." In other books it is sometimes used as though it might apply only to the species *homo sapiens;* at other times it is used as though it might apply to other animals. How "human" is an ant, a cat, a dog, a dolphin? If the author were asked to venture an opinion, he would probably say that the word "human" refers to a kind of relationship that can exist in the interaction of intelligent beings. This relationship helps determine their behavior toward each other, toward other beings and objects, and (perhaps necessarily) toward themselves. Cats and dogs often participate in this relationship, and so are partly "human." Dolphins may consider themselves to be very "human," as may any creatures from outer space that we might someday happen to meet, and, conceivably, it may become conventional to think of some *machines* as "human." (See the Exercises for this chapter; also see Chapter 9.)

**1–3.** The area of research that attempts to simulate the underlying processes involved in natural intelligence is known as *simulation of cognitive processes.* (See various entries of *Computers and Thought* in the Bibliography, cited as CT, for some introductory and early papers.) The coverage in this book is, again, primarily concerned with the extent to which machines can simulate the abilities of natural intelligence; only secondarily is the simulation of cognitive processes considered. However, it should be pointed out that some AI researchers view their work as being directed toward both goals—the subjects are certainly not mutually exclusive. Also, for the sake of exposition, we shall occasionally describe the processes used by intelligent machines in "personalistic" or "mentalistic" terminology, as though they were really similar to the cognitive processes used by people (or more exactly, as though they were similar to the cognitive processes that people often *describe* as being the ones they use: "I just had an idea," "My model didn't include that," "That was my concept also," "I've got a plan." See the discussion in note 7–1.

**1–4.** Turing's test is discussed in greater detail in the paper by Colby, Weber, Hilf, and Kraemer (1971).

## EXERCISES

*1–1.* Read Descartes and see if you can determine whether he thought machines could reproduce themselves.

*1–2.* Two other introspective philosophers were Montaigne and Pascal. What do you think their attitudes would have been toward artificial intelligence? How about Jefferson, Marx, Archimedes, and Einstein?

*1–3.* What do you think intelligence is?

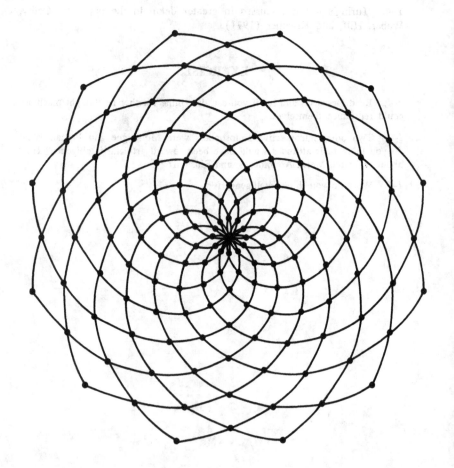

The Archimedean sunflower.

# 2

# MATHEMATICS, PHENOMENA, MACHINES

## INTRODUCTION

This chapter investigates in detail some of the mathematical background applicable to artificial intelligence. (The reader who wishes to commence the study of artificial intelligence research itself should turn to Chapter 3.) It presents a somewhat condensed discussion of automata theory, the branch of mathematics dealing with the nature of machines, since the way in which mathematics can be used to describe the operation of machines is essentially the way it can be used to describe natural phenomena in general. Thus, automata theory is a foundation for artificial intelligence (AI) research. It helps define the generality of a study that relies on computer programs to describe the phenomenon of intelligence.

In addition to discussing machines, the nature of mathematics itself will be discussed, with reference to the question, "Are there some things mathematics cannot describe completely?" It is argued in an informal, yet mathematical way that the answer is yes. There are limitations in the method of artificial intelligence research because it is based (as is all science) on mathematics and the capacities of mathematical descriptions. These limitations say nothing definite about whether AI research will succeed, only that it might not. The final discussion considers some very specific limits to the computational abilities of machines.

## ON MATHEMATICAL DESCRIPTION

A mathematical description of something consists of a finite set of statements (axioms) that utilize a finite set of undefined terms, together with a finite set of rules that govern the derivation of new statements from the axioms and from previously derived statements. Such a collection of statements is called a mathematical *system,* or *theory,* and the concept is that any statement, either given or derivable, is a true statement concerning the thing described by the theory. A mathematical theory may thus enable one to use a finite number of statements to describe something about which an infinite number of statements (those derivable under the theory) are true.

For example, the mathematical theory of Euclidean geometry gives us certain axioms or postulates concerning the undefined concepts of "point," "line," "plane," "between," etc.; the "thing" described by this theory is a "geometry," consisting of interrelationships existing among lines, points, planes, circles, spaces, etc.

The ingredients of a mathematical theory, then, are the following:

1. A set of basic words (e.g., "point," "line," "between," "distance," "x," "y," "not," "implies," "for all,") that refer to different objects, relations between objects, variables, logical connectives, quantifiers, and so on. These are the *undefined words* or *symbols* of the theory.
2. A set of basic sentences made of these basic words. These basic sentences are the *axioms* or *postulates* of the theory.
3. A set of logical rules, also made of these basic words, that tells us how to derive new sentences from the ones we are given.

Now, it is the essence of mathematical theories (note 2–1) that each of these sets be *finite;* the object described by the theory may be infinite, but the theory that describes it must be finite. In other words, the fact that there is a mathematical way of describing some object means that it is *finitely describable.*

This does not imply the converse, that if a thing is finitely describable it is therefore mathematically describable. It would take us too far afield, however, to consider this converse proposition (known as Church's thesis, or Turing's thesis) in detail (note 2–2). Since our interest is in mathematics and science, henceforth consider the phrases "finitely describable" and "mathematically describable" to be synonymous.

A mathematical description of something is thus a possibly infinite yet finitely describable set of sentences, each of which states something about the thing being described. If the thing (note 2–3) is infinite and yet finitely describable, then, intuitively, there are "patterns" which hold throughout the thing, and these patterns form the basis of our mathematical description. Thus, the Frontispiece figure to this chapter shows a collection of dots which could be infinitely extended. The entire collection, so extended, would be an "infinite thing." Yet the entire collection can be finitely described (see Exercise 2–1) because a "pattern" exists in the placement of the dots.

However, the simple existence of patterns in something does not guarantee that the thing is finitely describable: There may be an infinite number of patterns, none of which can be predicted from the others, each pattern adding its own infinite set of parts ("dots") to the thing.

So there are three possibilities that may hold if we are asked to describe something in a mathematical way: The thing may be finite, in which case presumably it is finitely describable (note 2–4); the thing may be infinite and yet finitely describable; the thing may be infinite and not finitely describable.

If the third possibility is the one that actually holds, then in fact we shall never be able to describe completely all of the thing in question. Rather we shall always be making discoveries like "there's another dot my description doesn't predict," or (perhaps) "oops, there's another subatomic particle. . . ."

As an indication (note 2–5) that there may be some things that cannot be finitely described, consider the following argument:

Assume we had a mathematical theory that would enable us to finitely describe the real numbers; that is, each sentence derivable in the theory would be a finite description of a real number, enabling the decimal expansion of that number to be computed accurately to as many places as desired. It is the nature of mathematical theories as we have described them that they may imply only a *countable* number of statements. But the real numbers are an uncountable set. Thus, no mathematical theory could enable us to derive a finite description for each real number; there must always be some real numbers that are not finitely describable.

All this explanation is by way of describing our notion of mathematical description. A good example of the usefulness of this type of description is the scientific method itself.

The scientific method is basically a way of selecting mathematical descriptions of the universe. To use the method, one develops several different mathematical descriptions of the known universe or of some part of the known universe (some set of "phenomena" in the universe; see the next section): To each of these descriptions there is a corresponding set of predictions that it makes about the rest of the universe; one rejects those descriptions that can be found by experiment to make false predictions or which make the same predictions as do other "less complicated" descriptions.

The scientific method has had many successes and therefore the usefulness of making and studying mathematical descriptions of things is well founded. Still, whenever one is called upon to consider a previously unstudied phenomenon, one *cannot* be entirely sure that it can be explained by the predictions of one's current mathematical descriptions of the universe. The reason for this is simple: There is no proof (note 2–6) that the universe is either a finite or an infinite thing. If one assumes it to be an infinite thing, one can never be sure in a finite amount of time whether mathematical descriptions have been developed to account for all the patterns that hold throughout it.

With this in mind, a person who is concerned with developing mathematical descriptions of the real world should understand that he might be engaged in an endless undertaking. It could be the case that there are an infinite number of phenomena in the universe, none of which can be predicted from a knowledge of other phenomena in the universe. It could even be the case that some phenomena in the universe are themselves not finitely describable.

On the other hand, it could be true that the universe *is* finite, or at least finitely describable.

What this has to say for our study of intelligence is simply that our success is not guaranteed. Current scientific theories do not all describe the universe as being finite. The caveat concerning the possible existence of undescribable phenomena must be heeded: There is no scientific guarantee that natural intelligence can be finitely described, either by our current scientific theories or by any mathematical description that could ever be developed—it may simply not be finitely describable.

For this reason we should take care to refer to the field we are studying as *"artificial* intelligence research." As will be seen in subsequent sections, the notion of "machine" corresponds to that of a "finitely describable phenomenon." Since it is an open question whether natural intelligence is a phenomenon that can be finitely described, we expect that "intelligent" machines will simulate some of the abilities of natural intelligence, but whether they will have them all remains unknown. Certainly, the evidence available suggests that intelligent machines will eventually have many abilities that are currently limited to natural intelligence.

# THE MATHEMATICAL DESCRIPTION OF PHENOMENA

## Time

With all the preceding conjectures in mind, let us see how it is that mathematics can be used to describe "phenomena," or "processes"; that is, things that happen in reality.

First of all, let us list names for some phenomena that are generally believed to exist. (See Exhibit A.) These are things people often talk about in the belief that they happen in the real world. Not all are necessarily things that can be described mathematically.

### EXHIBIT A

the playing of a game    chemical reactions    the evolution of species    thought processes    nuclear reactions    a person feeling emotion    waves traveling through a medium    cellular growth of organisms    crystal formation    sexual reproduction    a candle burning    a person living    a person dying    a stone falling to the ground    a bird flying    the motion of a weight on a spring    the formation of public opinion    conversion of energy from one form to another    dreaming    flipping a coin    the operation of a computer program    weather

To a mathematician looking at Exhibit A, perhaps the most immediate thing he would find common to all its elements is that each element involves "time"; each of these things may be said to happen as a "sequence of situations."[1]

Upon further inspection, the mathematician would discover that each of the phenomena named in Exhibit A can happen in a variety of ways. For some phenomena the variety is greater than for others. In

---

[1] An interesting and difficult open question is whether the automata-theoretic description of phenomena presented throughout this book is in conflict with relativistic findings concerning simultaneity. The reader interested in pursuing this question is invited to see Waksman (1966).

his desire to be general, he would say that the name of a phenomenon refers to the set of different ways in which it can occur.

A third thing the mathematician might note about Exhibit A is that it is possible for some phenomena to be made up of others; this chapter overlooks ways of describing this mathematically,[2] though as an example, it might be noted that "cellular growth of organisms" seems to be made up of "chemical reactions."

These observations are the essence of the mathematical approach to the description of phenomena. Mathematically, an *occurrence* of a phenomenon is viewed as a sequence of situations, and the phenomenon itself is viewed as being the set of all possible ways it can occur. A phenomenon is described by a mathematical theory of all ways in which it can occur; such a theory might describe it as either being made up of, or a part of, other (describable) phenomena.

The first ingredient in the mathematical description of a phenomenon is the specification of a time scale $T$ and of a set $X$ of all possible situations. We may take $T$ to be some subset of the real number line; for the moment we can leave $X$ unspecified. If $X$ is the set of all possible situations, then an *occurrence* $\theta$ is a function that associates to some of the elements $t$ of $T$ unique corresponding elements $\theta(t)$ from $X$. A phenomenon is a set of such functions $\{\theta_1, \theta_2, \cdot \cdot \cdot\}$, each representing an occurrence. A complete description of a phenomenon is, then, a description of its possibly infinite set of occurrences: The assumption that one can find a mathematical description for a given phenomenon is equivalent to the assumption that its set of occurrence functions is finitely describable. Since some occurrences of a given phenomenon might conceivably possess an infinite number of "details" (say, in the number of times at which situations are defined, or in the number of "true statements" about any particular situation), we may accept as a finite description any finite rule that allows us to compute these occurrence functions to an arbitrary accuracy. That is to say, we accept descriptions that are "effectively" true.[3]

---

[2] For this reason, although the theory of phenomena outlined in this chapter is adequate (to illustrate the limitations and generality of the mathematical approach), it is not an especially efficient way of describing anything other than very simple phenomena. Many approaches have been made toward developing a more efficient way of describing complex, "real-world" phenomena: Chapter 8 discusses briefly the possible formalizations for "parallel processes" and "hierarchical systems"; in Chapter 3 there is a brief discussion of logical systems for describing real-world situations and their interrelationships (causality, etc.).

[3] Thus, we shall ignore descriptions that describe "strictly noncomputable functions" (see Exercise 2–3), but we can accept descriptions that describe undecidable systems.

EXAMPLE 2–1. MOTION OF A WEIGHT ON A SPRING. In the simplest case of this example, where the motion of the weight is entirely vertical, one can describe any possible situation by a single real number, representing how far the spring is extended or compressed from its rest position. Thus, the set $X$ of all possible situations is described by the real-number line. Which particular occurrence of the phenomenon happens is dependent on such things as the mass of the weight, the damping factor of the spring, the initial position of the spring and weight, the spring constant $k$, etc.; the graph of any given occurrence function will generally look like Fig. 2–1. The phenomenon, or class of all possible occurrence functions, can be described by a single equation whose variables represent the factors given above.

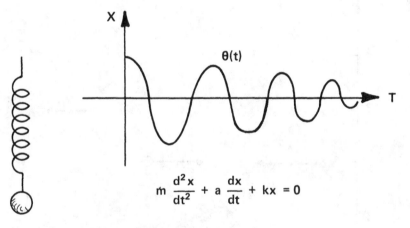

$$m \frac{d^2 x}{dt^2} + a \frac{dx}{dt} + kx = 0$$

Figure 2–1. An old friend to the physics student.

The considerations presented in the preceding section, on mathematical descriptions in general, still hold for the specific case of mathematical descriptions of phenomena. There may be phenomena that are not finitely describable. On the other hand, given only finite samples of the occurrences of a (possibly infinite) phenomenon, there is no way to prove that the phenomenon is not finitely describable—the most one can prove about it is that one's efforts to describe it have so far been unsuccessful. (Of course one might also prove that one's efforts to describe it have "so far" been successful.)

## Types of Phenomena

"Things that happen" may often be distinguished from each other by the nature of their occurrence functions. One[4] of the basic classifications defines three types of phenomena: *discrete, nondiscrete,* and *continuous.*

A phenomenon is *discrete* iff[5] each of its occurrences is a *step function.* A function $\theta$ is a step function iff it is constant or undefined throughout any closed interval $[t,t']$ except for a finite number of "jump discontinuities." Specifically, let $[t,t']$ be any closed interval of $T$ (possibly a point, if $t = t'$): Then there exist a finite number of points $t_k$, such that

$$t \leqq t_1 < t_2 < \cdots < t_n \leqq t'$$

and $\theta$ is either constant or undefined on each open subinterval $(t_{k-1}, t_k)$. Figure 2–2 gives an example of a step function.

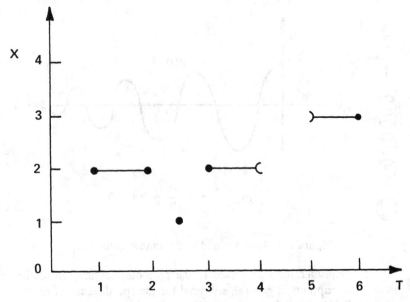

Figure 2–2. An occurrence of a discrete phenomenon.

---

[4] Some other classifications are determinacy versus nondeterminacy, periodicity versus nonperiodicity, etc. It should be kept in mind that these classifications are really being applied to descriptions of phenomena, not to phenomena themselves: For example, certain phenomena (electrons) can be described as being either discrete (particles) or nondiscrete (waves).

[5] "Iff" denotes "if and only if."

An equivalent definition of discrete phenomena is the following: Within an occurrence $\theta$ define an *event* to be an interval of time (closed or open or semiclosed) on which $\theta$ is constant. Then a discrete phenomenon is one such that each of its occurrences can be represented as a sequence of events, in which any event is either "terminal" or "next to" another event. An event is said to be terminal if no event follows it in time. One event is said to be next to another iff no event occurs between them in time.

A phenomenon is *nondiscrete* iff it is not discrete. Thus, a nondiscrete phenomenon has at least one occurrence in which there is a situation that is followed as closely in time as one chooses to look by mutually different situations.

A phenomenon is *continuous* iff it is nondiscrete, and for any occurrence (and for all $t$, $t'$) the difference between the situation that happens at time $t$ and the situation that happens at $t'$ tends to zero as the difference between $t$ and $t'$ tends to zero. This definition, of course, is meaningful only in cases where it is possible to establish a definition of "difference" that can be applied to the possible situations.

Throughout this book we shall primarily discuss discrete phenomena. Our reason for this is that by choosing the time intervals between situations to be suitably small, one can find occurrences of a discrete phenomenon that will match, to an arbitrary closeness, the occurrences of any nondiscrete phenomenon. Consequently, if there exists a finite description for a nondiscrete phenomenon, then there also exists a finite description for a discrete phenomenon that approximates it as closely as one wishes. (We can merely use the nondiscrete description to calculate the values during the appropriate discrete events.)

Let $A = \{\theta_1^A, \theta_2^A, \cdots\}$ be a nondiscrete phenomenon, and $B = \{\theta_1^B, \theta_2^B \cdots\}$ be a discrete phenomenon. If $A$ is continuous, then $B$ *matches* $A$ to a closeness $\delta$ if there exists a number $\epsilon$ greater than zero such that, for all $t$, $t'$, if $|t - t'| < \epsilon$, then

$$|\theta_i^B(t) - \theta_i^A(t')| < \delta$$

If $A$ is noncontinuous, then $B$ matches $A$ to a closeness $\delta$ if for all $t$ such that

$$\theta_i^B(t) \neq \theta_i^A(t)$$

there exists an $\epsilon$ such that $|\epsilon| < \delta$ and

$$\theta_i^B(t + \epsilon) = \theta_i^A(t + \epsilon)$$

It is always possible to find a discrete phenomenon that will match a given, finitely describable, nondiscrete phenomenon. Similarly, if $A$ is a discrete phenomenon, then $B$ *simulates* $A$, for all $i$ and for all $t$, if $\theta_i^A(t)$ being defined implies that $\theta_i^B(t) = \theta_i^A(t)$. If $B$ simulates $A$, then the occurrences of $A$ are reproduced exactly within the occurrences of $B$; an occurrence of $B$ may, however, contain situations that do not happen within the corresponding occurrence of $A$. If $B$ matches $A$, then the occurrences of $B$ reproduce those of $A$ in an approximate sense[6]; thus, if $B$ matches $A$, we shall also say that $B$ "simulates" $A$, approximately.

We shall see below that it is possible to construct a tool—a universal digital computer—that can reproduce exactly the occurrences of any mathematically describable discrete phenomenon. By suitably programming a fast enough digital computer, one can simulate any finitely describable phenomenon, regardless of whether it is discrete or nondiscrete or continuous. If intelligence is a finitely describable phenomenon, then it can theoretically be simulated on a (fast enough, big enough) computer.

## Discrete Phenomena

The preceding section gave a definition for discrete phenomena. The fact that there is a sense in which one can approximate any nondiscrete phenomenon to an arbitrary degree, using discrete phenomena, gives sufficient reason to investigate the subject of finite (mathematical) descriptions for discrete phenomena. What we desire is some way of characterizing all such descriptions. We shall see that this characterization is provided by automata theory.

In this respect the main thing to note is that we can describe any step function by a *string*, or sequence of symbols, provided we adopt an appropriate notation. Let us see how this could be done, using Fig. 2–3 as an example.

---

[6] These definitions describe "real-time" matching and simulation: They can be broadened to include notions of relative speed. Also, the equality sign can be taken to mean something like "is isomorphic to." However, it should be noted that even though a discrete phenomenon $B$ may match a nondiscrete phenomenon $A$, in general (and even if A is continuous) the set

$$\{t \mid \theta_i^B(t) = \theta_i^A(t)\}$$

will be of measure zero. Moreover, if $A$ is noncontinuous, then the set of points (in time) at which $B$ will be "close" (within a given $\delta$) to $A$ will in general be of measure zero; an expression that describes this set of points is

$$\{t \mid \exists \epsilon (|t' - t| < \epsilon \Rightarrow |\theta_i^A(t') - \theta_i^B(t)| < \delta)\}$$

Thus, the ability of discrete phenomena (machines) to "simulate" arbitrary nondiscrete, noncontinuous phenomena is somewhat limited.

In Fig. 2–3, the first event (note 2–7) is the happening of situation 2, which starts at $t = 1$ and ends at $t = 2$. Thus, we make the beginning of our descriptive string

2,1+,2+

The next event is the happening of situation 1, which occurs "during the instant" $t = 2.5$. The descriptive string now becomes

2,1+,2+,1,2.5+,2.5+

And so we continue, using minus signs whenever an event starts "immediately after" (note 2–8) some time $t$ or ends "immediately before." The final descriptive string would be

2,1+,2+,1,2.5+,2.5+,2,3+,4−,3,5−,6+

In general, any step function can be represented by such a descriptive string. If the function is defined only on a bounded time interval, then its descriptive string will be of finite length, even though the total number of points for which the function is defined may be infinite; for example, the descriptive string for Fig. 2–3 is finite, although the step function is defined for an infinite number of values of $t$ (note 2–9). Likewise, if the step function does not have a beginning or does not have a terminal event, then its descriptive string will be infinitely long.

Since any step function can be represented by a descriptive string, any set of step functions can be represented by a set of descriptive strings. Thus, to finitely describe the occurrences of a discrete phenomenon, one need only be able to finitely describe a certain set of strings: If the set is finite, we could simply list all its strings[7] (provided none of them is infinite), but what if the set of descriptive strings is infinite?

The answer to this problem lies in the following analysis: Even though the set is infinite, we can assign a number $1, 2, \cdots$ to each string in the set and proceed to talk of the first descriptive string, the second descriptive string, and so on (note 2–10). Then, if we can find a finitely describable rule that computes for each $n$ the $n$th descriptive string, we will in effect have found a finite description for the phenomenon. Thus, we can transfer our efforts from the finite description of discrete phenomena to the finite description of *functions*. Any discrete phenomenon is capable of being represented as a function that associates a unique descriptive string to each natural number. And, since any natural number can be represented by a string (of finite length), we are therefore concerned with finding finite descriptions of functions that map one set

---

[7] In practice, there are limits to the size of finite sets that can be enumerated (see note 2–4 and Chapter 3). Such sets are called "finite, effectively infinite."

of strings (those representing natural numbers) into another set of strings (those representing step functions).

The mathematical theory that deals with functions that map one set of strings into another set of strings is *automata theory;* a general way of characterizing functions of this sort is through the use of *Turing machines.* Automata theory is basically concerned with studies on the nature of Turing machines, its underlying hypothesis[8] being that this is the nature of all discrete machines; the *concept of machine is to be identified with that of "finitely describable phenomenon."* In this chapter we are concerned with some of the simplest types of machines. Automata theory discusses the abstract nature of machines, but it can include such aspects of real-world machines as their cost and probability of error.

Briefly, a Turing machine is composed of a finitely describable black box and an infinite, or potentially infinite,[9] tape (Fig. 2–3). The

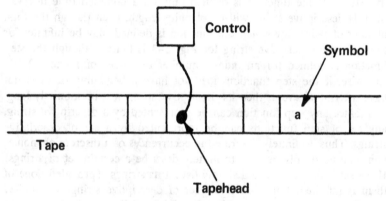

Figure 2–3. A Turing machine.

tape is divided into squares, each of which has a symbol (possibly the "blank" symbol) printed on it. The black box contains two subcomponents, a control and a tapehead; the tapehead is capable of scanning and writing symbols on one square of tape at a time, and of moving the tape either to the right or the left, all under instructions given to it by the control. The tapehead sends to the control the information as to what symbol it happens to be scanning, and the control decides on the basis of that information and a finite "memory" what actions it should instruct the tapehead to perform.

---

[8] Again, Church's thesis or Turing's thesis.

[9] By *potentially infinite* is meant that there is someone nearby ready to add more squares to the tape if necessary.

Although this may seem like a very simple type of machine with very limited capabilities, such is not the case. In fact, all evidence available to date indicates that Turing machines are capable of computing any finitely describable, computable function that maps one set of strings into another set of strings. There exist certain Turing machines which, given a suitable program, are capable of simulating the computations of any Turing machine. It can be shown that a Turing machine can effectively derive all provable theorems in any given mathematical theory. Indeed, Turing machines are capable of simulating[10] the phenomenon of self-reproduction. Therefore the rest of this chapter is devoted to a discussion of some results from automata theory.

# Finite-State Machines

Of all the elements of a Turing machine, the only one that requires mathematical formalization is the control: We need to specify more exactly how it is able to make decisions, what its memory is, etc. We now give a general definition of that class of machine which may serve as a control in a Turing machine; the machines in this class are usually referred to as finite-state automata.

> DEFINITION 2–1. A *finite-state machine,* or finite automaton, is a quintuple, $M = (Q,X,Y,\delta,\lambda)$, where:
> $Q$ is a finite set, the set of *states;*
> $X$ is a finite set, the set of *input* symbols;
> $Y$ is a finite set, the set of *output* symbols;
> $\delta: Q \times X \to Q$, the *next state* function;
> $\lambda: Q \times X \to Y$, the *next output* function.

Any quintuple of sets and functions satisfying this definition is to be interpreted as the mathematical description of a machine that, if given an input symbol $x$ while it is in state $q$, will output the symbol $\lambda(q,x)$ and go to state $\delta(q,x)$. (The two functions $\delta$ and $\lambda$ together are often referred to as the *transition function* of the finite state machine.)

Thus, a finite automaton is a machine that can exist in a finite set of states, where the particular state it is in at any given moment depends upon the inputs it has received and upon its previous states. The set of states in an automaton serves as its "memory": The only information that an automaton has concerning its past operation is the current state it is in; at least, this is the only information it can use in deciding its

---

[10] Discreetly . . ." (See Chapter 8 for a discussion of self-reproducing machines.)

next state and its next output when it is given an input symbol. Some examples would be instructive at this point.

EXAMPLE 2–2. A PARITY-CHECKING MACHINE. This machine has only two states; the machine will accept any finite string of zeros and ones; its output at a given moment will be the word "even" if the string it has so far received has an even number of ones, and "odd" otherwise, provided it starts in the "initial state" $q_0$. Let $Q = \{q_0, q_1\}$, $X = \{0, 1\}$, $Y = \{$"even," "odd"$\}$, and define $\delta$ and $\lambda$ by the following tables:

| $\delta$ | $q_0$ | $q_1$ |
|---|---|---|
| 0 | $q_0$ | $q_1$ |
| 1 | $q_1$ | $q_0$ |

| $\lambda$ | $q_0$ | $q_1$ |
|---|---|---|
| 0 | "even" | "odd" |
| 1 | "odd" | "even" |

For example, $\delta(q_0, 1) = q_1$; $\lambda(q_0, 1) =$ "odd".

The reader should verify for himself that this machine does what it is supposed to do, provided it is started in state $q_0$.

Actually, the use of tables to define the functions $\delta$ and $\lambda$ is rather clumsy and inefficient; if we were dealing with larger, more complicated machines, it would be very difficult to understand just what they were doing. It is customary to use a certain type of drawing, called a *state-transition diagram,* to describe a finite automaton. Figure 2–4 gives a state-transition diagram for the parity checker.

In such a diagram each state is represented by a *circle;* each tran-

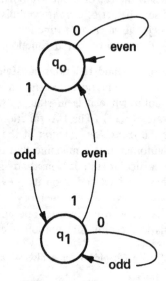

Figure 2–4. Parity checker.

sition between states is represented by an *arrow;* the input symbol causing the transition appears at the tail of the arrow, while the corresponding output symbol is inserted in the middle of the arrow.

Another good example of a finite automaton is a machine that adds
two binary numbers, provided they are suitably encoded into a string.

EXAMPLE 2–3. A BINARY ADDER. Let $Q = \{$"nocarry," "carry"$\}$,
$X = \{00,01,10,11\}$, $Y = \{0,1\}$, and let the functions $\delta$ and $\lambda$
be given by the state-transition diagram in Fig. 2–5. To add two
binary numbers, say 1101 and 10101 (decimal 13 and 21,
respectively), we first reverse them so that they are expressed
with their least significant digits first: 1011 and 10101. Next
we add sufficient zeros to them to make both strings be of the
same length and end in zero: 101100 and 101010. Finally, we
encode the two strings into a single string, whose symbols come
from the set $X$, by taking the first symbols of each string and
replacing them by their corresponding ordered pair, taking the
second symbols and doing the same, and so on. The string we
obtain is

   11,00,11,10,01,00

If we feed this string into the binary adder, then the sequence
of outputs that we get is

   0 1 0 0 0 1

This is the reverse of the binary number 100010 = thirty-four.

These two examples illustrate that finite-state machines do have
some computational ability and that they can be used in at least two
slightly different ways. The first example shows that it is possible to use
an automaton as an *acceptor* for a certain set of strings: If we replaced
its output set by "true" and "false," respectively, then the parity checker

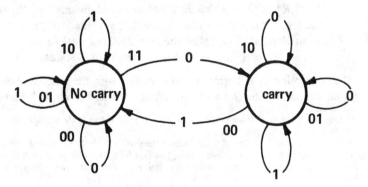

Figure 2–5. Binary adder.

would output "true" after the input of any finite string that contained an even number of 1's; that is, it would *accept* the set of all such strings.[11] The binary adder, on the other hand, illustrates that we can use finite-state machines to represent some of the functions that map one set of strings into another set of strings.

However, there are many functions that no finite-state machine can compute: One such function is multiplication. The reader might try his hand at designing a finite-state machine to multiply any two numbers. The basic reason it cannot be done is that the operation requires saving the complete information about each of the two numbers, and this requires either an infinite number of states or an infinite tape.

In fact, finite-state machines are only the building blocks of automata theory; they represent the simplest type of machine, one in which the future of an occurrence can depend on only a finite number of different "past histories," or states.

# Turing Machines

## Simple Turing Machines

Let us now return to our original discussion of Turing machines. The reader will recall that these were described as the most general type of discrete machine; so far as anyone knows, any function that can be computed can be computed by a suitable Turing machine.

> DEFINITION 2–2. A *Turing machine* (Tm) is an ordered quintuple, $T = (Q, X_b, P, q_0, F)$, where:
>
> $Q$ is a finite set of *states;*
>
> $X_b$ is a finite set of *tape symbols,* one of which is the *blank symbol b;*
>
> $P$ is the *next-move* function, a mapping from $Q \times X_b$ to $X \times \{L, O, R\} \times Q$ in which $L$, $O$, $R$ are symbols meaning "go to the left," "stay at the same place," "go to the right";
>
> $q_0$ is an element of $Q$, called the *start state,* or *initial state;*
>
> $F$ is a subset of $Q$ and is called the set of *final states.*

The operation of a Turing machine begins with the machine being in $q_0$ and examining the leftmost symbol of a string[12] from $X_b$* that is

---

[11] This acceptor is also a decider; that is, it rejects those finite strings not belonging to the set it accepts.

[12] If $A$ is a set, then by $A$* we denote the set of all finite strings whose symbols are elements of $A$. Thus, $\{a,b\}* = \{\epsilon, a, b, ab, ba, aa, bb, aaa, bbb, aba, bab, aab, baa, \ldots\}$, where $\epsilon$ denotes the *empty string,* which does not contain any symbols.

printed on some of the squares of its tape (every other square of the tape contains a blank symbol). The next-move function $P$ determines what symbol the tapehead prints on the square it is examining, whether the tapehead moves left or right one square or remains at the same square, and what state becomes the new state of the control.

The next-move function $P$ can be finitely described, and there is no difficulty in considering the control of the Turing machine to be a finite-state machine. Thus, the only essential difference between Turing machines and finite-state automata lies in the fact that a Turing machine is able to store its output on a potentially infinite tape and refer to it later. This single difference (note 2–11) is enough to enable Turing machines to be used as acceptors for a class of sets much larger than that of those accepted by finite-state machines, and it is enough for Turing machines to be able to compute a class of functions much larger than that of those which can be computed by finite-state machines. The sets that can be recognized (i.e., accepted) by Turing machines are the *recursively enumerable* sets; the functions that are computable by Turing machines (henceforth Tm-computable) are the *partial-recursive* functions.

Another way of stating Church's thesis or Turing's thesis is to say: *Any computable function can be represented as a partial-recursive function.* So far, every general definition of "finitely describable, computable functions that map one set of strings into another set of strings" has been shown equivalent to the definition for Turing machines.

EXAMPLE 2–4. A UNARY DOUBLER. We define a simple Turing machine that will produce a string of 1's of length $2m$ if it is given an input tape containing a string of 1's of length $m$. This is a computation that cannot be done (for all $m$) by any finite-state machine. Let

$$Q = \{q_0, q_1, q_2, q_3, q_4\}$$
$$X_b = \{b, 0, 1, A\}$$
$$F = \{q_0\}$$

and let the next-move function $P$ be defined by Table 2–1. If this machine is started on a string of the form

$$\ldots bb111 \ldots 11bbb \ldots$$

$$\underbrace{\qquad\qquad\qquad}$$

$$m \text{ 1's}$$

TABLE 2–1.   A Unary Doubler

| $Q$ | $\times$ | $X_b$ | $\rightarrow$ | $X_b$ | $\times$ | $\{L, O, R\}$ | $\times$ | $Q$ |
|---|---|---|---|---|---|---|---|---|
| $q_0$ | | 1 | | 0 | | $R$ | | $q_1$ |
| $q_1$ | | 1 | | 1 | | $R$ | | $q_1$ |
| $q_1$ | | 0 | | 0 | | $R$ | | $q_1$ |
| $q_1$ | | $b$ | | $A$ | | $R$ | | $q_2$ |
| $q_1$ | | $A$ | | $A$ | | $R$ | | $q_2$ |
| $q_2$ | | 1 | | 1 | | $R$ | | $q_2$ |
| $q_2$ | | $b$ | | 1 | | $R$ | | $q_3$ |
| $q_3$ | | $b$ | | 1 | | $L$ | | $q_4$ |
| $q_4$ | | 1 | | 1 | | $L$ | | $q_4$ |
| $q_4$ | | $A$ | | $A$ | | $L$ | | $q_0$ |
| $q_0$ | | 0 | | 0 | | $L$ | | $q_0$ |
| $q_0$ | | $b$ | | $A$ | | $0$ | | $q_0$ |
| $q_0$ | | $A$ | | $A$ | | $0$ | | $q_0$ |

at the leftmost 1, it will eventually halt in state $q_0$ (here the initial state is also the halting state), and its tape will hold a string of the form

$$\ldots bbA000\ldots00A\underbrace{1111\ldots11}_{m \text{ 0's}}\underbrace{111bbb}_{2m \text{ 1's}}\ldots$$

TABLE 2–2.   Operation of the
Unary Doubler

1.  (0) . . . $bb\overset{\downarrow}{1}11bb$ . . .

2.  (1) . . . $bb0\overset{\downarrow}{1}1bb$ . . .

3.  (1) . . . $bb01\overset{\downarrow}{1}bb$ . . .

4.  (1) . . . $bb011\overset{\downarrow}{b}bb$ . . .

5.  (2) . . . $bb011A\overset{\downarrow}{b}b$ . . .

6.  (3) . . . $bb011A1\overset{\downarrow}{b}bb$ . . .

7.  (4) . . . $bb011A1\overset{\downarrow}{1}bb$ . . .

8.  (4) . . . $bb011\overset{\downarrow}{A}11bb$ . . .

9.  (0) . . . $bb011\overset{\downarrow}{A}11bb$ . . .

10. (1) . . . $bb010\overset{\downarrow}{A}11bb$ . . .

11. (2) . . . $bb010A\overset{\downarrow}{1}1bb$ . . .

12. (2) . . . $bb010A1\overset{\downarrow}{1}bb$ . . .

13. (2) . . . $bb010A11\overset{\downarrow}{b}bb$ . . .

14. (3) . . . $bb010A111\overset{\downarrow}{b}bb$ . . .

15. (4) . . . $bb010A111\overset{\downarrow}{1}bb$ . . .

.
.
.

Table 2–2 shows the first 15 steps of the operation of the unary doubler. The eighth entry in the table, for example, is

8.  (4) ... $bb011\overset{\downarrow}{A}11bb$ ...

which means that in this step the machine is in state $q_4$, scanning the $A$ in the string $011A11$, with an otherwise blank tape. The reader should be able to continue the table and verify that the machine will reach a step

(0) ... $bbA000A111111bb$ ...

and that no further changes will take place on the tape. (It is a simple matter to add extra states that get rid of the output zeros.)

The main thing to be learned from Example 2–4 is that a Turing machine typically manages to surpass the limitations of the finite automaton by using "dummy symbols" to store on its tape information about its past operation. In the example, the dummy symbols are 0 and $A$, where 0 serves to store the information that a certain unit has already been doubled, and the appearance of two $A$'s on the tape represents the information (for us) that the machine has finished its computation. The tape of a Turing machine is thus a very significant part of its memory.

## Polycephalic Turing Machines

The Turing machine concept described above is very cumbersome for use on any but the simplest problems. It is more common to consider "polycephalic" Turing machines, which possess several ($n$-dimensional) tapes, each with its own finite number of tapeheads (Fig. 2–6); this model comes closer to the actual structure of modern computers. The formalization for polycephalic machines is relatively easy to construct. The relevant things to consider are:

1. The number of tapes the machine uses, say $n$.
2. The dimensionality of each tape. We can let each tape be an $m$-dimensional grid ($m$ is variable), and have each tape square be specified by an $m$-tuple of integers (e.g., $<2,0,-5,1,-3,6>$). The dimensionality of the $i$th tape can be denoted by a function $\delta(i)$.
3. The (finite) alphabet $X^i$ used on each tape.
4. The number $r$ of tapeheads used by the machine on each tape. These can be denoted $T_1^i, T_2^i, \ldots, T_j^i, \ldots, T_r^i$ for the $i$th tape.

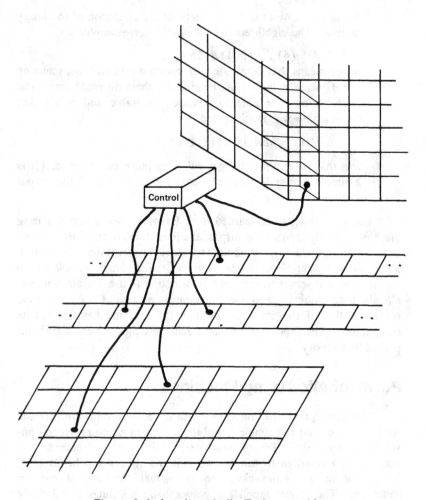

Figure 2–6. A polycephalic Turing machine.

5. What to do if two or more tapeheads are instructed to print on the same square at the same time. For each $i$, we can use a "dominance relation" $R^i$ that determines a unique "greatest" tapehead $T_j^i$ for any given set of tapeheads $\{T^{ki}\}$.

6. The set of states $Q$ for the control; also its initial state and its set of final states $F$.

7. The next-move function $P$, which for each tapehead $T_j^i$ maps $Q \times X^i$ into $X^i \times D^i \times Q$. We let $D^i$ denote the set of unit direction vectors for the $i$th tape; e.g., $<1,0,-1,1,01,0>$ is

a unit direction vector for a six-dimensional tape. We can assume that the tapeheads for each tape all start at the origin $<0,0, \ldots,0>$ of the tape.

The specification of (1) through (7) above then determines an individual polycephalic Turing machine.

The only advantage of polycephalic Turing machines over simple Turing machines is that they are more efficient to use: They are not more general with respect to the number or the nature of their uses. Any function that can be computed by a polychephalic Tm can also be computed by a suitable, ordinary Tm.

## Universal Turing Machines

One of the most surprising and important facts is that some Turing machines are capable of simulating the computations of any Turing machine. These machines are called "universal Turing machines"; the actual reason for their existence lies in two facts:

1. Any string containing only a finite number of different symbols can be "coded" as a unary string, consisting only of the symbols 1 and the "blank" symbol $b$.

2. Any Turing machine can be described by a finite string of symbols.

To show the first fact, note that a unique string of 1's can be assigned to each symbol in a set if the set contains only a finite number of symbols. Consequently, any string consisting only of symbols from that set can be represented by a string of the form "$\ldots bb1 \ldots 1b1 \ldots 1b1 \ldots 1bb \ldots$," consisting of a variable number of blocks of 1's, each of variable length, each block separated from the next by a single $b$.

To see the second fact, examine Table 2–1 and note that the total number of symbols used in the table is finite. Thus, the table itself can be represented as a (finite) string of quintuples, each of the form $(q,x,x,d,q)$. If one assigns a suitable unary coding to each of the symbols in Table 2–1, then any quintuple can be represented uniquely by a certain string of $b$'s and 1's; thus the table can be represented by a string of $b$'s and 1's in which certain substrings stand for quintuples. The unary string representing the quintuples of a given Turing machine $T$ will be denoted by $d_T$ and called the *descriptive string for the Turing machine* (not, in general, the same thing as a descriptive string for an occurrence of a discrete phenomenon).

The actual construction of a universal Turing machine $U$ is not very difficult, and the student should either try it for himself or consult one of the references on automata theory. For our purposes here it is simpler, and equally valid, to rely on the following description (see Fig. 2–7): $U$ works with two tapes, each acted upon by a single tape-head; the first tape contains the descriptive string $d_T$ for the Turing machine $T$ that $U$ is going to simulate, while the second tape contains a unary string $i$ representing (in the same code as that of $d_T$) the input to $T$; thus, no matter how many states or symbols $T$ may use, the machine $U$ will use only the symbols $b$ and 1 and a few dummy symbols of its own.

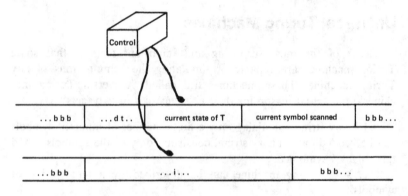

Figure 2–7. A universal Turing machine.

To simulate $T$, $U$ keeps a unary string representing the "current state" and the "current symbol scanned" of $T$ on its first tape, and it uses this information and $d_T$ to compute the corresponding actions it should take with respect to its second tape. In other words, $U$ does essentially the same thing a person does when he traces the operation of a given Turing machine on a given input tape: It merely keeps track of where $T$ is, of what state $T$ is in, and of what symbol $T$ is scanning, and it looks in a table $(d_T)$ to find out what actions $T$ would take; then it implements those actions on its own model of $T$'s tape.

A universal Turing machine, then, is one that can be "given a *program*" that enables it to simulate a Turing machine: In fact, a universal Tm is theoretically equivalent to a *general-purpose, discrete* (or *digital*) *computer,* and the program one gives a digital computer is analogous to a descriptive string $d_T$ for some Turing machine $T$. That is, a computer program is a descriptive string for a function $(T)$ that maps one set of strings (the possible inputs to $T$) into another set of strings (the possible outputs from $T$). Computers, then, are mechanisms for implementing finitely describable processes of symbol manipulation.

The fact that there are universal machines,[13] or computers, is very significant if we are investigating the behavior of machines in general. It enables us to conduct our investigations by referring to the behavior of a single machine as it is given various programs, rather than by building a new machine each time we want to observe a new behavior. In particular, it makes feasible a search for machines that simulate the abilities of intelligence. The work described in the following chapters would simply not be possible without digital computers. The reader who wishes to pursue the study of digital computers is invited to see the books by Bartee (1966), Bell and Newell (1971), Chapin, Mc-Cormick (1959), and Trakhtenbrot (1963). Papers and books relevant to the history of computers are those of Aiken (1937), Babbage (1864), Bernstein (1964), Burks et al. (1946), Bush (1945), Gardner (1958, 1970, 1971), S. Rosen (1971), Rosenberg (1969), Price (1959), Pylyshyn (1970), Shannon (1948, 1953), T. M. Smith (1970), and von Neumann (1951). The books by Arbib (1964, 1968, 1969, 1972) and Minsky (1967, 1969) are excellent introductions to the automata-theoretic nature of computers.

# LIMITS TO COMPUTATIONAL ABILITY

At this point the major purpose of this chapter has been satisfied, which was to show how it is that one can investigate finitely describable phenomena in general (and, especially, hope to simulate intelligence) by using computers.

It remains to complete the survey of those general limitations that can be placed upon the success of artificial intelligence research. We have already seen one such limitation, which is that the results of AI research must always be finitely describable: If natural intelligence is not a finitely describable phenomenon, then the best that AI research can do is to simulate some, but not all, of its abilities.

There is no scientific evidence that natural intelligence is not finitely describable—indeed, we have tried to show that there cannot be such evidence. On the other hand, there is some scientific evidence that natural intelligence is finitely describable; namely, the evidence that the brains of certain animals do "possess intelligence," plus the fact that these brains each contain finite numbers of cells. However, the evidence concerning the actual function and nature of brain cells is far from

---

[13] Not all Turing machines are universal.

final, and the exact way in which the intelligence of a brain is dependent upon its cells is still unknown (see Chapter 1). The most one can say is that the finite describability of true intelligence is likely but not proved.

Another general limitation concerning the properties of artificial intelligences can be derived: It can be shown that there are certain *unsolvable problems,* which cannot be solved by *any* machine, that is, by any finitely describable process; artificial intelligence research, then, can never produce a machine intelligent enough to solve one of these problems.

Before discussing one such unsolvable problem—the famous Halting Problem, first shown to be unsolvable by Turing—it is wise to note that there is probably no way any natural intelligence can be shown scientifically to be able to solve one of these problems. Certainly, unless Turing's thesis is false, no natural intelligence could ever give a finite description of a way to solve one of these problems.

The Halting Problem can be stated as follows: *For any Turing machine T, given any input tape i, tell whether or not T will eventually halt its computation.* By *halting* is meant that $T$ enters one of its final states $q_j \in F$, and prints a certain symbol (say, the halt symbol $H$) on a square of its tape; also, whenever $H$ occurs in a quintuple in the next-move function for $T$, the quintuple is always of the form $(q,x,H,0,q_j)$. (In particular we have $q_j,H,H,O,q_j$.) Thus, an outside observer, given a description of $T$, knows whenever he sees an $H$ appear on the tape of the Turing machine $T$ that $T$ is finished with its computation, and will do no more (significant) manipulation of its tape.

It can be shown that there is no Turing machine capable of solving the Halting Problem. That is, we can show that there does not exist a Turing machine $D$ which, given a description $d_T$ for any Turing machine $T$, and given any input tape $i$, will always compute in a finite time whether or not $T$ would eventually halt its computation if it were given the input tape $i$. There are many ways to go about proving this fact: One relatively simple way involves showing that if there were such a machine $D$, then one could use it as part of a larger machine (say, $E$) such that, given $d_E$ and an arbitrary $i$, $D$ would not be able to compute whether $E$ would ever halt. (See Minsky, 1967, for an exposition of this approach.)

Given, then, that there are problems no artificial intelligence can solve, it is natural to ask whether an artificial intelligence can be constructed so as to recognize these problems whenever they arise in the course of its operation, prove that they are unsolvable, and stop working on them. In fact, it can be shown that no Turing machine (thus, no artificial intelligence) is capable of recognizing *all* unsolvable problems.

For any mathematically describable problem-solving device there exists at least one problem that the device cannot solve, and cannot recognize to be unsolvable, provided the device is consistent (incapable of producing contradictory answers if given noncontradictory premises) and capable of doing simple arithmetic (addition and multiplication). This should not be taken to mean that if such a machine is confronted with such an unsolvable problem, it will never stop working on the problem, since the machine could easily be designed not to work on any problem past a certain time limit. Also, this limitation does not apply if the machine is allowed to be inconsistent—but, of course, with an inconsistent machine one cannot be sure that the answer the machine produces is correct. Whether this is true of human beings, whether there are problems that natural intelligence can never solve, and can never prove to be unsolvable, is an open question: It can be answered only in a scientific-mathematical way if it is shown that natural intelligence can be mathematically described—if it can be mathematically described, then problems of this sort probably exist.

These limitations on the generality of artificial intelligence, which have to do with the capacities of mathematical description and the existence of mechanically unsolvable problems, are both of a very theoretical yet vague nature. They really say nothing very concrete about the real-world capabilities of machines (or of people). We would do well, therefore, to investigate more specific limits on the computational abilities of machines. The remainder of this chapter is devoted to a discussion of the physical boundaries of the computational abilities of machines, and to establishing certain "conventions" regarding these boundaries, which are referred to (for illustrative purposes only—the boundaries are not exact) throughout the rest of this book.

To establish these conventions, note that there are three basic ways in which the description of Turing machines has, so far, been unrealistic[14]:

1. No real-world Turing machine can actually have an infinite tape, or even a truly "potentially infinite" tape; there are limits to how much "information" can be stored in a computer memory.
2. Any real-world Turing machine must conduct each of its actions (reading the tape, evaluating the next-move function, printing the tape, moving the tape) in a finite, nonzero time; there are limits to how fast a computer can operate.

---

[14] Ignore the fact that modern computers operate on a higher "level" than Turing machines (see the discussion of machine languages in Chapter 7).

3. Any real-world machine must conduct each of its actions with a nonzero "probability of error." Thus, in reading the tape there must be a nonzero probability that the Tm control will be incorrectly informed as to which symbol is actually on the tape square being examined by the tapehead. Similarly, there must be a nonzero probability that the next-move function will be misevaluated, etc. Thus, there are limits to the accuracy with which a given computer can operate.

Let us stress that, essentially, these same physical limitations apply to all real-world computers, not just to Turing machines.

The third limitation of machines means that real-world computers are actually *probabilistic* (perhaps *nondeterministic;* see Hopcroft and Ullman, 1969, and Manna, 1970b). In effect, any real-world machine is capable of errors in any computation it makes (so, in a sense, machines are inherently "inconsistent"). However, the inaccuracy of machines may often be minimized; in particular, it is often possible to build machines that are more "reliable" than their components, in terms of the accuracy with which they compute their respective functions. (The classic paper on this subject is that of von Neumann, 1956.) Although little will be said hereafter about the probabilistic nature of machines, a reasonable convention for modern-day computers is to assume that such a machine will normally make less than one error per billion read-evaluate-print-move cycles.

To discuss the memory-size limitation of computers, a brief but quantitative definition of the word "information" is needed. What does it mean to say one computer memory will hold more information than another? (Throughout this discussion we will be concerned only with the memory that corresponds to the tape of the Tm, not with the memory that corresponds to its finite-state control.) The qualitative answer is fairly simple: The amount of information a tape (memory) can hold is dependent on the number of squares that make up the tape and on the number of symbols that may be printed on each square.[15] Since the simplest tape is one for which each square may have printed on it only one of two symbols ("blank" and "1"; "0" and "1"; etc.),

---

[15] This is essentially the Shannon-Weaver (1949) concept of "information." A more intuitive approach to information would include some way of describing the probable causes, effects, and denotations of a given string of symbols. This is discussed more thoroughly in Chapter 7, but we may note here that there is still no clearly satisfactory formalization for the intuitive concept of information. Also, it is common to omit the "ceiling-function" and to allow information to come in noninteger quantities of bits.

it is customary to take this kind of tape as a standard. The number of squares that make up such a tape is referred to as the number of *bits* (*binary digits*) of *information* that it can hold. To find the number of bits of information that can be held by a given nonstandard tape, we must figure out how large a standard tape must be in order to store as many different strings of symbols as can be held by the nonstandard tape.

This is easily done. (Remember, any physical, real-world tape can be made up of only a finite number of squares.) Suppose each of the squares of the nonstandard tape is numbered successively: 1,2,3, . . ., $n$. Let the number of symbols that can be printed on square $i$ be $s(i)$—again, only one symbol may be printed on a square at a given moment. Then the product

$$s = s(1)s(2)s(3) \ldots s(n)$$

is the total number of different strings of symbols that can be stored on the given nonstandard tape. If $x$ is a real number, we define the *ceiling function* (see Knuth, 1969a) of $x$ to be the least integer that is greater than or equal to $x$. Denote the ceiling function of $x$ by the expression $\lceil x \rceil$. Thus, $\lceil 6.5 \rceil = 7$, $\lceil 4 \rceil = 4$, $\lceil -2.3 \rceil = \lceil -2 \rceil$, $\lceil 0 \rceil = 0$, etc. The reader may easily convince himself that the smallest standard tape that can hold as many different strings of symbols as those held by the nonstandard tape must have

$$\lceil \log_2 s \rceil$$

squares. We may therefore take this to be the amount of information (in bits) that can be held by the nonstandard tape.[16]

Modern computing systems make use of many different types of memory systems, each with its own characteristics. Some currently accurate conventions for the storage capabilities of these systems are: "core" memories may hold on the order of $10^7$ bits; "disk" memories may hold on the order of $10^8$ bits; magnetic tape memories may hold on the order of $10^9$ bits; optical (laser) memory systems currently in development may hold between $10^{10}$ and $10^{12}$ bits (see Damron et al., 1968; R. P. Hunt et al., 1970; Lohman et al., 1971). It should be noted that the *access time* necessary for a computer to determine what symbol is stored at a given position ("square") in a memory will, in

---

[16] Of course this notion of information does not really depend on whether the "tape" is actually made up of squares, on whether it is one-, two-, or *n*-dimensional, or on whether "symbols" are "printed" or "stored" in some other manner in the "squares" of the tape, etc.

general, increase with the size of the memory. Thus, the access time for a core memory is generally on the order of $10^{-7}$ second, whereas for an optical memory it is generally on the order of a second (see Chapter 8, "Hierarchical Systems").

Probably the conventions used most often throughout this book are those pertaining to limitation 2, that is, the speed with which a computer can operate. The basic actions performed by a modern computer are, in analogy to those performed by a Turing machine, "read location(s) in memory," "perform logical or arithmetical operations," "store result(s) in memory," "access new location(s) in memory." The performance of this sequence of operations corresponds to a *cycle* of the operation of the computer; in general, for each cycle of operation, the computer processes one machine instruction (i.e., evaluates one instance of the next-move function). It should be emphasized that, for most of the symbol-manipulation procedures in which we are interested, a typical computer will usually have to process several machine instructions to complete each step of the procedure (how many depends upon the program, the collection of machine instructions, that is being used to describe the procedure). We shall have occasion to make use of several different conventions for the speed with which the steps of a procedure can be carried out by a machine—each convention we use will pertain to a different type of machine. These machines, and the corresponding conventions, will be referred to as follows:

|   |   |
|---|---|
| *conventional* | 1 microsecond/step |
| *attainable* | 1 nanosecond/step |
| *theoretical serial* | $10^{-12}$ nanosecond/step |
| *theoretical parallel* | $10^{-88}$ nanosecond/step or $10^{-104}$ years/steps |

Again, these are rough estimates. Their accuracy and meaning will now be discussed.

*Conventional.* Modern computers process about 10 million instructions per second. It is estimated that, with optimal programming, the average step involved in the type of nonnumerical computations we are investigating (those that "simulate intelligence") might require ten machine instructions; probably this is conservative. For example, generating a successor to a chessboard configuration might, with extremely good machine-language programming, be done in 1 microsecond. Using the "conventional" time estimate will give the student a rough indication of the best speed he can expect a current computer to achieve in performing a given procedure.

*Attainable.* Some integrated-circuit chips have been synthesized which are small computers and memories. These chips typically have operation and access times on the order of nanoseconds. Using circuitry and computer chips specifically designed for a given procedure, it is conceivable that the steps of that procedure might be performed at the rate of 1 nanosecond/step. Should the time required for complete execution of a procedure be very large, using the "attainable" estimate, the student may conclude that current technology is not capable of building a machine to perform the procedure. (However, it should be noted that coherent optical systems may eventually be used to perform logical operations at rates on the order of one picosecond ($10^{-12}$ second) per operation (see Culver and Mehran, 1971).

*Theoretical Serial.* Bledsoe (1961) used quantum theoretical considerations to derive the minimum access time of a serial digital computer (in which all information is passed through a central processing unit) with a density less than or equal to 60 gm/cm$^3$. He obtained the figure $10^{-21}$ second = $10^{-12}$ nanosecond. Therefore $10^{-12}$ nanosecond/step is taken as the best speed with which a serial computer could perform the steps of a given procedure. It seems likely that this speed of computation is completely beyond the bounds of any anticipated technology.

*Theoretical Parallel.* Bremermann (1967) computed the maximum rate at which information can be processed in a universe of $10^{73}$ protons, and he obtained $7 \times 10^{103}$ bits/year. This estimate, in the form of $10^{-88}$ nanosecond/step or $10^{-104}$ year/step, is used as the maximum speed with which the steps of any given computational procedure can conceivably be performed. It is useful simply as a "clincher" to establish whether a procedure is completely beyond the bonds of computation.

There are real-world problems for which the only procedures we can describe that would yield exact solutions cannot be carried out: The performance of these procedures is beyond even the "theoretical parallel" bound to computational ability. (One such problem is the game of GO; see Chapter 4.) However, it should be emphasized that for most problems there are several procedures for arriving at (perhaps partial) solutions, and it is possible for some to be within bounds and others to be out. Similarly, it is usually possible to describe a given procedure with several different programs, some of which may be more quickly executed by a given machine than others.

Because so little is known about the functioning of the human brain, it is difficult to compare its physical limitations with those of computers. The consensus seems to be that the brain has a larger memory than that of the computer but that it performs its logical operations (whatever they are) much more slowly (on the order of milliseconds/operation). The slowness of the brain's operation seems to be relatively unimportant if we consider the complexity of its structure and the fact that it is highly parallel; these attributes probably account for its evident ability to perform extremely complex logical operations at about the same speed with which it performs more simple reflexes.

## SUMMARY

We have seen that there are limits to the things that computers can be used to simulate, to the problems they can be used to solve, and to the procedures they can perform. However, our knowledge of these limits and of natural intelligence is not sufficient to determine whether the attainment of a general artificial intelligence is within the bounds of computational ability. AI researchers still do not have enough evidence to decide whether machines can be made as intelligent as human beings.

### NOTES

**2–1.** In fact, we have here described what might be called simple mathematical theories. We may define a general mathematical theory to be such that its three sets are finitely describable. At any rate, the object described by the theory is still finitely describable.

**2–2.** The proposition that if a thing is finitely describable it is therefore mathematically describable is generally taken as a postulate of the philosophy of mathematics, since it has not been proved mathematically. Mathematics seems incapable of supplying or handling a nonmathematical definition for the concept of "finitely describable." The evidence so far is clear, however, that all mathematical ways of formalizing the concept of "finitely describable" are equivalent.

**2–3.** From the mathematician's point of view, the thing is often identified with the set of sentences describing it. One could say in this sense that the natural numbers do not exist separately from the axioms that generate the set of sentences describing them. The notion of mathematical description is an *effective* notion, something like "approximation": One can derive as many of the truths about the thing described as one likes, though one cannot necessarily derive *all* such truths, in a finite time.

**2–4.** Actually, this is not quite true. One can extend the arguments of Chaitin (1966, 1969) to prove that there exists a finite set $A$ with, say, $10^{200}$ elements, such that any finite description of $A$ requires at least as many elements (production-rules; see Chapter 7) as there are in $A$. That is, the smallest finite description of $A$ is the enumeration of the elements in $A$. However, the actual enumeration of the $10^{200}$ elements in $A$ is physically impossible. That is, the set $A$ is finite, but is "practically infinite, practically nondescribable."

The proof for the existence of such a set would, of necessity, be nonconstructive. Existence proofs of this type are not considered valid by some mathematicians (note 2–5).

**2–5.** (A set is countable if its elements may be put into one-to-one correspondence with the natural numbers 1,2,3, . . . . The set of sentences derivable within a given mathematical theory must be countable, since for each $n$ there can be only a finite number of sentences of length $n$ or less).

This conclusion was first drawn from the work of Georg Cantor. Naturally, it aroused much controversy, and there are many mathematicians today who disagree with Cantorism (see Kac and Ulam, 1968, pp. 12–14). In particular, there is no unanimous viewpoint among mathematicians as to the proper rules for reasoning about "infinity" or even, for that matter, as to the existence of infinite things (see Benacerraf and Putnam, 1964). Thus, the argument concerning the existence of mathematically nondescribable numbers is not a *proof*, especially if one does not grant the a priori validity of the infinity concept. Hilbert (in Benacerraf and Putnam, 1964, p. 136) argued that the results of scientific investigation have given no evidence for the existence of infinite things. The viewpoint in this book is that scientific investigation has given no incontrovertible evidence concerning either the existence or nonexistence of infinite things.

It should be pointed out that some scientists have disputed the completeness of mechanistic reasoning, using quantum theoretical arguments (see, e.g., Elsasser, 1969).

**2–6.** Certain aspects of the universe are, according to current scientific theories, described as being finite. According to relativity theory, there is a maximum possible velocity, that of light, although certain phase velocities can be greater. Albert Einstein suspected that the spatial size of the universe might be bounded, and estimated a figure for its radius.

**2–7.** The definition of discrete phenomena does not require that an occurrence have a "first" event. Even so, it is possible to make a descriptive string, with a beginning, for an occurrence with neither a first nor a last event. (How?)

**2–8.** Of course no one can observe precisely that an event starts "immediately after" some time $t;$ what this means is that there are different step functions that describe occurrences which appear to be the same. (Examples?) This also applies, of course, to "immediately before," and "during the instant."

**2–9.** This seems, incidentally, not to be the case with most nondiscrete functions. For these, the best one can usually do is to approximate the value at a given point to an arbitrary closeness in a finite number of steps. Even in the case of Fig. 2–2, "the weight on a spring," we really deal with a type of approximation: The differential equation that describes the class of all possible occurrence functions does have exactly one solution for any given assignment of values to its variables $(m,a,k)$, but in order to evaluate that solution and to see where the weight will be at a certain time $t$, we usually have to compute certain functions (sine, cosine, etc.) that yield approximations. Typically, a finite description of an occurrence of a non-discrete phenomenon will give exact information at a finite number of points (values of $t$) and information that is an arbitrarily exact approximation at an infinite number of points. Something of the reverse holds for discrete phenomena: A finite description will often give exact information about an infinite number of values of $t$, of which at most only a finite number are "approximate"; for example, $2,_\pi+,4+$.

**2–10.** The observant reader may object that surely one cannot represent phenomena that have a nondenumerable number of occurrences by descriptions which yield a denumerable number of occurrences. To (partially) answer this objection, consider Example 2–1, "the weight on a spring": This phenomenon may presumably have a nondenumerable number of occurrences. However, the set of occurrences one can actually *compute,* using its description (the differential equation), is denumerable, for three reasons:

    *a.* Each computable occurrence is specified by listing the values for the variables $m$, $a$, $k$, and the accuracy with which one wishes to evaluate the equation.

    *b.* Each of these values must be finitely described, and the finitely describable numbers are a countable set.

    *c.* The countable product of countable sets is countable.

Thus, although the description of the phenomenon applies in an "ideal" sense to an uncountable number of occurrences, it actually describes only a countable set.

**2–11.** *One of the subplots of Kurt Vonnegut's novel, The Sirens of Titan,* is relevant: The hero is part of an army trapped on Mars. Most of the soldiers in the army have radio receivers implanted in their brains and are remote-controlled by a person who has decided (for reasons extraneous to this discussion) to have them invade Earth. The hero manages to discover what is happening, despite the fact that he has a radio receiver implanted in his own brain, and he writes a letter to himself describing everything he knows about the invasion. After hiding his letter, his dislike for the army is found out. Surgical officers in the army erase a great deal of his memory, but after he returns to duty he discovers his letter. Reading it, he is able to replenish his memory and begin again. This cycle repeats several times.

## EXERCISES

**2–1.** (a) Find a finite description for the set of points that are the intersections of the 24 Archimedean spirals, having equations of the form

$$r = \pm(\theta + (k\pi/6))$$

where $k = 0,1,2, \ldots, 11$. (b) Find a finite description for the set of points formed by the analogous intersection of 24 exponential spirals, with equations of the form

$$r = \pm(e^{**}(\theta + (k\pi/6)))$$

2–2. Construct a next-move function for a "unary multiplier," which, given an input string

$$\ldots bbb \underbrace{11 \ldots 1}_{m} b \underbrace{111 \ldots 1}_{n} bbb \ldots$$

consisting of a string of $m$ 1's followed by a string of $n$ 1's, produces on its tape an output string containing $mn$ 1's.

2–3. (a) Show that any Turing machine can be represented by a natural number (an integer greater than zero). (b) Give a finite description for a function $f$ mapping the natural numbers into [0,1], such that $f(n)$ cannot be computed for any $n$ by any Turing machine.

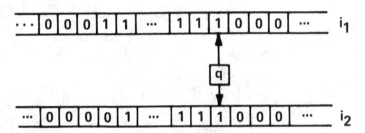

2–4. Consider a simple "polycephalic" Turing machine which has two tapes, $i_1$ and $i_2$, each of which is filled completely by zero's except for a single block of 1's. Let the blocks of 1's on the two tapes be right-justified, as indicated above. Find the simplest possible next-move function that will enable an outside observer to determine whether or not the number of 1's on tape $i_1$ is greater than or equal to the number of 1's on tape $i_2$, assuming that he cannot observe the state $q$ of the Turing machine.

2–5. In 1962 there were on this planet about 55,000 scientific journals publishing about 1,200,000 articles per year; there were also 60,000 scientific books and 100,000 other research reports issued per year (in the United States, scientific and technical publications have doubled in bulk approximately every 20 years since 1800). Estimate the size, in bits, of a computer memory capable of storing (a) all scientific publications produced in 1962, and (b) all scientific publications produced as of the present. Assume

  30 pages per article
  300 pages per book
  100 pages per research report
  60 lines of print per page
  70 symbols per line

and assume each symbol can be any of 128 different characters. How fast must one add to such a memory, to keep it up to date?

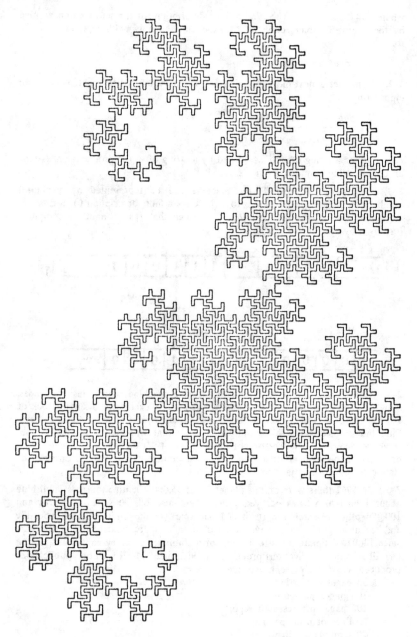

Dragon maze. (Courtesy of D. Ingalls, Xerox Palo Alto Research Center.)

# 3
# PROBLEM SOLVING

## INTRODUCTION

The opening sections of this chapter present a brief overview of the directions currently being taken by artificial-intelligence (AI) research and of the subjects that will be covered in subsequent chapters. The third section of this chapter describes some of the ways that AI researchers have formalized the concept of "problem." In the succeeding sections general problem solvers and reasoning programs, state-space problems, and heuristic search theory are discussed. Planning, learning, and reasoning by analogy are then introduced briefly. The final section is concerned with models, the "problem of problem representation," and the levels of competence that have been attained by artificial intelligences.

## PARADIGMS
### General Approaches

Speculations on the possibility of a search for mechanical intelligence were originally put forth by several individuals—including Alan Turing, John von Neumann, and Norbert Wiener—during the years 1943–1950; however, it was not until electronic digital computers became generally available in the early 1950s that experimental research in artificial intelligence could begin. Rapid progress in AI research did not occur until "symbolic processing languages," such as IPL and LISP,

67

were developed (note 3–1). To date there have been several thousand papers published on the subject of artificial intelligence. However, AI research is still in its embryonic state, and we cannot yet decide what its final form will be. Thus, this book must serve both as an introduction to a vast body of literature and a commentary on what appear to be the central topics discussed in that literature.

Kuhn (1962) discussed the importance of *paradigms* in the development of scientific investigations. (A paradigm is a general model of something that is found to be useful for investigating that thing.) AI researchers have developed many paradigms for artificial intelligence. Chapter 1 has already discussed one, which is represented by Turing's test: Artificial intelligence research is concerned with building machines that can perform tasks which people would ordinarily say require the "intellectual abilities" of a human being.

# Environments

Another way of viewing AI research is to see it as *an effort to design machines that are capable of existing on their own in environments produced by the real world*. An "environment produced by the real world" (or a *real-world environment*) is not necessarily our own environment. A mechanical intelligence might, for instance, operate in an environment consisting of "all published scientific works." Intuitively, an environment produced by the real world is always changing and does not have a known, complete description or prediction. We expect such an environment to exhibit regularities, or "patterns," and we expect a machine that operates in such an environment to encounter "problems." A machine operating successfully in a real-world environment will have to develop and represent internally its own "knowledge" of that environment. It may have to discover largely on its own the problems it needs to solve and the patterns it needs to recognize. If we design the environment ourselves, then many of these problems and patterns may be presented to the machine automatically (as with question-answering and fact-retrieving machines; see Chapter 7). Even if we do not design the environment, we may still know enough about it to give the machine automatic procedures for locating relevant problems and patterns. At any rate, the machine will have to be able to solve the problems and to perceive the patterns that it encounters.

Throughout this chapter we shall have much to say about the general nature of machines that are capable of existing in real-world environments. It is convenient to say that a machine which is capable of operating successfully in a real-world environment displays an *aptitude* for that environment. Also, reference is often made to environments simply as "problem domains" or "problem areas."

## Aptitudes

A closely related paradigm for AI research is to see it as being concerned with *frameworks for the engineering of mechanical aptitudes.* By this we mean that AI research can be viewed as an attempt to develop the computers and other hardware, programming languages, and human expertise necessary to design machines with aptitudes for specific real-world environments. This viewpoint springs from the recognition that some procedures (machines) will appear to be intelligent in some environments and unintelligent in others. Rather than search for a procedure that will be intelligent in all (or even many) environments, AI researchers may look for a framework (computer, language, expertise) within which to design procedures that can be tailored for intelligence in specific environments. Although AI researchers pursuing this paradigm are concerned with developing intelligent machines, they are more concerned with finding programming languages and computers that will facilitate the development and description of a wide variety of different intelligent machines, each with its own aptitude for solving problems in a real-world environment (some machines may have many of the aptitudes possessed by others). Many investigators have worked within this paradigm, too many for us to identify at this time all those who have made important contributions. Chapters 6, 7, and 8 are, in effect, a discussion of the work that has been done using this paradigm.

The idea of mechanical aptitudes is a valuable one, whether or not we seek a general framework within which to design them. Most AI researchers have not aimed directly at the goal of constructing completely intelligent machines, able to display intelligence at a human level. Rather, most work in artificial intelligence has been devoted to the machine simulation of *specific* intellectual abilities (giving machines specific aptitudes) such as the ability to play games or the ability to prove mathematical theorems. There are three basic reasons for this approach:

First, the theoretical and practical knowledge necessary to do really general work was (and is) extremely limited. There is no adequate guideline that can tell us in any detail how to build machines with a truly general artificial intelligence.

Second, one of the best ways to acquire this sort of information is to make a thorough comparison of human and machine abilities in limited problem areas or environments. The precise nature of the difficulties involved in AI research may show up more clearly if we confine our early inquiries to the simulation of specific aptitudes possessed by natural intelligence. Hopefully, many limited attempts at machine intelligence will eventually provide better grounds for generalization.

Finally, there is always pressure for some immediate results, both to solve current and practical problems—such as character recognition or assembly-line balancing (see, for example, Tonge, 1963)—that do not require general mechanisms for artificial intelligence, and to establish by experiment a likelihood that the more general attempts will eventually succeed.[1]

The specific machine aptitudes that have received the most investigation by AI researchers are problem solving, game playing, pattern recognizing, theorem proving, and language understanding. Two facts concerning "specific" machine aptitudes should be emphasized: First, there are *levels of generality* in the aptitudes that machines may possess. Thus, one procedure may have an aptitude for playing a specific game, such as Chess, and another procedure may have an aptitude for playing many different games; procedures with a "specific" aptitude for playing many different games are said to be *general* game-playing procedures. Similarly, a program with an aptitude for solving many different problems is called a "general" problem solver (it is not required that the program be able to solve *all* problems, or even that the problems it is able to solve be especially difficult). AI research has so far had only limited success in developing general problem solvers, general game players, general pattern recognizers, and general theorem provers. No procedures have yet been developed which we could fairly say are "general language understanders" (note 3–2).

The other fact that should be emphasized concerns the *interdependence of aptitudes*. Throughout this book we shall see many ways in which machines with one general aptitude must have other (perhaps less general) aptitudes. Thus, it can be shown that general game players must have an aptitude for pattern recognizing (see, e.g., Banerji, 1969), and general pattern-recognizing programs must have an aptitude for language understanding (see Chapters 5 and 7).

## Evolutionary and Reasoning Programs

The term *general artificial intelligence*, as it is used here, refers loosely to a machine (procedure) that has aptitudes for general problem solving, general game playing, general theorem proving, general pattern recognizing, and general language understanding, and also

---

[1] We discuss the practical uses and effects of general artificial intelligence more thoroughly in Chapter 9.

has aptitudes enabling it to display all other kinds of intelligent behavior normally exhibited by people. Again, no one has yet been successful in giving a machine the aptitudes corresponding to general artificial intelligence. There have been primarily two types of approach to the goal of achieving a general machine intelligence, one of which is an evolutionary approach, the other being what is called (following McCarthy) the "reasoning program" approach. These approaches are not mutually exclusive, but as yet they have not been combined: Thus, one can imagine reasoning programs that might change their rules of inference and "evolve," and one can imagine interrelating reasoning programs that would form a "self-organizing" whole, which might itself be a reasoning program (see Chapter 8).

The evolutionary programs, such as those written by Friedberg et al. (1958, 1959) and Fogel et al. (1966) and suggested by Holland (1970) and Campbell (1960), are programs that produce, select, and modify subprograms according to their ability to perform various tasks. There is no reason in theory why evolutionary programs might not eventually be used to produce a general artificial intelligence, but as yet the evolutionary approach has had little success.

The reasoning-program approach is an attempt to develop a single program capable of perceiving facts about its environment, of drawing conclusions from facts, of discovering an adequate means for the expression of facts, of formulating its own goals and strategies, and acting according to them—a program that would, in short, be a rational entity. The most well-known example of this approach is probably the "General Problem Solver" of Newell, Shaw, and Simon (1963), which might be described as a preliminary investigation of the rational process. McCarthy (1963a,b; with Hayes, 1968a) took a somewhat different approach to the same goal, concentrating particularly on what sort of internal language (means for expressing facts) would be the best for a reasoning program.

This chapter discusses the work of McCarthy, Newell, Shaw, Simon, Ernst, Nilsson, Amarel, Hewitt, Fikes, Pohl, and others, relevant to the construction of reasoning programs and to giving machines a "specific" aptitude for general problem solving.

# PARADIGMS FOR THE CONCEPT OF "PROBLEM"

## Situation-Space

What is a problem? Perhaps the best answer AI researchers can give is that the real-world nature of "problems" still has not been either fully formalized or fully investigated. We can, however, describe two basic models, or paradigms, for the concept of "problem," which would have to be included in any fully general formalization.

Our first general paradigm for "problem" is the *situation-space* model. A problem presented in this formalization consists of an initial situation, a set of possible situations, and a set of possible actions, together with a specification of how the various situations can be produced from each other by different actions, and the specification of a final, desired situation, or *goal;* the statement of the problem might also include a specification of certain situations to be avoided. A *solution* to a situation-space problem, then, is any sequence of actions that leads from the initial situation to the desired situation, and avoids the undesired situations.

Several additions should be made to this model for "problem" if we are to insure some generality in its application to the real world. First, we should allow the situations of a given situation-space to be *partially-specified;* that is, we should not require in general that a *complete* description be obtainable for any given situation (though for the simpler problems so far considered in AI research such descriptions are usually available); rather we might allow a given situation to be described by a set of sentences, each presenting a fact about the situation from which new sentences may possibly be derived. The set of sentences describing a given situation may be incomplete; that is, one may not be able to answer all conceivable questions about the situation. Again, the result of applying an action to a given situation will not necessarily be a completely specified situation. In the same vein, the goal to be obtained by solution of the problem may be only partially-specified.

Also, in full generality we would not require that the result of applying an action to a given situation necessarily be a unique situation, or even a unique partially-specified situation. That is, we should allow actions to be "nondeterministic" in their consequences, sometimes yielding one partially-specified situation out of a set of partially-specified situations.

Finally, a *solution* to a situation-space problem may in general be partially-specified; that is, the solution may be described as dependent on various contingencies that cannot be completely determined in advance. For example: "If $X$ should become a factor then do $Y$," "If $Z$ should happen, then formulate a new solution." Thus, the solution may in general be a *plan,* or *strategy,* not a specific string of actions. The various actions that might be included in a given solution should include "looking for a new solution"; "discovering more information about relevant situations"; and "interrupting one's actions, not doing anything."

A good example of a real-world problem that is partially-specified is McCarthy's Airport Problem: The problem consists in going from one's home to the airport. Two basic actions are available, *driving* and *walking*. To solve the problem, one starts at home, walks to one's car, and drives to the airport. However, in reality one cannot specify completely and invariably all of the details of the situations and actions that may occur in solving the problem; so, a single string of actions cannot be produced which is a guaranteed solution. One could, for example,

> Break one's leg going to the car.
> Have a flat tire while driving to the airport.
> Misread a highway-direction sign and get lost.
> Run out of gas or have engine trouble.
> Come to a roadblock or a detour.

A machine attempting to solve the Airport Problem could run into similar difficulties, yet these are all obstacles a general intelligence could surmount (though in doing so it might need to enlist the aid of other intelligences). The nature of this problem's difficulty lies in the partial-specification of its situations, actions, and solutions. This is true of most problems in the real world.

## System Inference

Our other paradigm for "problem" is the paradigm of *system inference*. Problems in this paradigm may take many different forms of representation, all of them theoretically equivalent, though a machine working within this paradigm might sometimes find the use of one representation to be more efficient than the use of another. Various forms of "system inference" would respectively require a problem-solving machine to be capable of inferring:

1. A function $f$ from a set $A$ to a set $B$, given examples of the function's values for a subset of $A$.
2. A relation $R$ within a set $X$, given a description of $X$ and a set of examples (positive or negative) of the way $R$ holds throughout $X$.
3. A grammar for a string language $L$, given a set of sentences that belong to $L$ and a set of sentences that do not.
4. A mathematical theory, given a set of propositions that are true within the theory, and a set that are not.
5. A Turing machine $T$, given a sample of its behavior on a set of input strings.

(Of course this list is not exhaustive.) The inference, or *system*, proposed by a problem solver as a solution to one of these requirements will typically be a finite description of a function, relation, string grammar, mathematical theory, or Turing machine.

The generality of this paradigm as a model of mathematical problems should be strongly suggested, but there may be some doubt as to its relevance to the real world. To help insure this relevance, we should allow the evidence for a given system-inference problem to be partially-specified and also allow the solutions (i.e., systems) proposed by the problem-solving machine to be partially-specified. Again, the machine should have some language for representing its knowledge of a given inference problem, and it should have some way of determining information that will help it decide among the various systems it might infer as a solution to a problem. It will often be the case that a machine will be able to infer several systems consistent with the evidence it has been given. However, we would not require that it be able to *derive* its inference(s) from the given evidence, nor even necessarily that it be able to *prove* that its proposed solutions are consistent—nevertheless, a system-inference machine should be able to defer to experience and not make an inference once it has recognized evidence that refutes it. Also, a system-inference machine should be able to detect, or try to detect, that its evidence is self-contradictory, and it should usually tend to propose increasingly better solutions.

An intuitive example of a real-world system-inference problem is the problem of *invention:* That is to say, given a description of some task to be performed (peel potatoes), find a description of an object that will perform the task (draw a blueprint for an automatic potato-peeler). The task to be performed can be corresponded to a function that maps situations into situations; the description of the task can be corresponded to a description of the function values on certain inputs; and the invention produced by the problem-solving machine can be corresponded to a finite description of a function (a program for a universal Turing machine) that performs the task. An efficient mechanical inventor should use what might be called the "principle of economy of invention": Do not design an invention to depend on other, unachieved inventions if you can help it (note 3–3).

Of course no one has built a "general invention-making machine," but the possibility is clearly in line with the notion of general artificial intelligence.

Actually, each of these paradigms for the concept of "problem," the situation-space model and the system-inference model, is equivalent to the other: It is likely that any problem which can be stated in one paradigm can be stated in the other, and that each of these models is merely a different way for representing the same underlying idea about the general nature of problem-solving ability. Still, we should emphasize again that neither paradigm has yet been completely formalized or investigated as regards its application to "problems of the real world." Finally, we should mention that for any problem, there are essentially *two levels of solution:* The first level is to prove the *existence* of a solution to the problem, and the second level is to *construct* the solution itself. Polya (1945) presented an excellent introduction to the nature of problems and their solutions, and gave attention to some aspects of real-world problems.

# PROBLEM SOLVERS, REASONING PROGRAMS, AND LANGUAGES

## General Problem Solver

The rest of this chapter will be concerned with computer programs that are capable of solving problems stated in the situation-space paradigm. Programs that work with problems stated in other paradigms are discussed primarily in Chapter 7. We shall see the situation-space paradigm used in Chapters 4 and 6 by programs which play games and prove theorems. In this section we are concerned with two questions: First, what should be the nature of a machine that would be a general problem-solver for problems of this type? Second, how should a machine of this type be designed to operate in a real-world environment similar to our own?

One example of a fairly general program for solving situation-space problems is the General Problem Solver (GPS) program of Newell, Shaw, Simon, and Ernst (1963 et seq.). GPS made use of an elementary language for the description of situation-space problems. That is, GPS was capable of accepting descriptions of *objects* and *operators* ( = situations and actions) and of accepting information that a certain object was the initial, or given, object and that a certain object was the desired object, or goal. The GPS language contained what might be called the first degrees of partial specification: One could specify to GPS that a *class* of objects (e.g., "any expression without an integral sign") was to be the goal, and the program could decide that some objects would be considered partial solutions. This was done with the use of *difference operators,* which were capable of detecting various types of differences between objects. The differences were themselves also treated as objects, and GPS could define subgoals of "changing the difference" between two objects. Thus, GPS would seek to minimize one difference at a time between two objects, and it usually was given an ordering for the various differences: Minimizing one difference could be considered more important than minimizing another.

GPS used the same problem-solving technique (referred to as *means-ends analysis* by its authors) on every situation-space problem it was given; the technique comprised three essential steps:

1. Evaluating the difference between the current situation and the goal.
2. Finding an operator that typically lowers the type of difference found in step 1.
3. Checking to see if the operator found in step 2 can be applied to the current situation; if it can, then apply it, else determine a situation required for the application of that operator, and establish it as a new (sub) goal; then go to step 1.

GPS was applied to many different simple problems, such as the Missionary-Cannibals Problem (see the last section) and the Tower of Hanoi (see the Exercises). It was also shown to be able to prove relatively simple theorems in mathematical theories; its authors were able to describe the resolution principle of J. A. Robinson (see Chapter 6) within their formalization for operators and objects. On all of these (fairly simple) problems, GPS was successful, though usually it was not as fast in producing answers as were special programs designed to solve the individual problems.

In several respects, GPS was not a fully general problem solver. In the first place, GPS could not produce a *plan* or *strategy* as its solution; the only solution GPS could produce would necessarily be a specific sequence of actions that would lead to the desired goal. Also, GPS could be applied only to problems that could be completely specified, where the various actions, objects, differences, etc., could be exactly described for the given problem. Thus, GPS was completely dependent on the ability of its programmer to produce a suitable representation for the problem.

As an example, GPS was given the famous Seven Bridges of Konigsberg Problem (see Fig. 3-1). The problem is to go over each of the seven bridges once and only once and return to the point from which you started. This problem was shown to be unsolvable by Euler in 1736, using certain topological considerations. When given the problem, GPS tried the same paths repeatedly and eventually gave up,

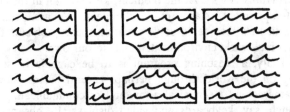

Figure 3–1. The seven bridges of Konigsberg.

unable to achieve a solution because it could not look at the problem in a general way: It could not develop a partially specified solution, or strategy, and then prove whether the strategy would work, nor could it prove theorems about the problem or its solutions. Since GPS could not invent Euler's "topological considerations," it could not prove the puzzle to be unsolvable.

Of course most people couldn't do this either, or at least not right away; otherwise the problem would never have become famous. Usually, the first thing a person will try is a GPS-like search. However, a person can stop such a search if it seems to be fruitless, and can try to reason about the problem itself.

All of which is to say that GPS was highly "representation dependent," more so than a truly general problem solver would be.[2] We should expect a representation-independent problem solver to be capable of:

1. Inventing new representations for a given problem, if it cannot solve the problem using the ones it has.
2. Discovering facts, and perhaps proving theorems, about representations and problems, their interrelations, etc.
3. Asking for, and looking for, help in the outside world.

Each of these abilities would be necessary to a problem solver that functions in the real world.

## Reasoning Programs

Following McCarthy and Hayes (1968a), let us label general problem solvers that work within the situation-space paradigm, and which possess independence from representations in this sense, as *rea-*

---

[2] This criticism also applies to the more recent general problem solvers such as FDS (Quinlan and Hunt, 1968), MULTIPLE (Slagle and Bursky, 1968), and REF-ARF (Fikes, 1970). These programs are each capable of solving a variety of different problems, but they are all highly representation-dependent.

*soning programs* (RP's). At the moment, RP's are still in the conceptual, "thought-experiment" stages of development. We are primarily concerned with RP's that could solve situation-space problems that might occur in a real-world environment similar to our own.

Basically, a reasoning program is to be capable of sensing and operating on the world through perhaps several means, such as television cameras and mechanical arms, and of communicating with people through, say, keyboards and video displays. Its observations at a given moment may be stored internally in several forms: Pictures, for example, might be stored as matrices, lists, or other data-structures. However, any data stored by the RP is ultimately to be described, within the RP, by sentences in a general language for the representation of phenomena. RP should be capable of proving theorems about phenomena, stated within this language, and of deciding what actions to perform on the basis of these theorems; its "phenomena language" should be capable of describing the actions it can perform, as well as the situations it can observe, and of describing interrelations between them. The language should be capable of describing hypothetical situations and actions, of designating some as desirable and others as not. Finally, the phenomena language should be capable of describing representations of problems, as well as problems themselves: RP should be capable of reasoning about its representations as well as with its representations, as described above.

A language is essentially a way of representing facts. An important question, then, is what kinds of facts are to be encountered by the RP and how they are best represented. It should be emphasized that the formalization presented in Chapter 2 for the description of phenomena is *not* adequate to the needs of the RP. The formalization in Chapter 2 can be said to be *metaphysically adequate,* insofar as the real world could conceivably be described by some statement within it; however, it is not *epistemologically adequate,* since the problems encountered by an RP in the real world cannot be described very easily within it. Two other examples of ways of describing the world, which could be metaphysically but not epistemologically adequate, are as follows:

1. The world as a quantum mechanical wave function.
2. The world as a cellular automaton. (See Chapter 8.)

One cannot easily represent within either of these frameworks such facts as "Today is my programmer's birthday," or "I don't know what you mean," or "San Francisco is in California," or "Ned's phone-number is 854–3662."

If we use human languages as an example, we can identify several things an RP language should be able to express very easily.

*Causality.* The language should enable RP to express various forms of causality relationships between situations and phenomena: "fire causes smoke."

*Temporality.* The language should be able to express that one situation precedes another, that one situation follows another immediately, that one situation may precede another, etc. "Harry will get home by the time John does."

*Ability.* The language shoud be able to express such notions as *"X can do Y"* (perhaps with appropriate modifiers; e.g., "if *X* is given certain knowledge"; thus, a person can open any combination safe, if he knows its combination).

*Relevance and Plausibility.* The language should make it possible to express the notion that certain situations or problems are relevant to each other, or may be relevant to each other, though perhaps not in any known way. The language should also include the possibility of expressing the plausibility and relevance of sentences: "These are all the sentences necessary to describe the problem"; *"X* is analogous to *Y";* "These sentences are plausible."

*Possibility and Probability.* The language should be able to express notions of indeterminacy and undecidability and, if necessary, treat them mathematically.

*Knowledge and Certainty.* The language should enable RP to express that something is known: "John knows Bill's phone-number"; "John knows how to find Bill's phone number"; "Someone here may know what time it is."

*Desirability and Undesirability.* The language should enable RP to denote situations (and perhaps actions) as being desirable or undesirable.

*Equivalence and Denotation.* RP should be able to express several different types of equivalence, such as "The morning star is the evening star"; "The velocity is 50 mph"; $"X^2 = X \cdot X."$

*Existence.* RP should be able to say that some things exist differently from others: *"X* is a solid"; *"Y* is an expression of information."

*Suppositionality or Hypotheticalness.* RP should be easily able to state that some of the statements it is using are "advanced for the sake of discussion" (see Carnap, 1947, 1950; Quine, 1955–1964; Hintikka, 1962, 1969; and Rescher, 1964, 1967).

This list is only illustrative; many more examples could be added, and each example could be treated in much greater detail. It is also true that these examples overlap each other; for a more thorough treatment, the reader should see the paper by McCarthy and Hayes (1968a).

One final thing to note on the subject of reasoning programs is that the language used by an RP will typically be changed with time by the RP. We should expect in general that a reasoning program will find it necessary to define new words or to accept definitions of new words; some of these words will denote new situations, actions, phenomena, or relations—the RP may have to infer the language it uses for solving a problem. Our most important requirement for the initial language is that any necessary extensions to it be capable of being easily added to it. For a further discussion on the nature of languages and their use by machines, see Chapter 7. Predicate calculus has been suggested as a possible basic language for an RP, and Chapter 6 discusses computer programs capable of proving statements expressed in predicate calculus theories. In the final section of this chapter, discussion is continued on the subject of representation-independent problem solvers.

# STATE-SPACE (SITUATION-SPACE) PROBLEMS

## Representation

This section discusses the situation-space paradigm itself in some detail, since it is perhaps the most popular one used by AI researchers, and since there has been a considerable theory of problem solving, known as *heuristic search theory,* developed around it.

The situation-space paradigm has been given several (slightly) different formalizations; in the literature of AI research it is usually called the "state space" paradigm, which is the name originally given to it by researchers in the fields of *operations research* and *control theory.* In this discussion "situation-space" and "state-space" terminologies are used somewhat interchangeably, as defined below. The formalization presented is essentially that of Nilsson (1971), which gives an extensive coverage of heuristic search theory. Other formalizations are presented in Banerji (1969), Sandewall (1969), and Quinlan and Hunt (1968).

Anyone who wishes to understand the current directions of AI research should make an effort to understand the state-space paradigm. While the ideas involved are not very difficult, their presentation will go easier if we consider a simple example. Such an example is the Three Coins Problem, which is stated below. After reading the statement of the problem, the reader is urged to solve it—its solution is very straight-forward.

## Three Coins Problem

Given three coins arranged as in Fig. 3–2, make them all the same (i.e., either all heads or all tails), using exactly three moves. By a *move* in this case is meant flipping one of the coins over, so that if it is heads before the move, it becomes tails afterward, etc.

Figure 3–2. Initial state of the Three Coins Problem.

The Three Coins Problem can be easily stated as a state-space problem. A configuration of the coins is a state. The initial state, or *start*, is denoted by the expression HHT. The desired states, or *goals*, are TTT and HHH. For any given state there are three possible *operators:* "turn the first coin over"; "turn the second coin over"; and "turn the third coin over." A move corresponds to the choice of one of these operators, and a solution to the problem is a sequence of three moves that transforms the start into one of the goals.

Let us label the three operators as $A$, $B$, and $C$, respectively. Thus, $B$ applied to HHT yields HTT; we can briefly denote this fact by the expression

$$HHT \xrightarrow{B} HTT$$

Since $B$ applied to HTT yields HHT, we shall, however, write

$$HHT \xleftrightarrow{B} HTT$$

Given this notation, the diagram shown in Fig. 3–3 depicts the *state space* of the Three Coins Problem; that is, all the possible states and the result for each state of applying each of the possible operators to it. By tracing through the diagram, we see that one sequence of moves which solves the problem is "first $A$, then $C$, then $A$," or $ACA$ for short. The other solutions to the problem are $AAC$, $CAA$, $BCB$, $BBC$, $CBB$, and $CCC$; each of these leads to the goal HHH. (There is no way to go from HHT to TTT in exactly three steps.)

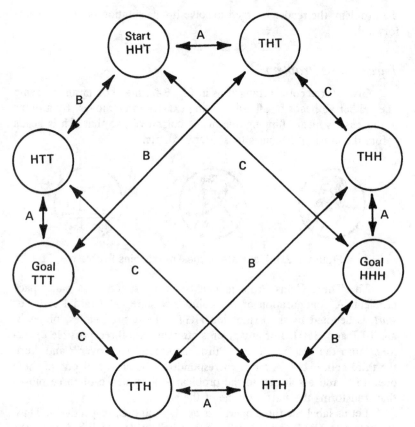

Figure 3–3. A state-space for the Three Coins Problem.

We shall consider a *state* to be a finitely describable mathematical object; in the Three Coins Problem each state was described by a string of three letters (e.g., HTH). Other ways in which states can be described include numbers, matrices, lists, graphs, sentences, sets, vectors, and trees. (Graphs and trees are defined below; the mathematical notion of "sentence" is discussed in Chapter 7.) A state could be infinite, but the fact that it has a finite description means we can discuss it logically, prove theorems about it, etc. However, throughout the rest of this book we shall be concerned only with finite states. From the computer's standpoint the description of a state is a data-structure (see Knuth, 1969).

Similarly, an operator is a finitely describable means of transforming one state into another state; there may be many ways of describing a given operator, and from the computer's standpoint an operator is a computational procedure.

A description of a state-space problem, then, is the specification of three things:

> $S$, a set of possible starting states
> $F$, a set of operators
> $G$, a set of desired states, or goals

A solution (or solution path) to a state-space problem is also the specification of three things:

> $s$, one of the possible starting states
> $g$, one of the desired states
> a finite sequence of operators that transforms $s$ into $g$

Thus, if $q$ is an operator and if we denote the result of applying $q$ to $s$ by the expression $q(s)$, and if $q_1, q_2, \ldots q_{n-1}, q_n$ is a solution to a state-space problem, then we have

$$g = q_n(q_{n-1}(\ldots q_2(q_1(s)) \ldots))$$

There may, of course, be many solutions to a given state-space problem $(S,F,G)$. We may consider a given $(S,F,G)$ state-space problem to be a collection of smaller state-space problems, each of the form $(\{s\},F,G)$, where $s \in S$—we shall say a procedure *solves* the $(S,F,G)$ problem if it is capable of producing a solution path for each of the corresponding $(\{s\},F,G)$ problems which has a solution.

The observant reader has probably noted that the definitions in the preceding paragraph make no mention of the sequence $q_1, q_2, \ldots, q_n$ consisting of three or any other prespecified number of operators. Yet we required in our informal statement of the Three Coins Problem that the solution use exactly three moves, that is, that $n$ be equal to 3. Can this sort of requirement be made within the framework of the definitions given in the preceding paragraph?

To see that it can, the Three Coins Problem is restated as follows: Let the initial state $s$ consist of the three coins, as in Fig. 3–2, and let $s$ also contain a "counter," initially set to zero. (The counter is to be capable of storing arbitrarily large numbers.) Denote the initial state $s$ by the expression (0,HHT). The three possible operators $A$, $B$, and $C$, which we can apply to an arbitrary state $(i,xyz)$, will now be respectively: "Turn coin $x$ over and replace $i$ by $i + 1$"; "turn coin $y$ over and replace $i$ by $i + 1$"; and "turn coin $z$ over and replace $i$ by $i + 1$." Finally, the set $G$ of goal states will contain two members: (3,HHH) and (3,TTT). The solutions to this statement of the problem are the same as our solutions to the previous statement.[3] On the other hand, the state space described by this statement of the problem, and shown in Fig. 3–4, is somewhat different from that of Fig. 3–3.

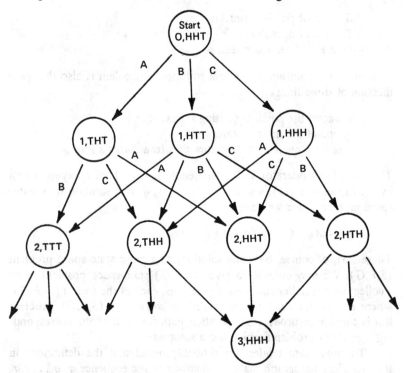

Figure 3–4. Another state-space for the Three Coins Problem.

It is possible to state many other problems within the $(S,F,G)$ format defined above (note 3–4). For some problems, especially those that place restrictions on the desired paths from start to goal, it is necessary to use "counters" or other devices. However, many problems can be stated rather simply within the $(S,F,G)$ state-space paradigm. This is true despite the fact that such problems will often have solutions that are very difficult to find. One reason for the popularity of the $(S,F,G)$ state-space paradigm within AI research is that it simplifies the problem of stating problems that often have very difficult, hard-to-find solutions.

---

[3] This is, incidentally, essentially the way the problem was stated to GPS, which solved it.

Of course the requirement that states and operators be finitely-describable objects is not entirely consistent with "problems of the real world." We can expect a problem solver in the real world to encounter things for which it does not have complete, finite descriptions. The statements and solutions of real-world problems, so far as the mechanical problem solver might be concerned, would still be finite descriptions, but they could be incomplete. A real-world mechanical intelligence might be able to make a statement like "There is an object on the road ahead, but I don't know what it is; I had better slow down and try to see what it is." In general, if the elements of a given problem are partially specified, we call them *situations, actions,* etc., whereas if they are completely described, we call them *states* and *operators,* etc. Thus, we distinguish between situation-space and state-space problems.

## Puzzles

None of the foregoing discussion is intended to deny, however, that state-space problems do occur in real-world environments or that the study of state-space problems can be of value to the study of situation-space problems. Many real-world problems can be expressed in the $(S,F,G)$ paradigm. A classic example is the Traveling Salesman Problem, which occurs in various forms in the scheduling of industrial production (see the Exercises). Formalizations for the situation-space paradigm are discussed in later sections of this chapter. It should be emphasized that many of the techniques being developed for the solution of state-space problems are directly applicable to situation-space problems. Thus, "games of strategy" are one general class of situation-space problems; Chapter 4 shows how the methods discussed in this chapter can be extended to game playing. The state-space problems considered in this chapter are essentially "one-person games of strategy"; these problems are also commonly called *puzzles*.

An example of a puzzle that is easily stated within the $(S,F,G)$ format, yet for which solutions are difficult to find, is the famous "15-Puzzle" (note 3–5). The puzzle uses a square tray adequate to hold 16 square tiles, in which 15 tiles are placed, each marked with a different number from 1 to 15. The space for the sixteenth tile is left empty; one configuration of the tiles may be changed into another configuration only by sliding a tile adjacent to the blank space into the blank space (this, of course, moves the blank space in the opposite direction). A "15-Puzzle Problem," or 15-Problem, is completely stated when we specify an initial configuration of the tiles and a goal configuration. Figure 3–5 shows a typical 15-Problem.

We can state a 15-Problem as an $(S,F,G)$ state-space problem as follows: A given configuration of tiles is a *state*. We shall denote each possible state by a 4 x 4 matrix, whose elements have values from 0 to 15. Thus, the start and goal states of the problem are indicated in Fig. 3–5. For a given state $s$ we denote the number in the $i$th row and

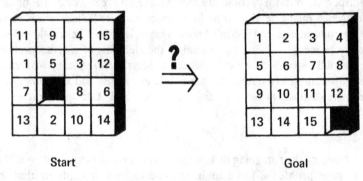

Start                                      Goal

Figure 3–5. A 15-Puzzle.

the $j$th column of its matrix by the expression $s_{ij}$. Thus, $s_{2,3} = 3$ for the start-state being considered. For a given state $s$, let $i_0$ and $j_0$ be the values of $i$ and $j$ such that $s_{ij} = 0$. We have $i_0 = 3$ and $j_0 = 2$ for the start state. Given this notation, we can describe four operators:

$A$. Replace $s_{i_0 j_0}$ by $s_{i_0,j_0+1}$  and $s_{i_0,j_0+1}$ by 0,   if $j_0 + 1 \leq 4$.
$B$. Replace $s_{i_0 j_0}$ by $s_{i_0+1,j_0}$  and $s_{i_0+1,j_0}$ by 0,   if $i_0 + 1 \leq 4$.
$C$. Replace $s_{i_0 j_0}$ by $s_{i_0,j_0-1}$  and $s_{i_0,j_0-1}$ by 0,   if $j_0 - 1 \geq 1$.
$D$. Replace $s_{i_0 j_0}$ by $s_{i_0-1,j_0}$  and $s_{i_0-1,j_0}$ by 0,   if $i_0 - 1 \geq 1$.

These correspond to moving the blank space "right," "down," "left," and "up," respectively. As is indicated in the description of the operators, an operator may not be applicable to a given state. However, for every state, at least two operators will be applicable. Part of the state space for the problem shown in Fig. 3–5 is shown in Fig. 3–6.

Altogether, there are $16! = 20,922,789,888,000$ different states in the state space of the 15-Puzzle. However, from any given starting state, only *half* of these states can be reached, using the operators $A,B,C,$ and $D$. The other 10½ trillion cannot be reached, regardless of the sequence of moves one tries (see Fig. 3–7). Computer programs have been written which are capable of solving the 15-Puzzle, that is, of finding a path between arbitrary start and goal states when such a path is possible and of recognizing start- and goal-state pairs for which there is no such

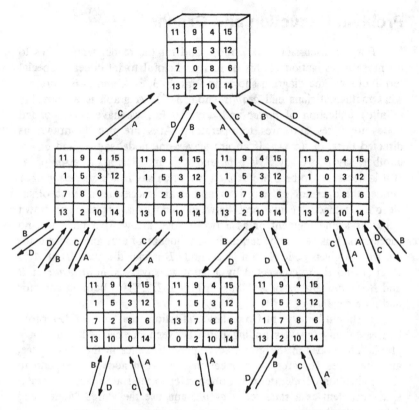

Figure 3–6. Part of the state-space for the 15-Puzzle.

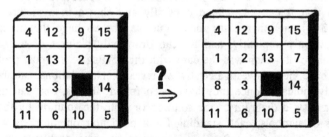

Figure 3–7. An unsolvable 15-Puzzle.

path. (One such program is discussed below.) However, so far as the author knows, no "general" problem-solving program (such as GPS, REF-ARF, and FDS discussed above) yet written is capable of solving the 15-Puzzle: Programs that can currently solve the 15-Puzzle are "special purpose." We shall return to this point later.

# Problem Reduction and Graphs

For the discussions that follow, and for the reader who wishes to do more investigation on his own, it is helpful to introduce a special terminology: The diagram shown in Figs. 3–3, 3–4, and 3–6 represent what mathematicians call *graphs* (note 3–6). A graph is a (possibly infinite) collection of *nodes* and *arcs;* so far, we have corresponded nodes to states and arcs to operators. Arcs are usually drawn as directed lines, or arrows. If an arc leaves one node, say *A,* and enters another, say *B,* we say *A* is a *parent* of *B* and *B* is a *successor* of *A.* If it is necessary to be more explicit, we often say *B* is a *successor* of *A* "under the operator *q*," etc. It *A* and *B* are successors of each other, we often replace the two arcs between them by a single *edge,* drawn either as a line segment or as a two-headed arrow. If a node has no successors, it is said to be *terminal.* A sequence of arcs and nodes leading from a given node *A* to a given node *B* is called a *path* from *A* to *B.* If *A* and *B* are connected by a path, we say *A* is an *ancestor* of *B* and *B* is a *descendant* of *A.* Thus, in Fig. 3–2, TTH is both an ancestor and a descendant of THT.

It should be evident from these definitions that an $(S,F,G)$ problem essentially involves finding paths between prespecified nodes in a graph. The nodes in the graph correspond to states in the state space, and the edges (or arcs, or connections) between nodes correspond to the application of operators. We often refer to the state space of a state-space problem as a *state-space graph,* and use the words "node" and "state" interchangeably.

For some state-space problems the state-space graph may be so small that it can be defined explicitly and shown in a picture (e.g., Fig. 3–4); in other cases the graph may be so huge that it can be defined only implicitly, and we can draw only pictures of very small portions of it—such was the case with the 15-Puzzle. In most problems that have been investigated by AI researchers, solutions can be explicitly indicated once they have been found; that is, one can usually either draw a diagram of the solution or state the solution by listing a series of symbols, each standing for a particular operator, as we did with the Three Coins Problem. However, in some problems even the solutions involve huge graphs and must be stated implicitly; some problems of this sort are games such as Checkers and Chess.

One of the most useful aspects of the state-space paradigm is that it can, in a sense, be applied to itself. Instead of identifying nodes by states and arcs by operators, we can identify nodes by *problems* and arcs by operators that change problems into other problems. We refer to finding a good path through a graph of this sort as a *problem-reduction* problem. The graph of a problem-reduction problem is known as an AND/OR graph, for reasons we shall learn in a moment.

Problem-reduction problems can be developed as a natural extension of $(S,F,G)$ state-space problems. To see how, consider the simplest possible $(S,F,G)$ state-space problem: What would it be?

Well, there are many extremely simple $(S,F,G)$ problems, but all are approximately of the same form. We can classify three types of trivial or primitive $(S,F,G)$ problems:

1. Problems of the form $(S, \{q\}, G)$, where there is only one operator available.
2. Problems of the form $(\{s\}, F, \{s\})$, in which no operator need be applied—more generally, problems of the form $(S,F,G)$ where $S \cap G \neq \phi$; that is, in which some start state is also a goal state.
3. Problems of the form $(S, \{\ \}, G)$ in which no operator can be applied and there is no start state that is also a goal state. The first two types of problem are trivially solvable; the last is trivially unsolvable.

Basically, the problem-reduction approach consists of finding operators that are capable of transforming complex $(S,F,G)$ problems into primitive $(S,F,G)$ problems. The particular operators one uses will depend upon the initial, complex $(S,F,G)$ problem, and it may often be very difficult to find good problem-reduction operators.

In general, an AND/OR graph contains two types of nodes: problem nodes and AND-nodes. These nodes are usually called *subgoals* when it is not necessary to distinguish them. The arcs connecting problem nodes to problem nodes, and problem nodes to AND-nodes, represent the application of problem-reduction operators. Those connecting AND-

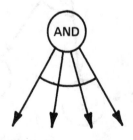

nodes to problem-nodes will be referred to as *and-links*. The *and*-links
from an AND-node usually subtend a circular arc, as shown here.

A typical, small AND/OR graph is shown in Fig. 3–8. A problem
node is said to be *solvable* if it is trivially solvable, or if any of its successor nodes is solvable. On the other hand, a problem-node is *unsolvable* if it is trivially unsolvable, or if *all* its successor nodes are
unsolvable. An AND-node is unsolvable if at least one of its successor
nodes is unsolvable; otherwise it is solvable.

Good examples of the problem-reduction approach are the "symbolic integration" problem solvers, such as SAINT (by Slagle, 1963) and

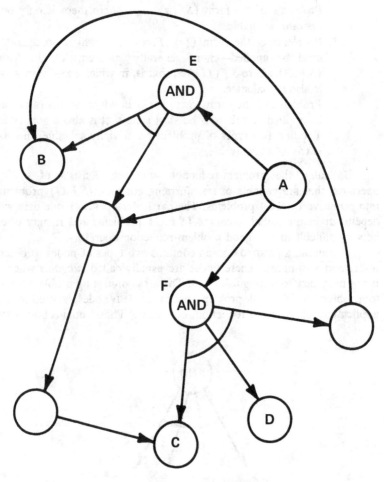

Figure 3–8. A small AND/OR graph.

SIN (Moses, 1967). These programs are capable of evaluating integrals such as

$$\int \frac{\alpha^4}{(1-X^2)^{5/2}} \, dx \qquad (3\text{--}1)$$

in a symbolic fashion similar (especially in the case of SAINT) to the way in which people go about solving such problems.

SAINT was constructed to use a table of trivial integral forms, such as

$$\int u^n \, du = \frac{u^{n+1}}{n+1} \quad (n \geq 1)$$
$$\int \sin u \, du = \cos u$$
$$\vdots$$

and it was given problem-reduction operators corresponding to various rules for the transformation of integrals, such as the integration-by-parts rule, the sum-decomposition rule, and certain trigonometric and algebraic substitution rules. For our purpose, it is not necessary to understand these rules or integral calculus.

When SAINT was given an expression like (3–1), it would attempt to reduce the expression to a combination of the trivial integrals in its table, by the proper application of its problem-reduction operators. In most cases its success at doing integration problems in this way was at about the level of a good first-year calculus student.

Figure 3–9 shows a portion of the AND/OR graph constructed by SAINT in its solution for the problem expression (1). The top part of the graph is similar to Figs. 3–3, 3–4, and 3–6. The trivial integral forms, or primitive problems, are the dark-bordered square nodes at the bottom of the figure. The first operator applied by SAINT was "trigonometric substitution"; this transformed the start node into the problem at node $A$ in the figure. Then SAINT applied two operators and obtained node $B$. Since SAINT estimated $B$ as being a difficult problem, the program went back to $A$, applied another reduction operator, and obtained node $C$. But $C$ also looked difficult, so SAINT went back to $A$ and applied a sequence of three operators, labeled "trigonometric identity," "trigonometric substitution," "algebraic identity," and obtained node $D$. Then SAINT applied the reduction operator "sum-decomposition," which transformed $D$ into *three* problems ($E, F,$ and $G$) linked together by an AND-node; the AND-node between node $D$ and nodes $E, F,$ and $G$ means that $D$ can be solved if $E, F,$ and $G$ can all be solved. Since $F$

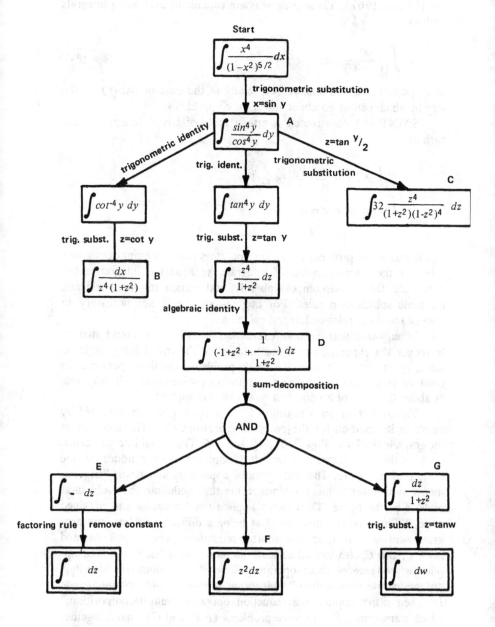

Figure 3–9. A portion of SAINT'S AND/OR graph for an integration problem. (Adapted with permission from Nilsson, 1971)

turned out to be primitive, or trivially solvable, $E$ and $G$ were quickly reduced to primitive problems.

Thus, SAINT deduced the following facts:

1. *Start* can be solved if $A$ can be solved.
2. $A$ can be solved if $B$, *or* $C$, *or* $D$ can be solved.
3. $D$ can be solved if $E$ *and* $F$ *and* $G$ can be solved.
4. $E,F$, and $G$ can all be solved.
5. *Start* can be solved.

Having proved that its initial problem could be solved, SAINT was able to construct the actual solution

$$\int \frac{X^4}{(1 - x^2)^{5/2}}\,dx = \arcsin X + \frac{1}{3}\tan^3(\arcsin x) - \tan(\arcsin x)^3$$

(3–2)

by first solving $E,F$, and $G$ and then undoing the sequence of substitutions it had used in going from *start* to $A$ to $D$. SAINT required about 11 minutes to solve this problem.

Notice that nothing has been said about how SAINT "estimated" the difficulty of problems. This subject is left for the next section; for the moment, we have concentrated on the nature of problem-reduction problems and AND/OR graphs.

The SIN program by Moses is more sophisticated than SAINT. SIN itself might be said to constitute a single reduction operator, which in most cases is capable of going directly from problem to solution without generating an AND/OR graph. SIN is capable of solving integration problems "at the difficulty approaching those in the larger integral tables" (Moses, 1967). For example, SIN can evaluate problem (3–1) in about 9 seconds (note 3–7); in doing so, it generates only two subgoals in contrast to the 13 required by SAINT for the same problem.

## Summary

We have seen two ways of stating problems that, in effect, ask the problem solver to find paths connecting prespecified sets of nodes in graphs. In many cases (including problem-reduction problems) the relevant graphs may be too large to store or generate completely by computer; this is probably true for most of the important problems a mechanical intelligence might be called upon to solve. How is it possible to solve problems of this sort? What enables a computer to avoid generating 10 trillion states of the 15-Puzzle, and yet still solve the problem?

# HEURISTIC SEARCH THEORY
## Need for Search

The field of AI research concerned with ways that computers can solve large state-space problems is known as *heuristic search theory*. In this section we introduce the reader to some of the central concepts of this field; more thorough discussions are provided by Nilsson (1971), Banerji (1969), Pohl (1970), and Michie (1971).

Given a finite description of a state-space problem, a computer can be programmed to generate the state space of the problem while checking to see if it has produced a solution. The *generation* process consists simply of producing finite descriptions (data-structures) for the nodes of the state space and for their connections to each other. With a large, difficult state-space problem, it will not be possible for the computer to generate descriptions for each of the nodes and connections between nodes of the state space of that problem. Rather, the computer may generate only a relatively small portion of that state space, and can check only that portion, to see whether it includes a path between nodes, which is a solution to the problem. With suitable programming, the computer may generate a portion of the state space containing on the order of $10^3$ nodes (for some problems it may be necessary and possible to generate a few orders of magnitude more; conversely, the "general problem solvers" discussed in this chapter typically may generate no more than 100 nodes), whereas the state space of a difficult problem may easily contain $10^9$ nodes. Thus, it is clear that the computer must be somewhat selective in the way that it generates the portion of a state space that it produces when trying to solve a state-space problem, if it is to be successful. Any procedure that a computer uses to *generate* a portion of the state space for a problem, and to *check* that portion for a solution, is said to *search* the state space, and is called a "search procedure." In this section we are interested in ways that search procedures can be designed to be "selective"; that is, ways they can be successful at finding a solution to a state-space problem without generating the entire state space of that problem.

Of course a search procedure might find a solution for a problem simply by randomly generating descriptions for nodes and their interconnections, but unless a large percentage of the paths through the state space of a problem happen to be solution paths, such a procedure will not generally be successful. Usually, what we desire in a search procedure is that it somehow be "systematically oriented" toward the problem it is being used to solve, in such a way that it can find a solution without generating the entire state space. A search procedure that is systematically oriented toward a problem will be said to embody *heuristic* (i.e., "serving to discover") *information* and will be called a *heuristic search procedure*. (Ways of achieving "systematic orientation" are discussed below.) If we can *prove* that a search procedure will always find a solution—if there is one—to any state-space problem $(\{s\},F,G)$ such that $s \in S$, then we say the search procedure is an *algorithmic search procedure* for the state-space problem $(S,F,G)$. It is possible for a given search procedure to be either heuristic or algorithmic or both or neither, with respect to some state-space problem $(S,F,G)$. Most often the search procedures used by problem-solving computers are heuristic, but not algorithmic; sometimes they are both (thus, a symbolic integration program using the Risch algorithm (note 3–7) would be heuristic and algorithmic, according to our definitions). Again, for large state-space problems, there is little value to a search procedure that is algorithmic but not heuristic, one that would solve the problem but might have to search the entire state space to do so.

Thus, "heuristic programming" refers to computer programs that employ procedures not necessarily proved to be correct, but which seem to be plausible. Most problems that have been considered by AI researchers are of the sort where no one knows any practical, completely correct procedures to solve them; therefore, a certain amount of proficiency in using hunches and partially verified search procedures is necessary to design programs that can solve them. So, by a *heuristic* is meant some rule of thumb that usually reduces the work required to obtain a solution to a problem. (Again, it may be possible to prove that the heuristic will *always* supply solutions to some set of problems, i.e., that it is algorithmic.) Clearly, much of the conscious thinking that people do is based upon the use of heuristics that have not been shown to be algorithms.[4] The realization of this fact and its incorporation in the design of computer programs was an important step in the development of artificial intelligence, signifying a recognition by AI researchers that intelligence is often exhibited in situations where one's understanding and knowledge are incomplete.

## Search Procedures

There are basically two methods of incorporating heuristic information about (i.e., "systematic orientation" toward) a state-space problem into a search procedure designed to solve that problem; these methods

---

[4] We might have a hard time proving this to a strict behaviorist. This is one of the places where the author invokes his "personalistic license," granted in note 1–2.

correspond to the use of "generator functions" and "evaluation functions." Our description of these methods will be facilitated if we examine in a little more detail the generation processes that may be used by search procedures.

The generation processes that AI researchers have investigated for state-space problems are made up of the following basic steps: First, a start node $s$ is given to the search procedure. This node corresponds to a finite description of a state and is stored by the computer as a data-structure.

Next, using the operators (in the set $F$ of the state-space problem), the successors to the start node are generated (i.e., a finite-description for each successor is generated). We denote by $\Gamma$ a procedure that calculates all successors to a given node. The process of applying $\Gamma$ to a node is known as *expanding* the node, or *generating* the successors to the node; thus, $\Gamma$ is often referred to as a *generator function*, or *generator*.

After a node is expanded, *pointers* are set up, leading back to the node from each of its successors. If a goal node is ever generated, then there will be pointers indicating a path from it back to the start node.

The successor nodes produced when a node is expanded are checked to see if one of them is a goal node. If no goal node is found, then the process of expanding nodes and setting up pointers is continued by expanding nodes that have been generated as successors. If a goal node is found, then the pointers that have been set up are used to trace a path back to the start node—the operators that were originally used by $\Gamma$ to produce the nodes along this path may be recovered and used to produce a solution path.

The various search procedures developd for solving state-space problems may be distinguished from each other on the basis of two criteria: how the process of expanding nodes and setting up pointers is continued, and the nature of their generator functions. A search procedure that expands nodes in the order in which they are generated, after generating all of them below a given node, is called a *breadth-first* search procedure. A search procedure that always expands the most recently generated node first is called a *depth-first* search procedure. Figures 3–10 and 3–11 show "snapshots" of the successive portions of a state-space graph that would be generated by breadth-first and depth-first search procedures. Both types of search procedure are examples of *blind search procedures* because the order in which they expand nodes is unrelated to the actual location of goal nodes in the state space (unless their generator functions incorporate heuristic information; see below). Thus, they are not heuristic search procedures.

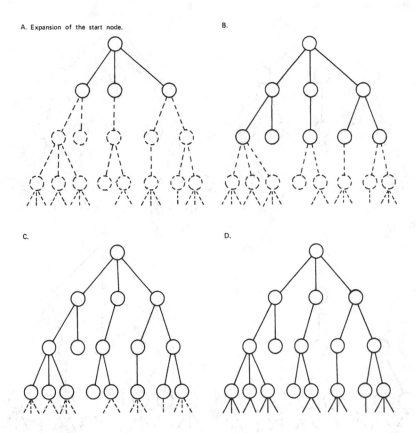

Figure 3–10. Snapshots of the search produced by breadth-first procedure. Dotted circles, ungenerated nodes; solid circles, generated nodes.

The breadth-first search procedure is algorithmic: If a path does exist from a given start node to a goal node, it will eventually be produced, using this procedure. It is possible for the depth-first procedure to search forever, going off in the wrong direction, without finding a solution path, even though one might exist. So, the depth-first procedure as stated here is not algorithmic. However, it may be modified to an algorithmic procedure by introducing the concept of the "depth" of a node (relative to the given start node): The *depth* of a node is zero if it is the start node, and is one plus the depth of its parent otherwise. A *bounded depth-first* search procedure is one which expands that previously generated, unexpanded node which has the greatest depth less

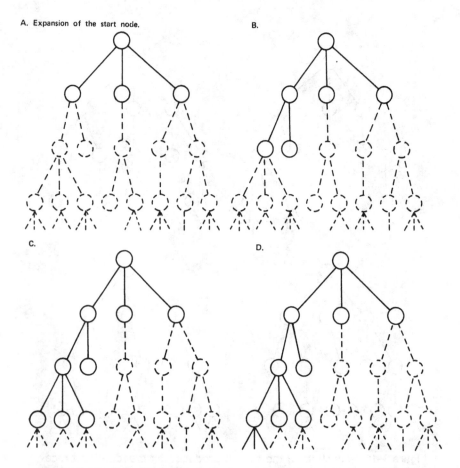

Figure 3–11. Snapshots of a search produced by a depth-first procedure. Dotted circles, ungenerated nodes; solid circles, generated nodes.

than the *depth (or level) bound l* established for the procedure. (If
there is more than one such node, it expands the one most recently
generated.) As illustrated by the snapshots in Fig. 3–12, a bounded
depth-first procedure generates nodes in a depth-first manner until it
reaches its depth bound; it then "backs up" and generates more nodes
in a different direction, etc. It is fairly simple to see how this idea may
be extended (essentially by allowing the depth bound to be systematically increased) to produce an algorithmic search procedure with a
basically "depth-first" nature.

Most heuristic search procedures are, in effect, modifications of the

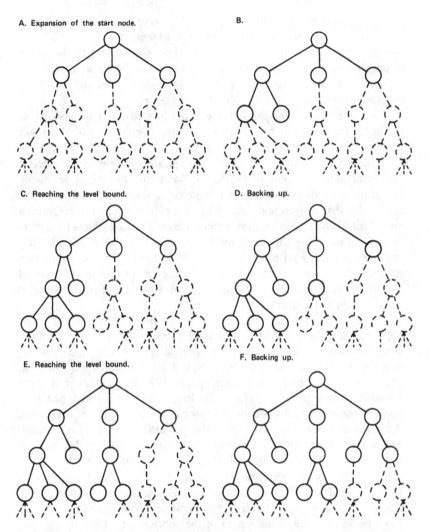

Figure 3–12. Snapshots of a search produced by a bounded depth-first procedure. Dotted circles, ungenerated nodes; solid circles, generated nodes.

bounded depth-first search procedure. As explained initially, heuristic search procedures rely on two methods, the use of generator functions and the use of "evaluation functions." A generator function may in-corporate heuristic information about a problem if it is designed to generate first those successors of the node to which it is applied which are most likely to lie on (preferably short) paths to some goal node. A search procedure that uses such a generator will tend to be "guided" toward a solution if it expands nodes according to the order in which they are produced by its generator. Thus, such a search procedure is "systematically oriented" by its generator toward searching the most promising portions of the state space first.

An *evaluation function* is some procedure that can be applied to the finite description of a node in a state-space problem and which will produce an estimate of the "value" of that node (the likelihood that the node lies on a path to a goal node). AI researchers have investigated a variety of different kinds of evaluation functions: For example, Slagle and Bursky (1968) designed an evaluation function that estimated the *probability* that a given node would be on a path to a goal node; Quinlan and Hunt (1968) used an evaluation function that constructed a *difference set,* measuring the (structural) differences between an arbitrary node and a given goal node; Samuel (1959, 1967) used an evaluation function that examined the important "features" possessed by a board configuration in checkers, to produce an estimate of the "strategic value" of the configuration (see Chapter 4).

The central results in heuristic search theory are those of Hart, Nilsson, and Raphael (1968) and Pohl (1970). Their results hold for evaluation functions that produce numerical estimates for the "values" of nodes in state spaces. By convention, if $f$ is an evaluation function, and $n$ and $n'$ are nodes in a state space, then we say that $n$ is more valuable than $n'$ if $f(n) < f(n')$; the lower the number assigned to a node by the evaluation function, the greater is the "value" of that node. An *ordered search procedure using the evaluation function $f$* is a search procedure that expands the previously generated, unexpanded node $n$ for which $f(n)$ is a minimum; if there is more than one such node $n$, then it expands the most recently generated one. The Hart-Nilsson-Raphael result may be stated as follows: For a given state-space problem $(\{s\}, F, G)$, let $g(n)$ be the depth of node $n$ from the start node; let $h(n)$ be an estimate of the length of the shortest path from $n$ to a goal node of the state space, and let $h_p(n)$ be the actual length of the shortest path from $n$ to a goal node. If for any node $n$ we have $h(n) \leqslant h_p(n)$, then an ordered search procedure using the evaluation function $f(x) = g(x) + h(x)$ will always find the shortest solution path for the state-space problem $(\{s\}, F, G)$, if there is a solution path at all. Furthermore, provided $h(x)$ is generally greater than zero, the ordered search procedure will often need to expand fewer nodes to produce its solution than would the breadth-first search procedure. Thus, such an ordered search procedure is both algorithmic and heuristic. If we relax the condition that $h(x) \leq h_p(x)$, the ordered search procedure will still be heuristic, but it may not be algorithmic. (More specific information about $h$ would be needed to determine whether it is algorithmic.)

## Search Trees

Actually, the presentation of heuristic search theory thus far is precisely correct only for problems whose state-space graphs have the nature of a "tree." A *tree* is a graph with the following characteristics:

1. The tree contains exactly one node that does not have a parent; this node is called the *root node*.
2. Every other node in the tree is a descendant of the root node.
3. Every other node in the tree has exactly one parent.

If the graph for a state-space problem $(\{s\}, F, G)$ is a tree, then the start node $s$ will, of course, be the root node of the tree: A tree that is a graph for a state-space problem is often called a *problem tree* for that problem. Figure 3–4 shows a portion of the problem tree for the Three Coins Problem.[5] A basic modification is needed to make an ordered search procedure, using the evaluation function $f(x) = g(x) + h(x)$, produce an optimal solution (in the sense of the preceding paragraph) when searching a state-space graph that is *not* a tree. The modification consists of providing the procedure with a means of "updating" its function $g(x)$; a node in a (general, non-treelike) graph may have more than one parent. Thus, we should define the "depth" of a node $n$ in a graph to be zero when it is the start node $s$; otherwise, we should define it as one plus the depth of its shallowest parent. The ordered search procedure may generate a node $n$ when expanding a node $n'$ with a depth of $d'$ and later generate the node $n$ again when expanding a node $n''$ with a depth $d''$. If $d'' < d'$, then the procedure should change its estimate for the depth of $n$, from $d' + 1$ to $d'' + 1$ (and it should make a similar change in its estimate of the depths of those nodes that are descendants of $n$).

Pohl (1970) presented similar results for an ordered search procedure using the evaluation function $f(x) = (1 - \omega)g(x) + \omega h(x)$ where $\omega$ is an adjustable parameter. (Pohl also discusses *bidirectional search procedures,* which are procedures that generate the state space of a problem outward from both the start and goal nodes; he concluded that such procedures are much more difficult to implement efficiently

---

[5] Problem trees are normally drawn upside down, with the root node at the top (see Knuth, 1969a, p. 307).

than are the ordinary "one-directional" searches we have discussed.)

It is often desirable to design the generator function used by a search procedure to have a "memory," and to generate the successors to a given node in a one-at-a-time manner that can be interrupted and resumed when necessary. If such a generator function incorporates heuristic information (thus tending to generate first those successors to a given node that have the greatest value), then a search procedure that uses it may search the most promising parts of the state space first, without needing to generate, store, or evaluate the less plausible nodes. And, if its first searches of the state space do not succeed in producing a goal node, the procedure may then reapply its generator function and search less plausible parts of the state space. Michie (1971) presented search-theoretical results for a general problem-solving program (GT4), which uses this type of generator function.

The general problem-solving programs we have discussed in this chapter (GPS, FDS, MULTIPLE, GT4—REF-ARF differs slightly, as we shall see below) are all programs that can accept a finite description for an arbitrary $(S,F,G)$ state-space problem, and which use that description to conduct a search through a portion of the state space of that problem. Their generality resides in the fact that they can accept finite descriptions for many different $(S,F,G)$ problems, and can often find solution paths for those problems. To a large extent, FDS and MULTIPLE are able to develop their own evaluation functions. The limitations of these problem solvers are due to two facts: They can search only relatively small state spaces; and they cannot develop a better finite description for a state-space problem than the one they are given—that is, they are not representation-independent.

Although we have not discussed applications of heuristic search theory to problem-reduction problems, much the same results can be obtained. It should be noted that the procedures for searching AND/OR graphs are essentially the same as those for searching the state spaces of two-person, nonchance, perfect-information games of strategy. (An AND node corresponds to a move belonging to one's opponent; an OR node corresponds to a move belonging to oneself.) The discussion in Chapter 4 is therefore relevant to search procedures for AND/OR state-space problems. However, for a thorough treatment of heuristic search theory as it applies to problem-reduction problems, and for a much more extensive discussion of the material in this section, we encourage the reader to see the book by Nilsson (1971).

# PLANNING, REASONING BY ANALOGY, AND LEARNING

## Planning

In its remaining two sections, this chapter discusses some important aspects of current AI research on problem solving. The topics discussed in this section are "planning," "learning," and "reasoning by analogy"; the next section discusses "models," the "problem of problem representation," and the "levels of competence" that have been attained by machine intelligences. These topics represent open questions for AI researchers, rather than established theories. Space does not permit presentation here in detail of the many viewpoints (and results!) that have been developed about these topics, although, because of the interdependence of aptitudes, they will be discussed in greater detail in subsequent chapters. At this point, only some brief summaries and references to the literature are presented.

A process that constructs and executes *plans* for solving problems is said to be a *planning* process. As emphasized throughout this chapter, many problems are partially specified, and for them there may not exist a single string of actions or operators that will always constitute a solution; therefore, the best, initial solution is often a plan. Plans for solving such problems may specify a wide variety of different actions, including "looking for outside help" and "making a new plan." Furthermore, these actions may be *conditional;* that is, a plan might include statements of the form "if $X$ happens, then do $Y$; otherwise do $Z$." Or, they may include *loops* and *recursion* such as: "Step1. Put money in the jukebox and punch-a-button; if it doesn't play what you want, then go to step1; otherwise, go to step2"; "If at first you don't succeed . . ."; "Move block $(x)$. If $x$ does not support anything, then pick it up and move it. Otherwise, for all $y$ such that $x$ supports $y$, first do move block $(y)$."

Again, the state spaces of some problems are extremely large, and the shortest solution-path for such a problem might be very long. A plan for such a problem might specify subgoals along the solution-path, and instruct the computer to search first for a path to the shallowest subgoal, and then for a path from that subgoal to the next, etc. (See McCarthy's San Diego Problem in Exercise 3–11.) This is often referred to as the "milepost" paradigm for plans.

AI research on plans and planning may be divided rather naturally into three categories: paradigms for the concept of plan; computer execution of plans; computer development of plans (which is what we should properly call "planning"). The degree of success attained by the research into these categories corresponds roughly to the order in which they are enumerated. Thus, a number of paradigms have been developed and many of them can be transformed into something that a computer can execute, but very little success has (so far) been obtained in having computers develop their own plans.

The preceding paragraphs summarize roughly the characteristics that have been proposed for "plan" by the various paradigms that have been developed; these characteristics may be condensed into something of a formal definition: A plan is a collection of procedures together with specifications for when those procedures should be used (i.e., "called"). Each of the various paradigms is a formalization of this idea in more detail for a specific problem domain. Perhaps the most extensive formalization is that provided by Hewitt (1968 et seq.), which is discussed further in Chapter 6. Other paradigms for "plan" are explored by Doran (1970) and Michie (1971).

Of course, there is a "strange paradox" here, because we have used the same words (in essentially the same phrases) to talk about the concept of "problem." Thus, a problem is a collection of procedures (operators) together with specifications for how they shall be used to construct a state-space graph, and information as to which paths in the state space are solution paths. The concepts of "problem" and "plan" may both be formalized by reference to procedures and their interactions with data-structures. Thus, one of the paradigms for "plan" is to see them as "nondeterministic" programs, whereas the REF-ARF general problem-solving program (Fikes, 1970) is designed to correspond problems with such programs. For a description and discussion of nondeterministic programs, see Manna (1970b). (A similar, but less well formalized paradigm for "plan" corresponds plans to "fuzzy" programs; see Zadeh, 1968.)

In passing, it should be noted that a program has been written which uses the "milepost" paradigm for plans in a procedure that evidently is capable of solving the problems of the 15-Puzzle. The program was written by Ashok K. Chandra of Stanford University at the request of John McCarthy. The "mileposts" or subgoals used by the program correspond to the correct placement of the blocks of puzzle in successive "gnomons" of the tray (see Fig. 3–13). The program conducts an ordered search through the state space of the 15-Puzzle, attempting first to correctly place the blocks in the outer gnomon, then the next outer, etc. Moreover, the search conducted by Chandra's program is bidirectional. At the present author's request, this program was run on

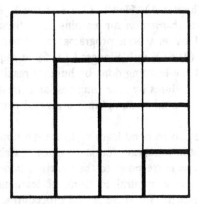

Figure 3–13. Gnomons.

a sample of about 300 randomly generated 15-Problems (start-node plus goal-node in the state-space), and found a solution-path for each problem that was solvable. Usually, the program required about 10 seconds to solve a given 15-Problem, and in doing so expanded less than a thousand nodes of the state-space.

## Reasoning by Analogy

The importance of designing problem solvers with an ability to "reason by analogy" has been stressed by a number of investigators. A number of basic kinds of analogies were identified by Kling (1971), discussed further in Chapter 6. The earliest program for "analogical reasoning" was that of Evans (1963). Winston (1970) presented an elegant formalization for the concept of "analogy" and showed how the results of Evans (1963) can be extended to three dimensions ("or more"; see Chapter 5). Becker (1969) discussed "semantic analogies," and Ramani (1971) presented a program that answers questions "by analogy."

## Learning

AI research has developed many paradigms for the concept of "learning." Learning for state-space problems may be formalized as a process that finds suitable evaluation functions and generator functions for an ordered search procedure. Samuel (1959, 1967) provided the classic treatment of this type of learning, in which the nature of an evaluation function for nodes in the game-tree of Checkers is changed by a checker-playing program, based on its previous experience with the game (see Chapter 4). Hewitt (1968 et seq.) presented the paradigm of *functional abstraction* for learning, and discussed some ways it might be utilized by PLANNER programs (see Chapter 6). McCarthy (1963), Minsky (1968) and Winograd (1968 et seq.) emphasized the fact that much of the learning done by humans results from their being taught various procedures by other humans, and stressed the desirability of incorporating "communication" into a general paradigm for learning (Chapter 7).

Since we may correspond learning to the development of the evaluation and generator functions for state-space search procedures, and since such functions correspond to the heuristic information these procedures may use, it is natural to think of learning as a process of "heuristic development" for these search procedures. In a sense, a program that modifies the evaluation function used by its search procedure is developing its own heuristics. However, it should be stressed that it is difficult (unless the program is "self-affecting"; see Chapter 8) to say that such a program is *really* developing its own heuristics: The process (program) by which heuristics are developed is itself a heuristic.

## MODELS, PROBLEM REPRESENTATIONS, AND LEVELS OF COMPETENCE

### Models

Throughout this book, and especially in Chapter 7, the role of "model-making" in artificial intelligence will be emphasized. In theorem-proving terminology (Chapter 6), a model is a particular interpretation of a statement, or of a set of statements. (A set of statements may have more than one model.) Any statement that is logically implied by a set of statements with a given model will hold true for that model, but any statement that does not hold true for the model cannot be logically implied by the set. This fact may be used as a device for recognizing non-derivable statements: A particular instance of a candidate statement may be compared against a model; if it is found to be false, then we know the candidate statement cannot be derived from the set of statements for which our model holds. A theorem-proving procedure may therefore be designed to reject automatically certain[6] statements it cannot hope to derive, if it has a means of developing models and using them for comparisons. Rather than statements, we may think of this

---

[6] . . . but not all, thanks to Godel (1931).

process as discarding possible successor nodes of a given node in a problem-reduction problem (see Nilsson, 1971). Gelernter (1959) presented a landmark testing of this concept in a program for proving theorems about plane geometry. For his program the use of models enabled, on the average, all but 5 out of 1,000 of the successors to a given node to be rejected.

# The Problem of Problem Representation

We have emphasized the desirability that general problem solvers be representation-independent; that is, capable of developing their own problem representations. No one has yet succeeded in giving representation independency to computers. However, Amarel (1968 et seq.) charted part of the basic mathematics necessary for such a task, and showed how a sequence of successively better problem representations for the Missionaries and Cannibals Problem (see the Exercises) can be produced, using the concepts of *macro-operators* and *macrostates*. Because of the fact that problems can be represented by programs, it is possible to treat the problem of problem-representation from a procedure-oriented point of view (see Hewitt, 1968 et seq.) in which the problem of developing (correct, improved) problem representations is equivalent to that of developing (correct, improved) programs (see Chapter 6). Again, we may see it as a problem of learning languages for problem description.

# Levels of Competence

Currently it may be said that AI research has produced the following "skillful" programs, which perform tasks with an aptitude that people normally correspond to that of a very practiced human intelligence:

Samuel's Checkers Player
Greenblatt's Chess Player
The symbolic integration programs of Slagle, Moses, and Risch
Feigenbaum's DENDRAL[7]
Wasserman's Bridge Bidder
Chandra's 15-Program

---

[7] DENDRAL is a heuristic program that infers the structure, of molecules from their mass spectrographs. Its performance compares favorably to that of graduate chemistry students (see Feigenbaum et al., 1971).

(The list is not exhaustive.) The aptitude of these programs for their tasks has been verified by direct comparison with the abilities of humans who are known to be "skillful" at performing the same tasks. Thus, Samuel's Checker Player is able to outplay all but the very best human Checker players. More modest claims must usually be made for the other "skillful" programs.

However, an important point should be noted: All these skillful programs are highly specific to their particular problems. At the moment there are no general problem solvers, general game players, etc., which can solve really difficult problems (e.g., the 15-Puzzle) or play really difficult games (e.g., Checkers or Chess) with a skill approaching that of human intelligence. And, it goes without saying, there are no general programs that can *learn* to perform these difficult tasks skillfully (note 3–2).

## NOTES

**3–1.** The IPL system was developed in 1956 by Allen Newell, J. C. Shaw, and Herbert Simon; LISP was developed in 1960 principally by John McCarthy. The importance of good programming languages to the development of AI research cannot be overestimated; AI research could not really get off the ground without IPL and LISP. It is doubtful that one of the most significant recent developments (Winograd's work on natural languages) could have been obtained without PLANNER and a similar programming language called PROGRAMMAR. Programming languages are discussed further in Chapters 6 and 7.

**3–2.** An empirically based theory that has been produced by many AI researchers is that the more general the aptitude possessed by a machine, the less efficient is its performance of the tasks that the aptitude enables it to perform. Thus, Newell, Shaw, and Simon noted that their General Problem Solver was less efficient at solving the problems it could solve than would have been programs specifically designed for solving each of those problems. This relation between generality and efficiency has been confirmed by the other general problem solvers mentioned in this chapter. However, there is room for doubting that the relation is a "real" one; perhaps it is possible to design general problem solvers that can learn to solve the problems in a given problem-domain more and more efficiently (for instance, the ability of people to learn to solve crossword puzzles) and, within a short time, approach the efficiency of problem solvers designed specifically to solve the problems of that domain.

**3–3.** The principle is basically that suggested by McCarthy (1956), namely, that "the enumeration of partial recursive functions should give an early place to compositions of the functions which have already appeared."

In this early paper, McCarthy suggested the system-inference paradigm, using the following argument: A problem should be something that has solutions. By a "well-defined problem" is meant one for which there is a definite test that verifies whether a proposed solution is correct. If a proposed solution is not correct, then the test may either reject it or not terminate, but if the solution is correct, then the test must always verify it in a finite, though possibly variable, time. Let us regard the test as being carried out by a Turing machine $T$, which, given as input a proposed solution (or description of a proposed solution), will output in a finite time an affirmative symbol $r$ if the proposed solution is correct. The statement of the problem then consists in a description of $T$ and the designation of $r$; a solution to the problem is any input string $i$ such that $T(i) = r$. A *general* problem solver is a machine that, given the description for the $m$th Turing machine $T_m$, will compute an i such that $T_m(i) = r$—if in fact there is such an $i$. Since, given $m$, one can construct the description of the $m$th Turing machine, a general problem solver can be said to compute a function $g$ on two inputs $(m,r)$ such that $T_m (g(m,r)) = r$; $g$, of course, is to be a partial function, not defined for all $m$ and $r$ (see also, McLamore, 1968).

**3–4.** An alternate statement for the state-space paradigm is given in Sandewall (1969). His formalization is briefly described here. To distinguish it from the $(S,F,G)$ formalization, let us call it the $(S',F',G')$ formalization.

The notion of *state* is the same as in the text; that is, a state is a data-structure. For the sake of simplicity, we shall call a collection (or set) of states a *particle* (physicists beware!), and say that the states in a particle *exist*. An *operator* is a computational rule, which can be applied to existing states in a particle to produce new states that will also exist in the particle. An operator can either change the states to which it is applied or it can remove some (perhaps all) of them from the particle, or it can simply add new states to the particle. A description of an $(S',F',G')$ state-space problem is the specification of three things: $S'$, an initial or starting particle; $F'$, a set of operators; and $G'$, a desired or goal particle. A solution to an $(S',F',G')$ problem is the specification of a finite sequence of operators and of how they are to be applied to $S'$ and its successors such that $G'$ will be produced.

Sandewall (1969) discussed various types of operators and formulated a theory of heuristic search for this type of problem. Also, he suggested that the proper representation for the possible ways of going from one particle to another is in terms of *lattices* rather than trees or graphs.

**3–5.** The 15-Puzzle was invented in 1878 by Sam Loyd (see Loyd, 1960) and was extremely popular during its early years, especially in Europe. Kasner and Newman (1956) reported that employers were forced to post notices forbidding their workers to play the puzzle during working hours. Loyd and others offered huge prizes to anyone capable of solving some of the unsolvable varieties of the puzzle. Some commentators at the time considered the 15-Puzzle to be a threat to society, attributing to it "untold headaches, neuroses, and neuralgias."

**3–6.** Equivalently, we can say a graph $G$ is an ordered pair $(N,R)$, where $N$ is a set of nodes $x, y, \ldots$, and $R$ is a set of binary relations $r_1, r_2, \ldots$ on the nodes. If $r_1(x,y)$ holds for a given $x$ and $y$ in $N$, then we say "$x$ connects to $y$ under the relation $r_1$."

As used here, "graph" is a generalization of one of the most common applications of the word: *graph of a function*. The graph of a function, say, $f: X \to Y$, is a pictorial description of the relation r such that $r(x,y)$ is true if $y = f(x)$.

**3–7.** Much of SIN's time advantage is due to the fact that it was run mostly in a compiled mode, whereas SAINT was run mostly in an interpretive mode. (For an explanation of these terms, see Knuth, 1969.) Moses estimated that his program is actually about three times faster than Slagle's.

More recently, Risch (1969) developed an algorithmic procedure for solving a wide class of symbolic integration problems; Risch's procedure does not need to generate a problem-tree and is guaranteed to always produce correct solutions (it might thus be said to display a "perfect aptitude" for its problem-domain). An introduction to the Risch algorithm and a summary of the current state of work on symbolic integration programs was provided by Moses (1971).

These programs, incidentally, are distinct from "numerical integration" programs, which typically compute numerical approximations to the values of definite integrals. Finally, SAINT is an acronym for "*S*ymbolic *A*utomatic *INT*egrator," and SIN stands for "*S*ymbolic *IN*tegrator."

## EXERCISES

In Exercises 3–1 through 3–10, and in Exercise 3–13, first solve the problems that are given. Next, make a list of the subproblems you considered while solving them. Discuss how a computer might be programmed to solve each of the given problems, and how each of the problems might be represented to the computer. If you find a state-space representation for a problem, estimate the size of the state space and try to identify heuristics and algorithms the machine could use to search it. If computer time is available to you, choose a problem and try to implement a computer program that can solve it.

**3–1.** Find your way out of the Maze of Dedalus.

*3-2.* (*The Missionaries-and-Cannibals Problem.*) Three missionaries and three cannibals are all on one bank of a river they wish to cross. They have a boat, which will hold two persons, but which can be rowed by one if necessary. If the cannibals ever outnumber the missionaries on a given bank, all the missionaries on that bank will be eaten. Otherwise, both parties will cooperate peacefully toward crossing the river. How can all the missionaries and cannibals be transported safely to the other bank? (b) Consider the general case in which there are $m$ missionaries and $n$ cannibals $(m \geq n)$, and in which the boat can hold $p$ persons, but requires at least $r$ persons to be rowed $(p \geq r)$.

*3-3.* (*The Confusion-of-Patents Problem.*)\* A certain patent attorney was astonished when he received the simultaneous allowance of five patents, for five separate clients, each of whom lived in a different city.

His astonishment turned to chagrin, however, when he learned what had happened to the patents. They had been received in his office on the same day, but because of an error made by a new clerk, they were sent out in wrong envelopes. Each client received a patent, but not his own.

The inventor of the steam shovel received the mousetrap patent, while the inventor of the latter found in his mail the papers that should have gone to Mr. Green. Mr. Blue received the patent for the rumbleseat awning. Mr. Black's patent was sent to Chicago; the patent that should have gone was sent to Boston.

Mr. Brown had the patent intended for New York. Mr White had Mr. Brown's patent. The non-refillable bottle patent was sent to Los Angeles; the inventor of the bottle received the patent of the Cleveland client, while in Cleveland the surprised client received a patent for an antisnore device.

Who should have received what where?

*3-4.* (*Traveling-Salesman Problems.*)

(a) For the map shown below, find the shortest path that starts at city A, visits each of the other cities only once, and then returns to A.

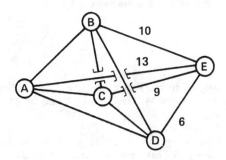

---

\* From Richard E. Fikes, "REF-ARF: A System for Solving Problems Stated as Procedures." *Artificial Intelligence Journal,* Vol. 1 (1970), pp. 27–120. Reprinted with permission.

(b) Find the path from start to finish, which passes once through all the nodes lettered *a* through *u*:

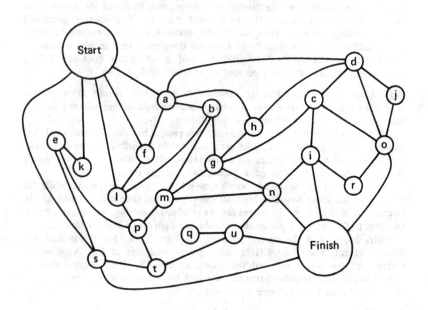

**3-5.** (*Crypt Addition.*) Assign a decimal digit to each of the letters in the words "send," "more," and "money," such that when the letters are replaced by the corresponding digits the following summation is true:

```
  send
+ more
------
 money
```

No digit may be assigned to more than one letter, and leading zeros are not allowed in the numbers formed by "send," "more," and "money."

**3-6.** (*Water-Jug Problems.*)
(a) Given a 3-gallon jug and a 4-gallon jug, how can precisely 2 gallons be put into the 4-gallon jug? There is a sink nearby, such that either jug can be filled from the tap and its contents can be poured down the drain. Also, water can be poured from either jug into the other, but the jugs themselves are the only measuring devices available.
(b) Given a 5-gallon jug and an 8-gallon jug, how can precisely 2 gallons be put into the 5-gallon jug? The conditions for this problem are the same as those in (a).

(c) Given a 5-gallon jug filled with water and an empty 2-gallon jug, how can precisely 1 gallon be obtained in the 2-gallon jug? In this problem, water may be either discarded or poured from one jug into another, but there is no source of water other than the initial 5 gallons. Again, the jugs themselves are the only measuring devices to be used.

*3–7.* (*The Monkey-and-Bananas Problem.*) A monkey is in a room where a bunch of bananas is hanging from the ceiling, too high to reach. In the corner of the room is a box, which is not under the bananas. The box is sturdy enough to support the monkey and light enough so that he can move it easily. If the box is under the bananas and the monkey is on the box, he will be able to reach the bananas. How can the monkey get the bananas (if he wants them)?

*3–8.* (*The Mutilated Checkerboard.*) Show that it is impossible to completely cover the "mutilated-checkerboard" with 1 x 2 tiles so that the tiles neither overlap nor stick out over the edge of the board.

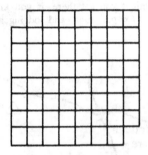

*3–9.* (*The Tower-of-Hanoi Problem.*) Initially three disks of different sizes, each having a hole in its center, are placed as shown in the diagram below, all of them about one of three pegs. It is desired to transfer their initial configuration to the third peg, moving them one at a time in such a way that only the top disk on a peg is ever the disk being moved and a larger disk is never placed on top of a smaller disk. How can this be done?

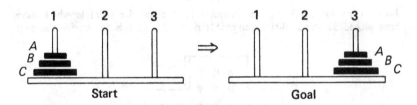

*3–10.* (*The Sliding Block Puzzle.*) Nine blocks are placed in a tray as shown below. (a) How many different configurations of the blocks may be obtained by sliding them about in the tray? (b) How many different configurations of the puzzle are there if configurations that may be obtained from each other by rotating or flipping the tray are considered to be the same? (c) Design a computer program that can explore the state-space of the sliding-block puzzle.

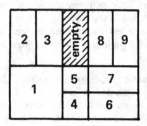

3–11.  (*The San Diego Problem.*) You have a road map for the area surrounding your present location; however, because the map was produced by the Super-
Duper gas-station chain, it shows only the roads in a 30-mile circle, the north and
east directions, the locations of two SuperDuper gas stations, and your present
location. Actually, you want to go to San Diego, which you know to be 400
miles to the south. How might you get there, if you know how to drive, have
a car, and sufficient money for gas, food, and lodging along the way?

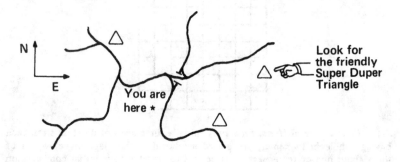

3–12.  Suppose we are given that nodes *B, C,* and *D* in Fig. 3–8 represent
trivially solvable problems. (a) What can be said about the solvability of node
*A?* (b) What if *B, C,* and *D* are unsolvable?

3–13.  (*Peg Solitaire.*) A board contains 33 standard-size holes in which have
been placed 32 standard-size, removable pegs. The goal is to remove the pegs

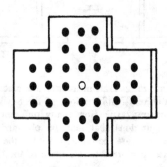

in such a way that a board is obtained which contains only one peg, placed in the (initially empty) center hole. Pegs may be removed only by "jumping" them, as in checkers; that is, a peg $A$ may be removed if and only if there is a peg $B$ next (left, up, right, or down) to it and an empty hole $C$ on the opposite side, and the removal of peg $A$ is accompanied by the placing of peg $B$ in hole $C$.

*3–14.* Why should a depth-first search procedure always expand the most recently generated node first?

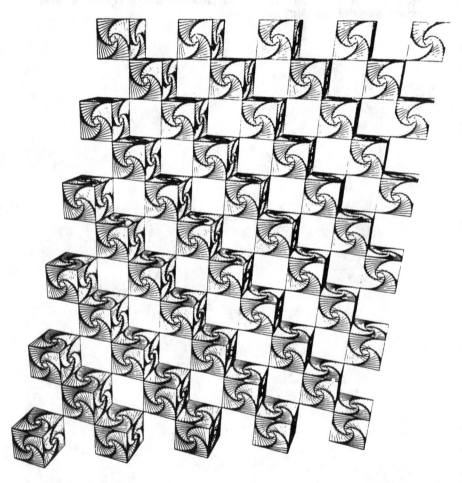

Checkerboard pattern. (Reproduced with permission, D. K. Robbins.)

# 4

# GAME PLAYING

## INTRODUCTION

In this chapter we investigate the ability of computers to play games. First the nature of the games that computers are able to play will be reviewed. Then the way in which computers may make use of heuristic search techniques in order to play these games will be described. These discussions will lead us to computer programs that are capable of playing Checkers, Chess, GO, Poker, and Bridge. The chapter concludes with a brief explanation of "general" game-playing programs.

## GAMES AND THEIR STATE SPACES

Some of the most important programs produced by AI research are those that simulate the human ability to play games: Games comprise a general class of problem concerned with reasoning about actions. They can be constructed with or without an element of chance involved, and they can be designed so as to specify that different players will have different information available to use in deciding how to play. Finally, games offer the possibility of a direct comparison between the abilities of machines and humans.

It is probably wise to remark that all games that computers can now play are of the type that is generally known as "games of strategy" because they possess well-defined rules and objectives for each player. Of course no claim is being made that these are the only games that exist. The reader is probably familiar with many games that do not

117

have objectives (and perhaps a few that do not have rules), which are also popular. Since the ultimate value of a game is the enjoyment one gets from playing it, games of strategy cannot be said to be the most valuable ones that people play. Still, they are the ones most easily identified with the use of intelligence, in its role as an ability to solve problems, and so it is natural that games of strategy should be studied extensively.

## Strategy

A *game of strategy* consists of a sequence of *moves,* each of which is an occasion for a *choice* between certain *alternatives,* made by one of the players of the game.[1] The *rules* of the game specify for each move which player makes the move and what his alternatives are. These rules are finitely describable, and are to be known to each of the players. What the rules will specify usually depends on the previous choices made in the game. For each move, only a finite number of alternatives are available. A complete sequence of choices (one that the rules define as terminating the game) is said to constitute a *play* of the game. In some games the rules will sometimes specify that the choice is to be made by *chance,* in which case the players are usually given a definite, or at least computable, probability distribution for the various alternatives. At each move, a player always knows completely what his alternatives are, but he may not know completely what choices have been made previously. If at each move every player knows completely all the choices that have been made so far in the game, it is said to be a game of *perfect information.* Finally, for each of the possible plays of the game, the rules specify a *payment*—which may be positive, negative, or zero —to be received by each of the players. The objective of each player is to maximize the payment he receives, by definition (if a player's payment is negative, then he is said to *make* the payment.)

These statements, of course, summarize only the logical, formal aspects of games of strategy, and say nothing about such questions as how a given game might be implemented physically, how the moves might be represented, etc. The computer programs discussed here accept symbolic descriptions of the choices made during a game, and when the rules require it, they produce symbolic descriptions of their own

---

[1] This paragraph comprises a brief summary of the basic definitions for "games of strategy" presented by von Neumann and Morgenstern (1944) in their "theory of games" (or, simply, "game theory"). The word "move" is used in their game-theoretic sense: "It's your move." In some games (e.g., Chess) it is common to use the word in an additional sense: "He moved the king's pawn forward two spaces."

"choices" (note 4–1). Other computer programs might be written to convert the symbolic descriptions produced by a game-playing program into a physical action, such as moving a pawn forward on a chessboard. Our concern in this chapter is only with computer programs that handle the "intellectual" aspects of game playing.

In general, a *strategy* is any set of rules that tells a player what choices he should make for all situations that might arise during the course of a game. A "good" strategy is one that guarantees its user will receive a high payment, or, in the case of games involving chance, it is one that provides for a high "mathematically expected" (in a sense, probable) payment. Given the complete description of a game, the theory of games provides a computational procedure capable of determining the correct strategies for all players, their best expectation in playing the game, etc.

This procedure depends, however, upon the enumeration of all strategies available to each of the players (including the strategies "chance" might use), which is something easy to describe but frequently difficult to perform. Thus, for many games, the number of strategies may be considered "effectively infinite," since any attempt to enumerate them all would require too much time. As we shall see below, this is true of the more difficult board games (Checkers, Chess, and GO) played by people. Yet people seem to be able to play these games fairly well (note 4–2). Throughout, this chapter emphasizes primarily the nature of the strategies that computers can use to play games and the extent to which computers can be enabled to select their own strategies. To pave the way for a discussion of this topic, let us present another, very similar formalization for "games of strategy."

## State Spaces

The brief description of games given above can be rephrased, using the terminology of the state-space paradigm for problems presented in Chapter 3. A game may be viewed as a state-space graph, together with a function associating some of the paths through the graph with payments (positive, negative, or zero) to be received by the players of the game. The nodes or states of the graph are descriptions of the *moves* or *situations* involved in the game; the arcs emanating from a given node (the operators applicable to a given state) are the alternatives associated with the corresponding move. Thus, a node in the state space of Checkers is a description of a legal configuration of pieces on a checkerboard, together with an indication of whose move it is. A node from which no arcs emanate (i.e., a terminal node) is one for which the game ends.

It is common to indicate that a certain player is to make a given move
(choose among the alternatives associated with that move) by drawing
the node for that move with a certain shape or shading that is different
from that used for the moves belonging to the other players. Terminal
nodes are usually drawn with the shape or shading of the player whose
move they would be, if the rules make a specification, even though they
do not have successors. Thus, Fig. 4–1 shows the state-space graph for
a simple game.

A game of strategy begins at the node in the graph labeled "start."
The person who has the starting move (player 3 in Fig. 4–1) in the
game chooses one of the available alternatives (one of the arcs leaving

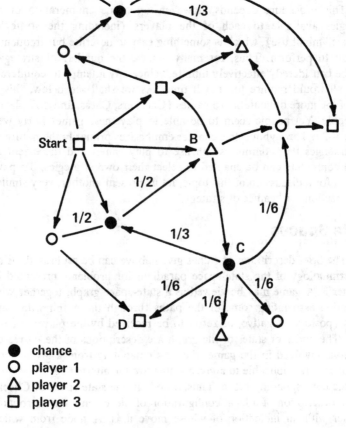

●  chance
○  player 1
△  player 2
□  player 3

Figure 4–1. The state-space graph for a simple game.

the start node) and "moves" the game along the corresponding arc to another node in the state space; say, *B* in Fig. 4–1. Node *B* represents the move by player 2, so player 2 chooses one of the available alternatives and "moves" the game along the corresponding arc to another node in the state space (for example, node *C*). Node *C* represents a move that is to be made by "chance"; the probabilities that chance will choose the various available alternatives are indicated by numbers next to the corresponding arcs (each number must be between zero and one, and their sum must equal one). The game continues in this fashion until it is moved to a terminal node (e.g., node *D*). A path through the state space that leads to a terminal node is known as a *play* of the game. If there are *n* players involved in a game, it is said to be an *n-person* game. If chance is involved in a game, it is said to be a *game of chance;* otherwise it is called a *nonchance* game. Thus, Fig. 4–1 shows the state-space graph for a three-person game of chance.

The strategic, "problematic" aspect of games of strategy arises from the definition of a *payment function,* which specifies that certain paths through the state space of such a game will yield payments to the players. By definition, a player in a game of strategy has the problem of trying to insure that he receives a high payment during the play of the game that actually occurs. Before the game starts, each player is assumed to have been given a complete, finite description of the state-space graph and of the payment function for the game. A player "acts strategically" when he makes a move after investigating the possible consequence of choosing the various alternatives, in light of what he knows about the game from the description of its state space and its payment function, and in light of what he knows about the path that has so far been taken through the state space of the game. The game is one of "perfect information" if all of the players always know what path has been taken; otherwise it is said to be a game of "imperfect information." Chess and Checkers are examples of perfect-information games. Double-blind Chess, or "Kriegspiel," is an example of an imperfect-information game (note 4–3). Bridge and Poker are also examples of imperfect-information games, since a person playing them usually does not know what hands are held by the other players. Bridge and Poker are also "games of chance," in contrast to Kriegspiel.

The payment function for a game of strategy can be very complex: However, it is *not* correct to assume that in a game of strategy each player is necessarily competing with the other players. In some games (e.g., Bridge) players must form teams, and each player must cooperate with those who are in his team. In other games (e.g., Poker) it may sometimes be strategically sound for two players to form a temporary alliance against another player. One can even devise games of strategy in which there is no competition between players at all (note 4–4).

However, most popular games of strategy do have some degree of competition involved in them. Many games can be described as *strictly competitive;* such games are also known as *zero-sum* games, since their payment functions specify that for any play of such a game the sum of the payments received by all players in the game must be equal to zero. In a strictly competitive game, no player receives a positive payment unless other players receive counterbalancing negative payments. Examples of zero-sum games are Chess, Checkers, Tic-tac-toe, Hex, and GO. In these games the payments that may be received are 1, 0, and −1 ("win," "draw," and "lose").[2] These games are also examples of two-person, nonchance games of perfect information.

For a computer program to play a game, it must be able to select a legal alternative whenever it is required to make a move. For it to play the game *well,* it must select alternatives that will tend to bring about plays of the game for which the payment-function awards the program a large payment.

There are essentially two ways a computer program can go about selecting desirable alternatives. We shall refer to them as the *local* and *global* approaches. The local approach has been fairly successful with a few difficult games (Kalah, Checkers, and Chess), although its success has diminished with the more difficult games. Except for games with very small state spaces, or during the "end plays" of very large games, it is generally not possible for the local approach to work perfectly, in the sense of always selecting the best available alternative. On the other hand, the global approach has had success with a few limited classes of relatively simple games (e.g., Tic-tac-toe, Nim, and Hex), but its techniques may eventually be extended to more difficult games.

A program that uses the global approach is designed to analyze the game as a whole. The computer might, for example, prove theorems about the game, using its description of the game and its past experience at playing the game. Such theorems might reduce the game to other, simpler games. This approach has been investigated by Banerji, Koffman, Amarel, Pitrat, and others. Programs that use it are discussed in the final section of this chapter.

A program that uses the local approach is designed to analyze a part of the state space of the game. Given a situation that is the program's move, the program can enumerate some of the paths through the

---

[2] It is often more convenient for a programmer to effectively give win, lose, and draw the values $\infty$, 0, and $-\infty$.

state space of the game which might result from choosing among the available alternatives. The program could be designed to analyze these paths, using its description of the payment function, and to select an alternative that "leads to a desirable set of paths." In the case of zero-sum games, the type of analysis the program must perform is known as a *minimax* analysis. The collection of paths through the state space of a game, which emanate from a given node in that state space, is known as the *game tree* below that node; the game tree below the start node of a game is often referred to simply as the game tree of the game.

Most AI research on game playing has been concerned with developing computer programs that use local analysis to play zero-sum, two-person, nonchance games of perfect information. The next section describes ways in which heuristic search techniques can be used to analyze such games locally. The remaining sections of the chapter discuss programs that have been written to play games. Some programs discussed play imperfect-information games of chance. For a more extensive yet simple treatment of classical (enumerative) game theory, see Williams (1954). The original book on the subject (von Neumann and Morgenstern, 1944) is highly recommended. Discussions of the state-space approach to the description of games and some examples of global analysis of games are given in Banerji (1969, 1970). Nilsson (1971) and Slagle (1971) present detailed formalizations of current applications of heuristic search techniques to game playing (i.e., local analysis).

# GAME TREES AND HEURISTIC SEARCH

## Game Trees and Minimax Analysis

In general, the techniques presented in Chapter 3 for searching graphs and trees in order to solve problems are applicable to the design of game-playing programs. In particular, the terminologies and concepts associated with game trees are similar to those for problem trees, as defined in Chapter 3. The basic difference between games and the puzzle-like problems already discussed is that, with a game, different nodes belong to different players and no player can completely control the path that is actually taken through the state space (note 4–5). Throughout this section we shall be concerned only with zero-sum, two-person, nonchance games of perfect information.

The *game tree below a given node* in the state space of a game is usually drawn with the given node at the top, as the *root* of the tree. The successors to the root node are placed immediately below it, and arcs are drawn from the root to each of its successors. The root node and its successors are known as the "top nodes" of the tree. The process is then repeated for each of the successors to the root node. Figure 4–2

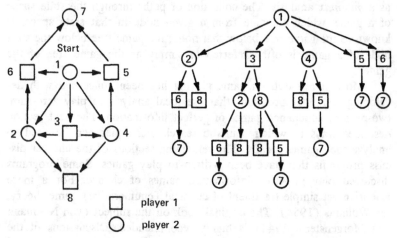

Figure 4–2: A state-space graph for a simple game and the corresponding game-tree.

shows a state-space graph for a simple two-person game, and the corresponding *game tree of the game* (i.e., the game tree below the start, or root, node). As may be seen from Fig. 4–2, it is possible for the "same" state-space node to occur in many places throughout a game tree. This is just another way of saying that there may be many paths connecting two nodes in a state space. Thus, "3,2,6,7" and "3,4,5,7" are two different paths connecting state-space nodes 3 and 7 in Fig. 4–2. The purpose of a game tree is to represent separately each of the possible paths through the state space of the game. Thus, each game-tree node represents a path through the state space of the game; that is, a *sequence* of state-space nodes. The expansion or *generation* of a game tree terminates with those nodes that do not have successors; a terminal node in a game tree is often referred to as a *tip* node. In Fig. 4–2 there are ten tip nodes in the game tree and only two terminal nodes in the corresponding state-space graph. The number of plays of a game is equal to the number of tip nodes of its game tree. Thus, there are ten plays for the game whose state-space graph is shown in Fig. 4–2.

It is evident from Fig. 4–2 that even a game with a small state space may have a large game tree and a large number of possible plays. Games like Checkers, Chess, and GO have game trees so large that they cannot be physically generated completely. (They also have large state spaces.) When it is not possible to actually count the number of plays of a game, the number may be approximated by using estimates of the average branching-factor $B$ and the average depth $D$ of the game tree. Thus, we estimate[3] that the game tree of Checkers has an average depth of 100 and an average branching factor of 6 (i.e., the average possible play of the game might run 100 moves and each move might have an average of 6 available alternatives): The total number of possible plays for Checkers is then

$$B^D = 6^{100} \cong 10^{78}$$

Similarly, it has been estimated that there are $10^{120}$ possible plays for Chess (Shannon, 1950a,b) and $10^{761}$ possible plays for GO (Zobrist, 1969).

Let us suppose for a moment that a player, whom we shall call *player 1,* actually could generate the entire game tree for *any* finite game, no matter how large, and discuss how he might select a strategy for a (zero-sum, two-person, nonchance, perfect-information) game such as Checkers, Chess, or GO. We shall give this person "infinite time and resources" and see what happens.

As stated before, each player's objective is to maximize the payment he receives during the play of the game that actually occurs. We are concerned only with perfect-information games. Thus, the player whose move it is knows exactly what path has been taken from the start node to arrive at the current situation in the game, and he knows exactly what the current situation is (e.g., what pieces are where on the checkerboard). Because he is given a description of the rules of the game, he knows exactly what alternatives he can choose to apply to the current situation and what situations will result. Because our player has infinite time and resources, he can generate the complete game tree below the given node, and can determine the payments associated with each of the plays emanating from that node. He will then have information something like that indicated in Fig. 4–3.

Nodes $B,C,E,F,$ and $G$ in the figure are tip nodes of the tree. Each of the tip nodes of the game tree identifies a different play of the game. Using the payment function, player 1 can calculate the payment specified for each play of the game, and he can consider this payment to be

---

[3] This is based on a conversation with Arthur Samuel and is only a very rough estimate. Another figure often given is $10^{40}$ plays (sometimes $10^{40}$ nodes in the game tree), based on an estimate in Samuel (1959). The higher estimate is used here.

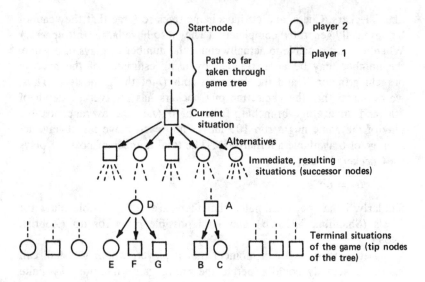

Figure 4-3. Player 1's maximum necessary game tree.

associated with the corresponding tip node of the tree. Thus, he might determine that the play represented by node $B$ would yield him a payment of $+10$, whereas the play represented by node $C$ would yield him a payment of $-2$. Consequently, he would know that if he and player 2 should take the path through the state space represented by node $A$ in the game tree, they would arrive at a situation in which it would be player 1's turn to move, and he would be able to select from two alternatives: one yielding him a payment of $+10$, the other yielding a payment of $-2$. Because each player's objective is to maximize the payment he receives, we say that the *value* of the path represented by node A is $+10$ to player 1 and $-10$ to player 2.

Similarly, player 1 could determine the payments associated with the plays represented by tip nodes $E,F,G$, respectively; he might calculate that play $E$ would yield him a payment of $+1$, play $F$ a payment of $-6$, and play $G$ a payment of $+9$. Because the game is a zero-sum game, he knows that $E,F,$ and $G$ will yield player 2 payments of $-1$, $+6$, and $-9$, respectively. Consequently, he would know that if he and player 2 should take the path through the state space represented by node $D$ in the game tree, they would arrive at a situation in which it would be player 2's turn to move, and player 2 would choose the alternative leading to play $F$. Thus, the *value* of node $D$ is $-6$ to player 1 and $+6$ to player 2.

Because he has "infinite time and resources," player 1 can continue to find the value of each node in the game tree below his current situation by "backing up" his evaluation of nodes from the tip nodes of the game tree according to the following rule: *The value of a given node to player 1 is the maximum of the values of its successor nodes to player 1.* Similarly, the value of a given node to player 2 is the maximum of the values of its successor nodes to player 2. Moreover, because of the zero-sum nature of the game, *the value of a given node to player 1 is the minimum of the values of its successor nodes to player 2,* and the value of a given node to player 2 is the minimum of the values of its successor nodes to player 1. If player 1 determines the values of all nodes in the game tree below his current situation according to these rules, he is said to have done a complete *minimax analysis,* or *evaluation,* of the game tree below his current situation. The value for a node that one obtains by performing a complete minimax evaluation is referred to as the *theoretical value* of the node.

All the game trees considered are finite, so player 1 will be able to generate the complete game tree and do a minimax analysis of its nodes in a finite time (he is given infinite time and resources simply because there is no a priori limit to how big the tree might be and how much time he might require—whatever time he does require, though, will be finite). He will then know the theoretical values of the successor nodes to his current situation. If player 1 chooses an alternative leading to a successor node that has the maximum theoretical value (to him) of all the successor nodes to his current situation, and if he continues to choose in this way whenever it becomes his turn to move, we say that he is playing "perfectly." If player 1 plays perfectly, then the best payment he can expect from the game, if player 2 plays perfectly, is guaranteed. If player 2 does not play perfectly, then player 1 will receive an even higher payment (note 4–6).

EXAMPLE 4–1. *How much time would it take an "attainable" machine to generate and minimax-evaluate the complete game tree for Checkers?* We have assumed $B = 6$ and $D = 100$. By the rule for trees, developed in Chapter 3, there are approximately $(B^{D+1})/(B-1)$ nodes in the complete game tree; thus, there are $(6^{101})/5$ = approximately $2 \times 10^{78}$ nodes in the game tree of Checkers. The "attainable" machine might generate the game tree at a rate of 1 node per nanosecond and then minimax-evaluate it at a rate of 1 node per nanosecond. There are $3.15 \times 10^{18}$ nanoseconds in a century, so the machine would require $(4 \times 10^{78})/(3.15 \times 10^{18})$ = approximately $10^{60}$ centuries to complete this procedure.

Example 4–1 illustrates that it really would be necessary to give player 1 "infinite" time and resources, at least by comparison with current scientific estimates for the lifetime of the universe ($<10^{11}$ years), if we expect him to do a complete minimax analysis of Checkers at an "attainable" rate. In general, it is not possible for a computer program to minimax-evaluate the complete game tree below a given node of Checkers unless that node happens to be very close to the tips of the tree. The Exercises at the end of this chapter show that the same results obtain for Chess and GO.

Even though a computer program cannot usually generate the entire game tree below a given node, it can still generate a portion of that game tree. In most of the possible situations (nodes in the state space) that might occur in games like Checkers, for example, the average node may have six successors, but of these six perhaps only three would be considered "plausible" or "reasonable" by a human Checkers player. If a program could be designed to generate only those successors that were "reasonable," that could do a minimax analysis on the resulting *reasonable game-tree,* and that could select the alternative below its current situation with the highest *reasonable evaluation,* it would still be able to play a very good game. We can estimate that Checkers has, on the average, three reasonable successors to each node and that the average reasonable play has a length of 40 moves. There are thus $3^{40} = 10^{19}$ reasonable plays of Checkers. Similarly, there are about $5^{50} \cong 10^{35}$ reasonable plays of Chess and $10^{100}$ reasonable plays of GO. Of course playing "reasonably" is not the same thing as playing "perfectly." If we had the complete evaluation of the Checkers (or Chess or GO) game tree, we might find that some nodes people currently consider "reasonable" have in fact very low theoretical values; conversely, we might find that some nodes people currently consider "unreasonable" have very high theoretical values.

> EXAMPLE 4–2. *How much time would it take an "attainable" machine to generate and minimax-evaluate the complete reasonable game tree for Checkers?* On the basis of $B = 3$ and $D = 40$, there are approximately $(3^{41})/2$ nodes in the complete, reasonable game tree. The "attainable" machine might generate the game tree at a rate of 1 node per nanosecond and minimax-evaluate it at the same rate. Thus, the machine would require about $(3.5 \times 10^{19})/(3.15 \times 10^{18}) =$ approximately 10 centuries, or a thousand years, to complete this procedure.

Again it is not practically possible for a computer to evaluate the complete, reasonable game trees of games like Chess, Checkers, and GO.

Instead, when a computer program attempts to select the best alternative, or successor, available for a given node in the game tree, it will (if it uses local analysis) generate only a *portion* of the reasonable game tree below that node, and will minimax-evaluate that portion to estimate the best immediate alternative. It will then output a description of that alternative (as being its "choice" for the move) and wait until it is required to make another move (estimate another alternative). The rest of this section considers how a computer program can generate a reasonable portion of a game tree and how it may analyze the portion that it generates.

## Static Evaluations and Backed-up Evaluations

In order to generate and analyze a portion of the reasonable game tree below a given node, it is necessary to judge the "reasonableness" of nodes in some way that is not dependent upon having judged many of their successor nodes. A *static evaluation function* is a method for estimating the value of a node which is not dependent on the values of the successors to that node. A good static evaluation function is one that tends to give estimates that agree with the true, theoretical values of the nodes in a game tree. Different games require different static-evaluation functions. In general, it is not possible to design a static-evaluation function that is *perfect* for a given game; that is, one that will estimate for each node a value that is equal to the theoretical value of that node.[4]

For our purposes, a static evaluation function is necessarily a computational procedure that can be applied by a computer to its description for any given situation that might occur during a play of the game. The function should yield for the situation a numerical value approximating that which would be obtained by analyzing the game completely. When applied to a given node (situation), the static evaluation function may take into account such things as the number of pieces one has, the positions of a game board one occupies, the number of successor nodes to the given node, or whether any successor nodes represent "captures."

The next few pages discuss how a game tree may be analyzed, or evaluated, given a static-evaluation function. As stated before, static-

---

[4] If one did have a perfect static-evaluation function, there would be no need to generate a game tree at all; instead, to determine the best arc from a given node, one would merely have to apply the static-evaluation function to each of the successors to that node, and then select an arc leading to a node with the maximum static evaluation. Thus, one would be playing perfectly, in the sense defined in the text.

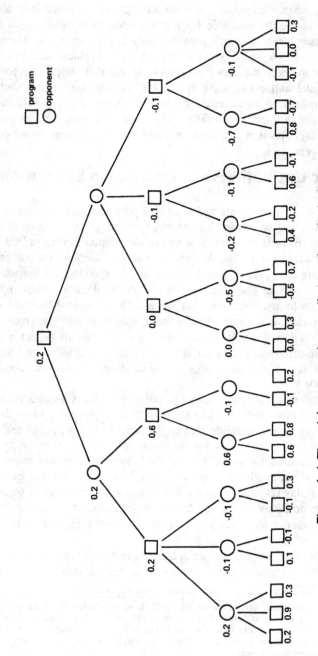

Figure 4-4. The minimax procedure applied to a hypothetical game tree.

evaluation functions may also be used to generate game trees that are reasonable portions of complete game trees. However, for purposes of explication, the discussion of techniques for generating game trees will be deferred to the end of this section.

Suppose, therefore, that a portion of the reasonable game tree below a given node has been generated; such a game tree can be evaluated by a computer program that makes use of minimax analysis and a static-evaluation function. The rules for the minimax analysis are as follows: If a given node is one for which it is the program's turn to move, its value is the maximum of its successors. If a node is one for which it is the opponent's turn to move, its value (to the computer) is the minimum of the values of its successors. The value of a tip node is its *static evaluation;* that is, the result of applying the static-evaluation function to it. Figure 4–4 shows a portion of a game tree for which the tip nodes have been assigned values according to some hypothetical static-evaluation function and the remaining nodes have been given values according to the rules of the minimax procedure. A value given by the minimax procedure to a node that is not a tip node is known as *backed-up value* for the node. To determine the backed-up value for a given node, one must first find the static evaluations of the tip nodes that are below it in the game tree, and then, using the rules of the minimax, "back up" evaluations until a value reaches the given node.

The accuracy of the backed-up evaluation of a given node (how close it is to the theoretical value for that node) is greatly dependent on the amount of the game tree below that node to which one applies the minimax procedure. Again, in general, neither the static evaluation nor the backed-up evaluation will be infallible indicators of the theoretical evaluation for a given node. However, the accuracy of the static evaluation will often be better for nodes near the tips of the complete game tree than it is for nodes near the root. Thus, the backed-up evaluation of a node near the root of the tree will tend to be more accurate than the static evaluation of that node, because the minimax procedure makes use of the static-evaluation function where it is most accurate (on nodes nearer the tips of the tree).

## The Alpha-Beta Technique

In practice, it is important to recognize that the nature of the minimax procedure makes it unnecessary to obtain evaluations for all nodes in the game tree when evaluating the top nodes. A method exists for determining whether the evaluation of a given node can affect the evaluation of nodes that are above it in the game tree. This method is known as the *alpha-beta technique*.[5] To see how it works, let us suppose that a game-playing program is given the task of evaluating the (portion of a) game tree in Fig. 4–4, and is proceeding to minimax from left to right. The tree is reproduced in Fig. 4–5, with the addition that certain significant nodes have been lettered.

The first step of the program is to obtain the static evaluations of nodes $A$, $B$, and $C$. These are found to be 0.2, 0.9, and 0.3, respectively, so the backed-up value of node $D$ above them is determined by the minimax procedure to be 0.2. The next step of the program is to obtain the static evaluation of node $E$, which is found to be 0.1. Consequently, we know that the backed-up value of node $F$ must be less than or equal to 0.1 (since the value of $F$ is the minimum of the values of its successors). Now, the value of node $G$ is the maximum of all values immediately below it because $G$ represents a situation in which it is the program's turn to move, and the program should take the choice that has the greatest evaluation. *This means that node F and all the nodes below it need not be considered further.* The reason is that the value of node $D$ has already been determined, and whatever the value of $F$ it is less than that of $D$. Similarly, when the program evaluates node $H$ as being $-0.1$, it knows that neither 1 nor any other nodes below it need be evaluated. Thus, the value of node $G$ is set at 0.2.

Next, the program evaluates situations $J$ and $K$ and sets the value of $L$ at their minimum, which is 0.6. Since the evaluation of $M$ is the maximum of the values of the nodes immediately below it, the value of $M$ is greater than or equal to that of 1; so, $M \geqq 0.6$.

The value of $N$, however, is the minimum of those of $G$ and $M$. But since $G$ has a value of 0.2, it is not necessary to evaluate any more of the nodes below $M$, and thus the value of $N$ can be set equal to 0.2.

The program continues in this manner to evaluate only those nodes in the tree that could change the values of the nodes above them. As an exercise, the reader may verify that, in order to determine the most desirable alternative, the program need continue developing evaluations only for those nodes labeled $P$ through $U$ in Fig. 4–5. The other nodes of the tree need not be considered at all.

The alpha-beta technique is essentially a process of "using common sense" to carry evaluations up the tree with a minimum amount of work. It can be proved that, with respect to a given static-evaluation function, the alpha-beta technique will always assign the same values to the top nodes of a given tree as would the minimax procedure. (A detailed

---

[5] This technique received its name from McCarthy, who, with his students, did research on it at MIT.

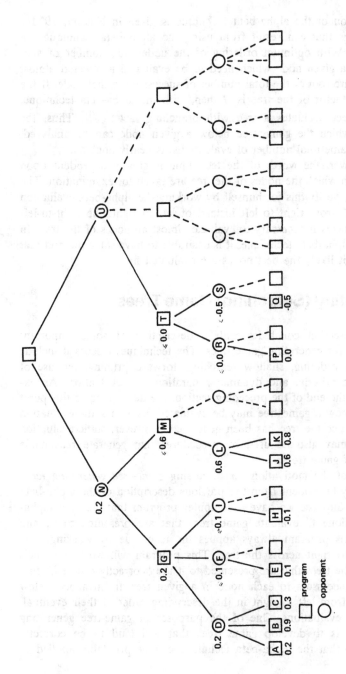

Figure 4-5. The application of the alpha-beta technique to the tree of Figure 4-4.

formalization of the alpha-beta technique is given in Nilsson, 1971.) The savings that can result from using the alpha-beta technique are enormous. With optimum ordering of the nodes, the number of successors to a given node which need to be evaluated is lowered almost to the square root of the total number of successors to that node. If the branching factor of the tree is $B$ then, with the alpha-beta technique, one in effect evaluates a tree with branching factor $\sqrt{B}$. Thus, the depth to which the game tree below a given node can be analyzed, using the same total number of evaluations, is nearly doubled.

However, the worth of the technique is greatly dependent upon the order in which the nodes of the tree are taken for examination. The reader may verify this for himself by working the alpha-beta evaluation of Fig. 4–5 from right to left instead of left to right; the right-to-left ordering makes it necessary to evaluate almost all nodes of the tree. In using the alpha-beta technique, it is desirable to have some method that will make it likely the best nodes are evaluated first.

## Generating (Searching) Game Trees

This section concludes with a description of some important techniques for generating game trees. The techniques discussed include plausibility ordering, shallow searching, forward pruning, the use of termination criteria, and dynamic generation and evaluation. As explained at the end of the preceding section, the description to this point relates to how a game tree may be evaluated, using a static-evaluation function, once the tree has been generated. However, static-evaluation functions may also be used by procedures that generate reasonable portions of game trees.

Part of the motivation for discussing *game-tree generating techniques* may be evident from the previous description of the alpha-beta technique. Suppose we have a computer program that uses the alpha-beta technique to evaluate game trees that are presented to it, and suppose this program always applies the technique by working, say, from left to right across the tree. This program will work most efficiently if the trees that are presented to it are "correctly ordered," that is, if the successors to each node in a given tree are arranged below that node from left to right in the descending order of their eventual, backed-up evaluations. One of the purposes of game-tree generating techniques is to develop game trees that will tend to be correctly ordered so that the alpha-beta technique can be profitably applied to them.

Of course there is no way to insure that a game tree is correctly ordered without having already performed the back-up, or minimax, evaluation that we wish the alpha-beta technique to replace. However, we can increase the likelihood that the tree will be correctly ordered if we make use of some less extensive technique that will give the nodes in the tree a "plausible ordering." Three types of *plausibility-ordering techniques* are generators, shallow search, and dynamic ordering. Dynamic ordering will be described at the end of this section.

A *generator* is a procedure that automatically produces first the most desirable alternatives (and the situations to which they lead) below a given situation and then produces less desirable alternatives, etc. Thus, a generator in Chess might be designed to first produce those alternatives that create situations in which the opponent will be in check or a piece will be captured. The nodes in a game tree can be given a plausibility ordering corresponding to the sequence in which they are produced by a generator.

A *shallow search* technique is a procedure that makes use of a static-evaluation function (not necessarily the same one used by the alpha-beta technique) to conduct a limited tree-generation and evaluation process below each of the nodes that are to be ordered. Thus, suppose a plausibly ordered game tree is being generated below node $A$ in Fig. 4–6; a shallow search technique might first generate the small portion of the tree below $A$, shown in Fig. 4–6. It would then apply its static-evaluation function to the nodes at the bottom of this tree and back up evaluations (probably using its own alpha-beta technique) to nodes $B$, $C$, and $D$. The "shallow evaluations" it obtained for $B$, $C$, and $D$ might indicate that they should be plausibly ordered $C$, $B$, $D$; a shallow search might then be done below $C$ to determine a plausible ordering for nodes $E$, $F$, and $G$. In general, when the game tree below node $A$ is generated, it is most profitable (for the overall application of the alpha-beta technique) that shallow search be used to plausibly order nodes near the top of the tree; it makes relatively little difference whether shallow search is used to order nodes near the bottom (i.e., near the tip nodes).

The other major purpose of game-tree generating techniques is simply to generate a reasonable portion of the complete game tree below a given node; this is in contrast to its purpose in making sure that the generated portion is plausibly ordered. The relevant techniques are forward-pruning and the use of termination criteria; each may be considered a special case of the other. A game-tree generating procedure employs *termination criteria* when it decides not to continue extension of the game tree it is generating, thus creating tip nodes in the game tree

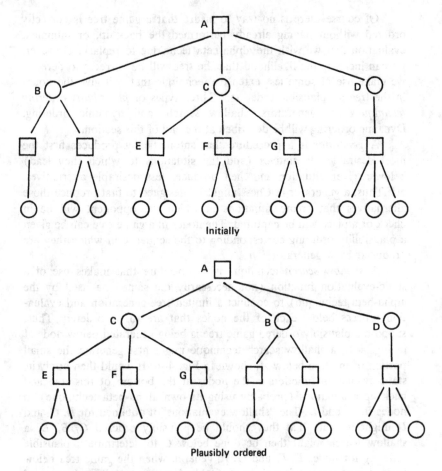

Figure 4–6. Using a shallow search technique to plausible order nodes
B, C, D.

it produces. Some useful termination criteria are "game over" (the tip node produced is actually a tip node of the complete game tree), "maximum depth," and "minimum depth." The maximum-depth termination criterion is employed by procedures that do not produce game trees having a depth greater than some preassigned value. Nodes at that depth below the root node automatically become tip nodes of the game tree that is produced. The minimum-depth criterion is used by procedures that do not produce game trees having a depth less than some preassigned value.

Game-tree generating procedures that employ the minimum- and maximum-depth criteria generally use other criteria that may override them. Thus, the minimum-depth criterion may be overridden by the "game over" criterion. Similarly, many procedures make use of a "dead position" criterion, specifying that a node will not be considered a tip node unless it is "dead"—what deadness means depends upon the game being played. Thus, a situation in Checkers in which there are jumps available is "live" (i.e., not "dead"); a situation in Chess in which someone is in check or there are captures available is live; a situation in GO in which there is a possible "ladder attack" is live. If the dead-position criterion is not satisfied, then the maximum-depth criterion will generally be overridden. Live nodes will always have their successors generated and evaluated (unless the program runs completely out of time or memory space). This is good because it is difficult to find static-evaluation functions that give accurate evaluations for live nodes.[6]

A game-tree generating procedure uses *forward pruning* when it decides not to continue generating successors of a node that might otherwise be considered. Thus, returning to Fig. 4–6, after having plausibly ordered nodes *B, C,* and *D,* a game-tree generating procedure might decide that node *D* is too implausible to merit further investigation; the portion of the complete game tree below *D* would therefore be "pruned" from the game tree produced by the generating procedure. The time saved by not generating or evaluating nodes below *D* can be used to search more deeply elsewhere. In *n-best forward pruning,* only the nodes below the *n* most plausible successors to a given node are searched (generated and evaluated); other successors are pruned. (Thus, the discussion of Fig. 4–6 might illustrate 2-best forward-pruning). In *tapered* n-best forward pruning, the parameter *n* is decreased as the depth of the given node increases. Again, the most plausible successors below the given node may be determined either by use of a generator or by conducting a shallow search. The time saved by forward pruning must be weighed against the chance that relevant portions of the game tree, which might otherwise be considered, will be pruned out. Thus, the shallow search below node *D* might have been misleading; there might have been very valuable nodes farther below.

We have now examined the basic techniques used by game-playing programs that use local analysis to determine how they should play zero-sum, two-person, nonchance games of perfect information. In this exposition the process by which such a program searches a game tree to determine its most desirable alternative has been separated into

---

[6] The dead-position criterion was first suggested for Chess by Turing (1953).

two parts: a generation procedure that produces plausibly ordered, reasonable portions of the game tree, and an alpha-beta technique that evaluates game trees that are supplied to it. In fact, the distinction made is an artificial one: A truly efficient game-playing program will conduct both procedures in a simultaneous, or *dynamic,* fashion. The alert reader will probably have already suspected that something like this should be done. After all, why generate nodes if the alpha-beta technique is later going to decide not to evaluate them?

A game-playing program is said to use *dynamic generation and evaluation* if it applies the alpha-beta technique as another part of its forward-pruning and plausibility-ordering techniques. Essentially, such a program will generate "plausible branches" of the game tree, using the results of the alpha-beta technique to guide their generation. The generation of a plausible branch will terminate when it reaches maximum depth and its tip node is dead. Evaluation is made of each node in a plausible branch as it is generated. After it is evaluated, the backed-up values of nodes above it in the tree are changed accordingly, using the alpha-beta technique. This may cause some nodes to be pruned from further consideration, or change the plausibility orderings of other nodes (*dynamic ordering*), or indicate that new plausible branches should be generated. (A formalization for dynamic search procedures is given in Nilsson, 1971). A game-playing program using dynamic search may approach a reduction of the branching factor of the complete game tree from $B$ to $\sqrt{B}$ (note 4–7).

# CHECKERS
## Checker Player

Samuel (1959,1967) wrote a computer program capable of playing Checkers at a championship level. The program is capable of beating all but the very best players, and once beat a Checkers master, Robert W. Nealey (see Fig. 4–7). This section discusses Samuel's Checkers Player.

Samuel's program conducts an alpha-beta tree search, using forward pruning. To insure the effectiveness of the alpha-beta technique, the Checkers Player does a shallow, breadth-first search to order alternatives according to plausibility. The overall tree search terminates whenever a node is at a maximum depth and the program judges it to be "dead" (i.e., there are no immediate jumps available). During a game it usually takes the Checkers Player less than a minute to perform

| Game move | Black (computer) | White (Nealey) | Game move | Black (computer) | White (Nealey) |
|---|---|---|---|---|---|
| 1 | 11–15 | | 28 | | 27–23 |
| 2 | | 23–19 | 29 | 15–19 | |
| 3 | 8–11 | | 30 | | 23–16 |
| 4 | | 22–17 | 31 | 12–19 | |
| 5 | 4–8 | | 32 | | 32–27 |
| 6 | | 17–13 | 33 | 19–24 | |
| 7 | 15–18 | | 34 | | 27–23 |
| 8 | | 24–20 | 35 | 24–27 | |
| 9 | 9–14 | | 36 | | 22–18 |
| 10 | | 26–23 | 37 | 27–31 | |
| 11 | 10–15 | | 38 | | 18–9 |
| 12 | | 19–10 | 39 | 31–22 | |
| 13 | 6–15 | | 40 | | 9–5 |
| 14 | | 28–24 | 41 | 22–26 | |
| 15 | 15–19 | | 42 | | 23–19 |
| 16 | | 24–15 | 43 | 26–22 | |
| 17 | 5–9 | | 44 | | 19–16 |
| 18 | | 13–6 | 45 | 22–18 | |
| 19 | 1–10–19–26 | | 46 | | 21–17 |
| 20 | | 31–22–15 | 47 | 18–23 | |
| 21 | 11–18 | | 48 | | 17–13 |
| 22 | | 30–26 | 49 | 2–6 | |
| 23 | 8–11 | | 50 | | 16–11 |
| 24 | | 25–22 | 51 | 7–16 | |
| 25 | 18–25 | | 52 | | 20–11 |
| 26 | | 29–22 | 53 | 23–19 | |
| 27 | 11–15 | | | White concedes | |

Figure 4–7. One of the program's early victories. (Samuel, 1959, 1967.)

its tree search and decide how it will move. Samuel's program is unique in that it is, to some extent, capable of developing its own static-evaluation function. The Checkers Player is capable of using and selecting a static-evaluation function that is a composite function of a set of *parametric functions* (this is described below). Together with the concepts presented in the preceding section, this description is sufficient to show how Samuel's program plays the game, once it has a good static-evaluation function. So, this section explains how the Checkers Player is able to achieve its evaluation function.

Samuel had three basic problems in constructing a good evaluation function. He had to determine the proper set of parametric functions; the proper type of composite evaluation function; and how the program's experience should influence it to modify its evaluation function.

The solutions for each of these three problems involved a considerable amount of heuristic programming. From the standpoint of bona fide machine learning, perhaps the most important thing would be to enable the program to create its own set of parametric functions, since these functions are an inherent limitation on its ability to play the game and since their specification by an outside source is a substantial hint as to the proper way of playing. The Checkers Player is not programmed to do this: All parametric functions are supplied in advance to its operation, and are carefully chosen for their relevance to the game. (A typical parameter is MOB (total mobility), equal to the number of squares to which the Player can potentially move, disregarding forced jumps.)

The second problem, determining the proper type of composite evaluation function, has been approached in two ways in different versions of the Checkers Player. The original approach was to let the evaluation function be a *polynomial* of the form $a_1t_1 + \ldots + a_nt_n$; that is, a weighted sum of the values of the terms $t_n$. (For example, $3t_1 + 5t_2$ is a polynomial function of $t_1$ and $t_2$: If $t_1 = 4$ and $t_2 = 7$, then the function has a value of $3 \times 4 + 5 \times 7 = 12 + 35 = 47$.) The greater the value of the polynomial, the more favorable one's evaluation of the configuration in question. This approach has the advantages of permitting an easy modification of the function, obtained by changing the weights $a_i$. The disadvantage comes from the linear nature of the polynomial and the fact that it is not really plausible to assume that the theoretical evaluation function can be linearly expressed in terms of the given parametric functions $t_i$. Samuel's original program overcame this to some extent with the use of two techniques: First, it was made possible to introduce new terms that were binary connectives of the previous ones (i.e., terms that corresponded to logical expressions of the form $(t_i \wedge t_j, \sim t_i \vee t_j$, etc.). A second, more recent technique was to divide the course of the game into six successive phases (determined primarily by the number of pieces on the board); in each phase a different polynomial could be used (one with a different set of terms and different coefficients for each term).

Another, more direct method of constructing a nonlinear evaluation function has been investigated more recently, and is the one currently used by the Checkers Player. The method consists of constructing a hierarchy of "signature tables" as follows: First, the possible values of the parametric functions are restricted; that is, some parameters are allowed to have only five values $(-2,-1,0,1,2)$ and the rest are allowed to have only three values $(-1,0,1)$. Next, six collections (called *signature types*) of parameters are chosen. Each signature type contains four elements, of which one is a five-value parameter and the rest are three-value parameters. (Some parameters may be included in more than one signature type.)

For each signature type, a signature table is to be constructed; this table lists an evaluation (either $-2,-1,0,1$, or 2) for every combination of values of the four elements. There are thus 125 entries in each signature table, every entry being $-2,-1,0,1$, or 2. (Actually, it is only necessary to include 63 entries in a given signature table, since the parametric functions are designed to be "symmetric" for each of the players. If $(1,2,-1,0)$ is listed in a given signature table as having an evaluation of 2, the evaluation of $(-1,-2,1,0)$ is automatically determined to be $-2$, and it is not necessary to list it in the table.)

To build the hierarchy, two second-level signature tables are constructed as in Fig. 4–8, each of which has a second-level evaluation (an integer from $-7$ to 7) for all possible combinations of values of the three first-level tables it describes. There are thus 125 entries in each of the second-level tables.

Finally, a third-level table assigns an evaluation to each possible combination of values from the second-level table.

To determine the evaluation of a given board-configuration, it is necessary to:

1. Determine the values of each of the parametric functions for that particular configuration.
2. Look in the six first-level signature tables and find the first-level evaluations of the configuration.
3. Look in the two second-level tables and find the second-level evaluations of the configuration.
4. Obtain the final evaluation by looking in the third-level signature table.

As a further improvement on the quality of these evaluations, a different signature-table hierarchy is used for each of the six phases of the game.

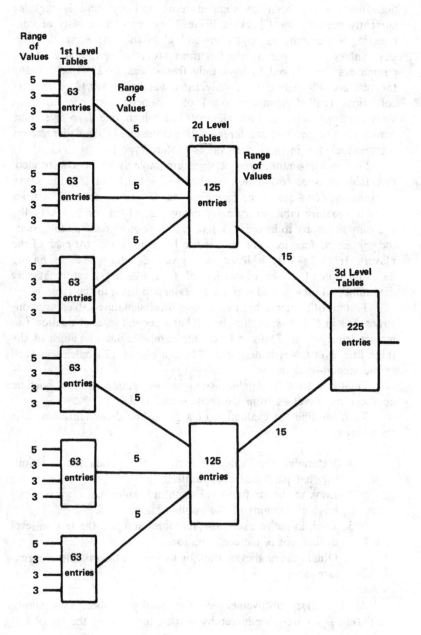

Figure 4–8. The signature-table hierarchy. (Samuel, 1967.)

## Learning

The remaining question is how these evaluation functions, whether polynomials or signature tables, are to be obtained from the game-playing experience of the program. The method for developing these functions constitutes the "learning ability" of the program.

Work on the Checkers Player has been primarily devoted to two ways of doing this, referred to as *rote learning* and *learning by generalization*. Rote learning can be accomplished by establishing a large file of those board configurations and their evaluations that are encountered during the course of the games the program plays. The establishment of this file eliminates the need to recompute an evaluation each time such a configuration arises, so it has the benefit of increasing the efficiency of the program (provided the search time through the file is kept low). Learning is effected as follows: If the program is

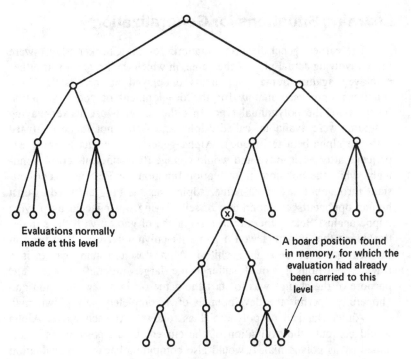

Evaluations normally
made at this level

A board position found
in memory, for which the
evaluation had already
been carried to this
level

Figure 4–9. The effect of "rote learning."

144 INTRODUCTION TO ARTIFICIAL INTELLIGENCE

presented with an arbitrary board configuration and asked to determine the correct choice for the next move, it will often find in the file some configurations that are descendants of the configuration in question. The evaluation for these descendants, before they were put in the file, was originally made in terms of *their* descendants, which might ordinarily be too far away from the original move to be investigated. The evaluation of the various alternatives for the move can thus be much deeper than the normal limits on computation would allow (see Fig. 4–9). The rote-learning method was particularly good at developing the Checker Player's opening and end games.

Learning by generalization is the technique that does most of the work in constructing the evaluation function, however. In the case of the polynomial type of evaluation function, the basic process is that of changing the coefficients for the various terms, whereas in the case of the signature-table function, the process is that of changing the various entries in the tables. These processes are accomplished in different manners, depending on the use of particular learning situations to which the program can be subjected.

## Learning Situations for Generalization

The earliest generalization situations for the Checker Player were those involving actual play of the game, in which the program was either employed against human opponents or played against itself. These situations were used mainly for the development of good evaluation functions of the polynomial type. In either case, two Checker-playing programs were available, called Alpha and Beta (not to be confused with the alpha-beta technique). Alpha generalized on its learning experience after each move and would change its coefficients correspondingly, while the polynomial evaluation function for Beta was kept constant throughout any given name. Alpha was the program used against human opponents; the condition of self-playing was effected by playing Alpha against Beta, generally in a sequence of games, with the stipulation being that if Alpha won a game its polynomial would be used in the next game by Beta also, while if Alpha lost too many games in a row its polynomial would suffer some large, random change. The purpose of the change was to start the game off in a new direction and (hopefully) permit the development of a completely new polynomial.

Alpha changed its polynomial as follows: At each move, Alpha would compute the evaluation of the current board position as determined by its polynomial. It would also compute a backed-up evaluation of the current board position, determined by looking ahead in the game tree and minimaxing backward from the tips of the tree, as defined in the preceding section.

At any rate, given the evaluation, immediate evaluation, and the backed-up evaluation, Alpha would adjust the coefficients of its evaluation function so as to make its new immediate evaluation of the configuration closer to that it had obtained by the look-ahead method.

The success achieved by this technique of "learning while playing" was significant, although somewhat time-consuming. It was particularly good at developing the middle-game performance.

## Book Learning

In the normal operation of the Checkers Player, time spans on the order of a minute are required for it to make the choice of a move. This results in a great deal of time consumption and makes it desirable that a faster method than "learning while playing" be found to accomplish the learning process.

The generalization technique was therefore explored in a third learning situation, referred to as *book learning*. Approximately 250,000 different board configurations, together with the moves recommended for them, were transcribed from the Checkers literature and stored on magnetic tape, and the program was structured so as to learn under their guidance. This learning situation was used for the development of both the signature-table evaluation function and the polynomial evaluation function.

The procedures in both cases were similar: Given a particular board configuration, the program would look at the various alternatives for the move and store each of their resultant configurations. One of the alternatives would be the book-recommended choice.

Next, in the case of the polynomial function, a table would be formed, listing each such resultant configuration against the values of each of the parametric functions when applied to it. Using the table, a count was made of the number of configurations for which a given parametric function had a value higher than it had for the book-recommended configuration; also counted were the number of configurations for which it had a value lower than that for the book-recommended configuration. These numbers were added to the cumulative totals $H$ and $L$ of that particular parameter for all the configurations that the program had so far considered; a coefficient $C$ for that parameter was defined to be the ratio $(L - H)/(L + H)$. This was the coefficient associated with the parameter in the polynomial evaluation function.

Roughly, the same thing was done in the development of the signature tables. These tables listed each resultant configuration against its values with respect to each of the signature types (i.e., against its *signatures*), and cumulative totals $D$ and $A$ were accumulated for each signature with respect to all the board configurations so far considered, using the rule that $D$ was increased by one for each signature of an alternative not recommended by the book, while $n$ (the total number of nonbook moves) was added to the $A$ total for each signature that corresponded to a book-recommended move. The *correlation coefficient* for a given signature, defined as $C = (A - D)/(A + D)$, was used as the entry for the signatures that occurred in the third-level table and, if the signature occurred in a lower level, it was adjusted to fit the values possible there.

## Results

The coefficient $C$ for a given parameter (or signature) serves as cumulative measure of the goodness of the parameter in predicting the book move. The book-learning technique worked well, especially for the signature-table type of evaluation function. After analyzing approximately 175,000 board situations, the Checkers Player was able to predict book-recommended moves with an accuracy of 48%, simply on the basis of its evaluation function, without doing any tree searching. In actual play the program follows book-recommended moves to a much greater extent because it uses tree-searching techniques.

These, then, were the fundamental heuristics behind the Checkers Player's approach to learning the game. Samuel's Checkers Player was one of the earliest major successes of AI research, being the first computer program to perform at a championship level in a difficult game of strategy. The program improved to the point where it could beat its own designer. It remains today one of the best achievements in game-playing programs.

# CHESS AND GO

## Chess

Shannon (1950 a, b) was one of the first to point out the impossibility of using exhaustive search to play Chess, and suggested that a terminating tree search should be used. Turing (1953) described a simple Chess program and suggested that termination of the tree search should be governed by whether or not the positions ultimately reached were "dead." (Turing defined a dead position to be one in which there were no immediate captures available.) Since then, Chess programs have been written by Gillogly, Bernstein, Bastian, Newell, McCarthy, and others. An article by Good (1968) describes a "Five-Year Plan" for the development of an expert Chess-playing program. Some of the ideas mentioned have been implemented, though many deserve further investigation. One of the best Chess-playing programs to date is that of Greenblatt, Eastlake, and Crocker (1967); it is usually referred to either as the "Greenblatt Chess Program" or as "Mac Hack Six."

In order to describe Greenblatt's program, some of the customary terminology used by Chess players is adopted: We shall refer to each of the various alternatives for moving pieces on a chessboard that a player can legally use in one turn as being *Chess moves* or, more simply, *moves*. In all other sections of this chapter the word "move" has been used in its (von Neumann-Morgenstern, 1944) game-theoretic sense, to denote a situation in which a player can choose among alternatives.

The tree search done by Greenblatt's Chess Player program is rather sophisticated, but it can be explained within the state-space paradigm. The possible board configurations, together with the Chess moves that allow one to go from one configuration to another, are the state space of Chess. Greenblatt's program utilizes heuristic information in evaluating both the states *and* the operators of the state space. When presented with an initial board configuration, the program employs a *plausible-move generator* to enumerate legal Chess moves (operators) possible from that configuration and to estimate the desirability of each move.

The plausible-move generator incorporates a large amount of heuristic information in the way it evaluates a given move. Basically, however, its evaluation of a move is a comparison of the positions and pieces attacked before the move, to those attacked after the move. Gains or losses resulting from blocking or unblocking pieces are taken into account, and factors are added to increase the evaluated plausibility of moves that attack certain weak spots (for example, pinned pieces). The evaluation also incorporates very specific heuristic information, such as: "It is bad to move pieces in front of center pawns on their original squares."

The moves are ordered according to the score they receive from the plausible-move generator, and some of them are selected for further consideration. The board configuration resulting from the first of these moves is calculated and the plausible-move generator is applied to it; the process continues to a preset depth, at which point an *evaluation function* is applied to the resultant board configuration. If there are many pieces in danger (*en prise*), the plausible-move generator is applied again and the analysis is carried down another level of the tree. Otherwise, the evaluation function returns a value for the configuration, dependent upon a comparison of the pieces held by each of the players, how much their pieces have changed since the initial configuration, the presence or absence of certain "pawn structures," the safety of the kings, the extent to which the two sides control the center, and the number of plausible captures that can be made from the position. (Plausible captures are investigated in a manner similar to that for plausible moves.)

Thus, the tree search of Greenblatt's program terminates at a depth dependent upon the configurations themselves and the extent to which there are or are not pieces *en prise* (see Turing's "dead" position idea, described in the section "Generating Game Trees"). Similarly, the width of the tree search is *tapered* (see the second section of this chapter) so that at successive levels of the tree the number of plausible moves from each configuration considered for further investigation is 15,15,9,9,7, . . . (all levels below the fifth have a branching factor of 7).[7] However, the width at any level can be expanded if there is heuristic information that an important move (a check, for example) is being ignored. The alpha-beta technique is used throughout the generation of the game tree so that the investigation of many plausible moves is obviated. (It is estimated that the use of the alpha-beta technique reduces the amount of computation by a factor of 100.) Also, the program avoids considering the same board configuration twice by maintaining a table of those configurations it has already encountered and evaluated. Finally, the program contains a table of "book openings," which provides it with the moves recommended by human experts for board configurations that often occur during the beginnings of Chess games.

In 1967 the program was given a tournament rating of about 1,400. (The mean of all United States tournament players is about 1,800; the mean of all Chess players, about 900.) In April 1967, the program won the Massachusetts Class D amateur trophy. The program has been continually improved and at present wins at least 80% of its games against nontournament players. In 1969, Good estimated that the program would play about 2,000 in England. The program is an honorary

----

[7] During nontournament play the program typically expands its plausible game tree with a constant branching factor of 6.

member of the United States Chess Federation, under the name of Mac Hack Six. Figure 4–10 shows Mac Hack Six winning the first game of tournament Chess to be won by a computer.

Mac Hack Six is not a "learning" program in the sense of Samuel's Checker Player. It is, however, one of the "skillful" programs so far produced by AI research (see Chapter 3). The level of skill of Greenblatt's program, relative to that attainable by humans, is probably not as great as that attained by Samuel's Checkers Player or Feigenbaum's et al. (1971) DENDRAL, but it is still considerable—with more development, Mac Hack Six may reach the master tournament level.

**White is Mac Hack Six; black is a human rated 1510**

| 1 | P–K4 | P–QB4 | | 12 | QxQP | B–Q2 |
|---|------|-------|---|----|------|------|
| 2 | P–Q4 | PxP | | 13 | B–R4 | B–N2 |
| 3 | QxP | N–QB3 | | 14 | N–Q5 | NxP |
| 4 | Q–Q3 | N–B3 | | 15 | N–B7ch | QxN |
| 5 | N–QB3 | P–KN3 | | 16 | QxQ | N–B4 |
| 6 | N–B3 | P–Q3 | | 17 | Q–Q6 | B–KB1 |
| 7 | B–B4 | P–K4 | | 18 | Q–Q5 | R–B1 |
| 8 | B–N3 | P–QR3 | | 19 | NxP | B–K3 |
| 9 | O–O–O | P–QN4 | | 20 | QxNch | RxQ |
| 10 | P–QR4 | B–R3ch | | 21 | R–Q8mate | |
| 11 | K–N1 | P–N5 | | | | |

Figure 4–10. First game won by computer in tournament competition: Game 3, Tournament 2, Massachusetts State Championship, 1967. (Greenblatt et al., 1967, reprinted with permission.)

# The Game of GO

Of all the various perfect-information board games described previously, GO is probably the most difficult (see the second section of this chapter). No really successful GO-playing program has yet appeared. However, Thorp and Walden (1970) investigated some of the logical aspects of the game, Zobrist (1969) described a program that plays a legal game and has "reached the bottom rung of the ladder of human GO players," and Ryder (1971) described a program that uses heuristic search techniques to play a "fair beginner's" game.

The *rules of* GO are fairly simple to state: The game is played on a 19 x 19 board (see Fig. 4–11) between two players, each of whom has an unlimited number of *stones,* the stones of one player being white and those of the other being black. The players alternate in making moves. In a given move, a player may place a stone on any unoccupied

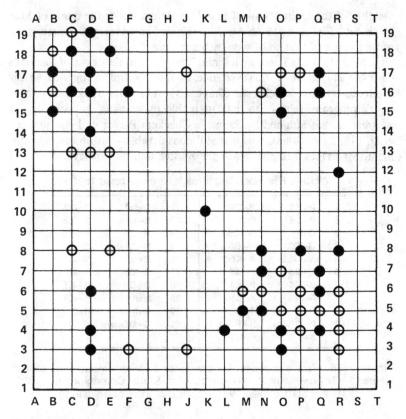

Figure 4–11. An illustration of GO. (Courtesy of E. Fiala and H. E. Sturgis, Xerox Palo Alto Research Center.)

intersection of the board (subject to two restrictions, described below) or he may pass. The game is over if the two players pass in succession. Stones of the same color which form a connected string lying along a row or a column of the board are said to form a *chain*. The *breathing spaces* of a chain are the empty intersections adjacent (by row or column adjacency; diagonal adjacency is not sufficient) to the chain. When a player places a stone on the board, he may not form a chain without breathing spaces, unless he is capturing. He may not capture a stone that has captured one of his stones on the preceding turn, unless he also captures one or more additional stones. Otherwise, if a chain has no breathing spaces, it is captured by the opponent. At the end of the game a player's payment is equal to the sum of the intersections surrounded by his stones plus the number of his opponent's stones that he has captured. The technique in capturing stones is to maneuver so that the chains of one's opponent have no breathing spaces.

Zobrist's program uses pattern-recognition techniques (see the next chapter) to aid its investigation of a given GO board configuration. It possesses 85 "templates," which are capable of matching configurations of stones already on the board, and either suggest places for the program to place its stone or suggest areas in which the program should conduct a limited look-ahead. When the program does look ahead, it does not perform an extensive tree search.

Ryder's program represents a departure from the strict alphabeta, heuristic tree-search techniques that have worked so well for Checkers and Chess, and comprises a unification with recent developments in pattern recognition and problem solving. At least two aspects of its operation are significant: First, it is designed to recognize recursively defined features of configurations of stones on the board. Second, the program is a goal-oriented plan for playing GO: It is capable of establishing and rejecting limited goals (e.g., "target captures") and of searching for move sequences ("tactics") that will lead to them. The recognition of recursively defined patterns has been investigated by Morofsky and Wong (1971), and by Hewitt (1968 et seq.).

GO is an extremely difficult game to play. It may be several years before a program can be written that will be "skillful" at playing the game, even at an amateur level comparable to the current Greenblatt Chess-playing program.

# POKER AND MACHINE DEVELOPMENT OF HEURISTICS

Waterman (1968) designed a language in which heuristics for Draw Poker could be expressed as sentences, and he attempted to construct a program that could select the appropriate sentences under the guidance of experience. Waterman's Poker-playing program, though perhaps not as well known as other game-playing programs, is one of the few such programs to differ significantly in its approach from the Checkers Player.

He distinguished between two types of heuristic, *heuristic rules* and *heuristic definitions*. A heuristic rule specifies an action to be taken and the type of situation that prompts taking the action. Heuristic definitions define terms that may occur in the statement of other heuristic rules or definitions. A heuristic rule in Poker, for example, could be a statement such as: "If the pot is high, call"; the term *high* could be defined with the use of a heuristic definition such as "the pot is high if it is greater than or equal to $B$," and the term $B$ also could be defined by a heuristic definition like "$B$ equals 1000."

Waterman's program works within the state-space paradigm for the statement of problems. Poker *states* (the "hand" one holds, the bids that have been made, etc.) are described by vectors. Given an input state-vector $\mathbf{v} = v_1, \ldots v_n$, the problem for the Poker-player program is to decide upon an output state-vector that is both legal according to the rules of Poker and desirable from the program's standpoint of trying to do well in the game. (Thus, a legal output state-vector may include a change in the program's current bid.) The Poker-playing program develops an ordered list of heuristic rules and definitions, which we shall call a *heuristic block,* that specifies an output state-vector for each input state-vector. (Waterman's program is an example of a program that develops subprograms. We discuss various aspects of this subject in Chapters 6 and 7.)

In Waterman's Poker player the general expression for a heuristic rule is of the form

$$(V_1, \ldots, V_n) \rightarrow (f_1(v), \ldots, f_n(v))$$

where each $V_i$ represents a set of values for corresponding variable $v_i$. Essentially, such an expression says: "Whenever the state vector $\mathbf{v}$ is such that $v_1$ is a member of the set $V_1, \ldots$, and $v_n$ is a member of the set $V_n$, the resultant state vector $v'$ is defined to be $v' = (f_1(v), \ldots, f_n(v))$." A heuristic definition is either an expression of the form "$A1 \rightarrow A, A = 2$," which means that an element $a$ is considered a member of the set $A1$ if it belongs to the set $A$ and if $a = 2$, or an expression such as "$X \rightarrow K1 + Y$," which means that $X$ is defined by the sum of $K1$ and $Y$.

The first step in executing a heuristic block is to compare the input state-vector with all the heuristic definitions until the most general description possible (in the heuristic terms that have been defined) of the state vector is obtained; this description can now be matched against the left-hand sides of the heuristic rules. The convention is adopted that the description of the state vector is to be compared with heuristic rules, in order from the top down, until a match is made, at which point the appropriate action is taken.

To illustrate, suppose the input state-vector is (3,2,4) and the heuristic block is as follows (where the asterisk means that any value is acceptable):

*Heuristic*
*Rules*

$(A1,*,B1) \rightarrow (*,X + v_2,*)$
$(A2,*,C1) \rightarrow (v_1 + 1,*,*)$
$(*,B2,*) \rightarrow (*,*,v_1 + 3)$

*Heuristic*
*Definitions*

$A1 \rightarrow A, A > 5$     $X \rightarrow K1 \times D$
$A2 \rightarrow A, A < 4$     $A \rightarrow a, a$ a member of $(1,2,\ldots)$
$B1 \rightarrow B, B \geq 2$     $B \rightarrow b, b$ a member of $(1,2,\ldots)$
$B2 \rightarrow B, B < 3$     $C \rightarrow c, c$ a member of $(1,2,\ldots)$
$C1 \rightarrow C, C = 5$

Comparison of the input with the heuristic definitions yields $(A2,(B1,B2),C)$ as the description of the state vector (which is to be read: "the input is in the situation $A2$ either $B1$ or $B2$, and $C$"). This description is compared with the left-hand sides of the heuristic rules; the first rule it is found to match is $(*,B2,*) \rightarrow (*,*,v_1 + 3)$, so the output vector is $(3,2,6)$. (Provisions can be made to establish constants that are fixed within the system, such as $K1$, or to allow variables and constants that can be updated, such as $X$ or $D$.)

Given this framework for the description and implementation of heuristics, essentially four operations can be applied to a heuristic block to produce a new block.

First, a given heuristic rule can be modified to match a vector **v** by enlarging some of the sets $V_i$ in the left-hand part of the rule expression. Second, such a rule can be modified by making one or more variables irrelevant (introducing an asterisk in the left-hand part of the expression), again to insure that it matches a given vector **v**. Third, if a rule is found to cause an error (i.e., if experience should indicate that there are situations for which it prescribes a wrong action), it can be modified so as to not match a given vector $(v_1, \ldots, v_n)$ and a rule below it can be modified to match it (in both cases by altering sets $V_i$ in the left-hand part of their expressions). Finally, an error-causing rule can be overridden by inserting a new heuristic rule directly above it in the heuristic block.

Before these operations can be applied, some questions need to be answered. The most obvious question is, "What is an acceptable output vector for the given input?" Two others are, "What sets are relevant?" and "How should they be changed?" For example, if one is given the information that (4,2,4) is an acceptable output vector, that $C1$ is a relevant set, and that $C1$ should be made to include more values, then one can determine that "$C1{\to}C,C{=}5$" is the heuristic which should be changed and that "$C1{\to}C,C{\geq}4$" is a heuristic definition which can be substituted in its place. In this example there is nothing further to be done: The input vector (3,2,4) is now represented as $(A2,(B1,B2),C1)$, and this symbolic description is matched by the second heuristic rule,

$$(A2,*,C1){\to}(v_1 +1,*,*)$$

with the result that (4,2,4) is the output vector.

How is the program to extract from its experience the answers to these questions? It is possible to supply this information from the outside, in which case one might say the program is being trained; Waterman investigated this approach and achieved a Poker-playing program that could play a better-than-average game. (See Table 4–1 for the rules used by Poker Player.) Waterman also investigated ways the program could infer the necessary information on its own, although his approach did not completely free the program from dependence on outside help. He was able to structure the program so that it could solve the first two questions and then, with the aid of a decision matrix given to it by the programmer, solve the third question. (Waterman's use of a decision matrix parallels Newell and Simon's use of a "difference table" in GPS.) Given this decision matrix, the program was capable of "learning" to play a fair game of Draw Poker, although its success at learning poker was not nearly as dramatic as the success of Samuel's program at learning checkers. Waterman's program is distinct from the game-playing programs discussed in previous sections in that it plays a game of "imperfect information."

The problem of designing a program that develops its own heuristics is still unsolved. Again, perhaps the only thing clear is that such a program would have to be guided by heuristics, and that it will eventually be necessary for AI researchers to consider the nature of heuristics that develop heuristics.

# BRIDGE

A recent program by Wasserman (1970b) is capable of bidding skillfully in the game of Contract Bridge. Bridge bidding is a significant intellectual task, involving imperfect information and requiring an ability to work and communicate with a partner. Wasserman's program

TABLE 4–1. Rules for Poker Used by Waterman's Program

---

Definitions of State-Vector Variables and Symbolic Values

| | |
|---|---|
| VDHAND: | the value of your hand |
| POT: | the amount of money in the pot |
| LASTBET: | the amount of money last bet |
| BLUFFO: | a measure of the probability that the opponent can be bluffed |
| POTBET: | the ratio of the money in the pot to the amount last bet |
| ORP: | the number of cards replaced by the opponent |
| OSTYLE: | a measure of conservative style by the opponent |
| OH: | the expected value of the opponent's hand |
| OB: | a measure of the probability that the opponent is bluffing |
| CS: | a measure of conservative style by the opponent |
| BO: | a measure of the probability that the opponent can be bluffed |
| LAP: | the largest bet possible without causing the opponent to drop |
| SB: | a small bet |
| MB: | a medium size bet |
| BB: | a large bet made in an attempt to bluff the opponent |
| BBS: | a small bluff bet |
| BBL: | a large bluff bet |
| OAVGBET: | the average bet made during a round of play |
| OTBET: | the number of bets made by the opponent during a round of play |
| OBLUFFS: | the number of times the opponent was caught bluffing |
| OCORREL: | a measure of the correlation between the opponent's hands and bets |
| OD: | the number of times the opponent has dropped |
| SW: | a sure-to-win hand |
| EC: | an excellent-chance-of-winning hand |
| GC: | a good-chance-of-winning hand |
| PC: | a poor-chance-of-winning hand |
| NC: | a no-chance-of-winning hand |
| K1 to K31: | constants |

| | | | |
|---|---|---|---|
| 1. a. | (SW P8 B5 * * * *) | → (* POT+(2×LASTBET) O * * * *) | call |
| b. | (SW * * * * * *) | → (* POT+(2×LASTBET) LAP * * * *) | bet |
| 2. a. | (EC P1 B5 * * * *) | → (* POT+(2×LASTBET) O * * * *) | call |
| b. | | P1 → P, P > K1 | bf |
| c. | | B5 → B, B > O | bf |
| d. | (EC * * * * * *) | → (* POT+(2×LASTBET) LAP * * * *) | bet |
| 3. a. | (GC P2 B5 * * OR1 *) | → (* POT+(2×LASTBET) O * * * *) | call |
| b. | | P2 → P, P > K2 | bf |
| c. | | OR1 → R, R = O or 1 | bf |
| d. | (GC P9 B6 * * OR1 *) | → (* POT+(2×LASTBET) O * * * *) | call |
| e. | | P9 → P, P > 15 | bf |
| f. | | B6 → B, B > 7 | bf |
| g. | (GC * B5 * * OR2 CS1) | → (* POT+(2×LASTBET) O * * * *) | call |
| h. | | OR2 → R, R = 2 | bf |
| i. | | CS1 → OCS, OCS > K3 | bf |
| | (GC P3 B5 * * OR3 *) | → (* POT+(2×LASTBET) O * * * *) | call |

## TABLE 4-1 (continued)

| | | |
|---|---|---|
| k. | $P3 \rightarrow P, P > K4$ | bf |
| l. | $OR3 \rightarrow R, R = -1$ | bf |
| m. (GC * * BO1 * OR3) | $\rightarrow$ (* POT+(2×LASTBET) SB * * * *) | bet |
| n. | $BO1 \rightarrow BFO, BFO > K5$ | bf |
| o. (GC P4 B5 * * * *) | $\rightarrow$ (* POT+(2×LASTBET) O * * * *) | call |
| p. | $P4 \rightarrow P, P > K6$ | bf |
| q. (GC P9 B7 * * * *) | $\rightarrow$ (* POT+(2×LASTBET) O * * * *) | call |
| r. | $B7 \rightarrow B, B > 10$ | bf |
| s. (GC * * * * * *) | $\rightarrow$ (* POT+(2×LASTBET) MB * * * *) | bet |
| 4. a. (PC * B5 * PB2 OR4 *) | $\rightarrow$ (* POT+(2×LASTBET) O * * * *) | call |
| b. | $PB2 \rightarrow PB, PB > 1$ | bf |
| c. | $OR4 \rightarrow R, R = O$ | bf |
| d. (PC * B5 * PB2 OR2 CS2) | $\rightarrow$ (* POT+(2×LASTBET) O * * * *) | call |
| e. | $CS2 \rightarrow OCS, OCS > K7$ | bf |
| f. (PC P6 B9 BO1 PB3 OR6 *) | $\rightarrow$ (* POT+(2×LASTBET) BB * * * *) | bet |
| g. | $P6 \rightarrow P, P < K14$ | bf |
| h. | $B9 \rightarrow B, B < 5 \wedge B \neq O$ | bf |
| i. | $PB3 \rightarrow PB, PB > 3$ | bf |
| j. | $OR6 \rightarrow R, R \neq -1$ | bf |
| k. (PC P5 B2 BO2 * * *) | $\rightarrow$ (* POT+(2×LASTBET) BB * * * *) | bet |
| l. | $P5 \rightarrow P, P < K9$ | bf |
| m. | $B2 \rightarrow B, B < K10$ | bf |
| n. | $BO2 \rightarrow BFO, BFO > K11$ | bf |
| o. (PC * B8 * PB4 OR6 *) | $\rightarrow$ (O * O * * * *) | drop |
| p. | $B8 \rightarrow B, B > 9$ | bf |
| q. | $PB4 \rightarrow PB, PB < 2$ | bf |
| r. (PC * B5 * * * *) | $\rightarrow$ (* POT+(2×LASTBET) O * * * *) | call |
| s. (PC * * * * * *) | $\rightarrow$ (* POT+(2×LASTBET) SB * * * *) | bet |
| 5. a. (NC * * * * OR4 *) | $\rightarrow$ (O * O * * * *) | drop |
| b. (NC * * * * OR2 CS3) | $\rightarrow$ (O * O * * * *) | drop |
| c. | $CS3 \rightarrow OCS, OCS > K12$ | bf |
| d. (NC P10 B9 BO1 * OR7 *) | $\rightarrow$ (* POT+(2×LASTBET) BBS * * * *) | bet |
| e. | $P10 \rightarrow P, P < 13$ | bf |
| f. | $OR7 \rightarrow R, R = 3$ | bf |
| g. (NC P6 B4 BO3 * OR6 *) | $\rightarrow$ (* POT+(2×LASTBET) BBL * * * *) | bet |
| h. | $P6 \rightarrow P, P < K14$ | bf |
| i. | $B4 \rightarrow B, B < K15$ | bf |
| j. | $BO3 \rightarrow BFO, BFO > K16$ | bf |
| k. (NC * B5 PB1 * *) | $\rightarrow$ (* POT+(2×LASTBET) O * * * *) | call |
| l. | $PB1 \rightarrow PB, PB > K17$ | bf |
| m. (NC P7 B9 * * * *) | $\rightarrow$ (* POT+(2×LASTBET) O * * * *) | call |
| n. | $P7 \rightarrow P, P < K32$ | bf |
| o. (NC P7 B3 * * * *) | $\rightarrow$ (* POT+(2×LASTBET) SB * * * *) | bet |
| p. | $B3 \rightarrow B, B < K13$ | bf |
| q. (NC P6 B3 * * OR6 *) | $\rightarrow$ (* POT+(2×LASTBET) SB * * * *) | bet |
| r. (NC * * * * * *) | $\rightarrow$ (O * O * * * *) | drop |
| 6. | $SW \rightarrow H, H - OH > K18$ and $H \geq K19$ | bf |

## TABLE 4-1 *(continued)*

| | | |
|---|---|---|
| 7. | EC → H, H − OH > K18 and H < K19 | bf |
| 8. | GC → H, K20 < H − OH ≤ K18 | bf |
| 9. | PC → H, K21 < H − OH ≤ K20 | bf |
| 10. | NC → H, H − OH ≤ K21 | bf |
| 11. | OH → K22 − (K23 × OAVGBET × OTBET × OB) | ff |
| 12. | OB → (K24 × OBLUFFS) − (K25 × CS) | ff |
| 13. | CS → (K26 × OCORREL) + (K27 × OD) | ff |
| 14. | BO → (K28 × CS) − (K29 × OH) | ff |
| 15. | LAP → K30 − (K31 × BO) | ff |
| 16. | SB → random(1,5) | ff |
| 17. | MB → random(3,9) | ff |
| 18. | BBS → random(10,15) | ff |
| 19. | BB → random(8,14) | ff |
| 20. | BBL → random(14,20) | ff |
| 21. | H → VDHAND, VDHAND > 0 | bf |
| 22. | P → POT, POT > −1 | bf |
| 23. | B → LASTBET, O ≤ LASTBET < 21 | bf |
| 24. | BFO → BLUFFO, BLUFFO < O ∨ BLUFFO ≥ O | bf |
| 25. | PB → POTBET, POTBET ≥ O | bf |
| 26. | R → ORP, −1 < ORP < 4 | bf |
| 27. | OCS → OSTYLE, OSTYLE < O ∨ OSTYLE ≥ O | bf |

### Values of Constants K1 through K32

The values of the constants used in defining the production rules representing the heuristics for Draw Poker are given below.

| | |
|---|---|
| K1 = 40 | K17 = 4 |
| K2 = 22 | K18 = 27 |
| K3 = 1 | K19 = 376 |
| K4 = 9 | K20 = 10 |
| K5 = 5 | K21 = 0 |
| K6 = 30 | K22 = 6 |
| K7 = 1 | K23 = .05 |
| K8 = 6 | K24 = 1 |
| K9 = 23 | K25 = 2 |
| K10 = 7 | K26 = 1 |
| K11 = 10 | K27 = 2 |
| K12 = 1 | K28 = 8 |
| K13 = 1 | K29 = 1 |
| K14 = 21 | K30 = 5 |
| K15 = 4 | K31 = 1 |
| K16 = 20 | K32 = 8 |

Source: From Waterman (1968). Reprinted with permission.

achieves the level of human experts in partnership bidding and is estimated to be slightly more skillful at competitive bidding than is the
average duplicate Bridge player. The program is capable of bidding
skillfully according to four systems: Standard American, Goren,
Schenken, and Kaplan-Schweinwold (an ability few humans possess).
Figure 4–12 shows Wasserman's program bidding all four hands (independently) of a random deal of the cards.

In March 1969, the program's competitive bidding ability was
tested against two human players who had often played as partners,
one a Life Master having approximately 1,000 points, the other possessing nearly 100 points. The contest was conducted in two sessions,
with 15 hands being bid in each session. (Hands and scoring information were obtained from the American Contract Bridge League National
Tournament, held at Cleveland in March 1969.) The program won one
session and lost the other, being defeated overall by a score of 388.50 to
361.50.

Wasserman's program is similar to Greenblatt's Chess Player, and
Samuel's Checkers Player, in that it is designed to evaluate Bridge
hands, using features and procedures similar to those described by good
human Bridge players. Unlike Samuel's program, the Wasserman
Bridge bidder does not "learn" to improve its performance. Even so,
Wasserman's program is significant because it does perform a difficult
intellectual task.

# GENERAL GAME-PLAYING PROGRAMS

Ultimately, the most desirable game-playing program would be
one that could accept the definition of any game of strategy and which,
with practice, could learn to play the game with a skill comparable or
greater than that which people could develop in playing the game. At
the moment, the attainment of a *general game-playing program* is an
indefinite prospect. However, programs have been written that are
general with respect to certain specific classes of games. In this section a
brief description is given of the classes of games that have been investigated and the programs that are capable of playing them.

The first class of games are the *positional games*. These include
two-, three-, and $n$-dimensional Tic-tac-toe, Hex, Go-Moku (not to be
confused with GO), the Shannon switching games (e.g., Bridg-it), and
many others. Essentially, a positional game is defined by three sets, say,
$N$, $A$, and $B$. The set $N$ is considered to be a set of *positions; A* and $B$
each contain subsets of $N$. A positional game is played by two players,

NORTH

| S - | A | Q | 6 | 4 |
| H - | 10 | 9 | 6 | |
| D - | 5 | 3 | | |
| C - | Q | 7 | 6 | 5 |

WEST

| S - | J | 9 | 7 | | |
| H - | A | Q | 8 | 5 | 2 |
| D - | 10 | 6 | 2 | | |
| C - | A | 8 | | | |

EAST

| S - | 3 | | | | | | |
| H - | 3 | | | | | | |
| D - | A | K | Q | J | 8 | 7 | 4 |
| C - | J | 10 | 9 | 3 | | | |

SOUTH

| S - | K | 10 | 8 | 5 | 2 |
| H - | K | J | 7 | 4 | |
| D - | 9 | | | | |
| C - | K | 4 | 2 | | |

| SOUTH | WEST | NORTH | EAST |
|-------|------|-------|------|
| | PASS | PASS | 1 D |
| DOUBLE | REDOUBLE | 1 S | 2 D |
| 2 S | 3 H | PASS | 4 D |
| PASS | 5 D | DOUBLE | PASS |
| PASS | PASS | | |

Figure 4–12. A complicated and highly competitive bidding sequence. (Wasserman, 1970), reprinted with permission.)

who alternate in choosing elements from $N$ (once chosen, an element may not be rechosen). The first player tries to construct one of the sets belonging to $A$, and the second player tries to construct one of the sets belonging to $B$. The winning player is the one who first succeeds in constructing one of the desired sets. Positional games may involve elements of aggressive strategy, since one player may choose an element from $N$ that he knows the other player would like to choose.

| 1 | 2 | 3 |
|---|---|---|
| 4 | 5 | 6 |
| 7 | 8 | 9 |

To illustrate, the positions in two-dimensional Tic-tac-toe may be numbered as shown by the sketch. The set $N$ for Tic-tac-toe may thus be considered equal to $\{1,2,\ldots,9\}$, while the set $A$ and the set $B$ both contain the sets

$\{1,2,3\}$, $\{4,5,6\}$, $\{7,8,9\}$, $\{1,4,7\}$, $\{2,5,8\}$, $\{3,6,9\}$, $\{1,5,9\}$, and $\{7,5,3\}$

A player in the game usually indicates that he has chosen a position by placing his "mark" (which is either an $X$ or an $O$) on the position.

Positional games were formalized by Koffman (1967) and have been studied by many researchers, including Banerji (1970), Citrenbaum, Pitrat (1971), and Banerji and Ernst (1971). Programs have been constructed which are capable of accepting the definition of an arbitrary positional game and, with practice, of "learning" to play the game quite well. Koffman constructed a program that learns to recognize sets of important board configurations in 4x4x4 Tic-tac-toe, and which requires about 12 games before it starts beating its opponents. Koffman's program describes a given set of board configurations by means of a weighed graph. Fig. 4–13 shows a situation in 4x4x4 Tic-tac-toe from which player $X$ can force a win in six moves; Fig. 4–13b shows the sequence of moves that leads to the win; Fig. 13c shows the "winning paths" used in the force and their interconnections; and Fig. 13d shows the weight-graph representation for the situation. Figure

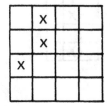

A. Winning situation

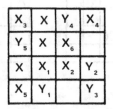

B. Sequence of moves which forces a win from A

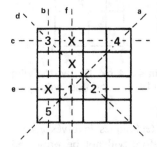

C. Analysis of B in terms of principal rows, columns, and diagonals it uses

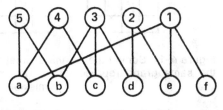

D. Graph representation for C

Figure 4–13. A winning situation in 4 x 4 x 4 Tic-tac-toe and its graphic representation. (Koffman, 1967, reprinted with permission.)

4–14 shows some other winning positions that have the same weighted graph representation.

Another general class of game that has received a great deal of study is the *nimlike* game, formalized by Berge in 1962. A given nimlike game consists of a directed graph and a counter, initially placed on one of the nodes of the graph. The graph of a nimlike game is required to have terminal nodes and it may not have "loops." Two players alternate in moving the counter from its position to an adjacent node along a directed arc. The first player to reach a terminal node wins.

Nimlike games have been studied by Berge, Banerji and Ernst (1971). Many techniques for decomposing a given nimlike game into smaller games (see "problem reduction" in Chapter 3) or for proving that the strategies of one game can be used for another, have been de-

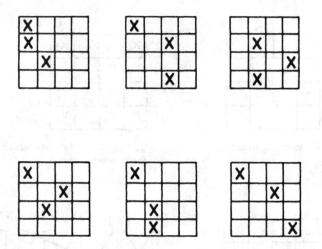

Figure 4–14. Some other winning situations in 4 x 4 x 4 Tic-tac-toe with the same graph representation. (Koffman, 1970, reprinted with permission.)

veloped. However, the description of these techniques involves a considerable amount of mathematics, and therefore will not be presented here.

The development of general game-playing programs is hampered by the fact that there is as yet no clearly satisfactory theory of what it means for two games to be "strategically isomorphic," or of how to find simpler games that are strategically isomorphic to a more difficult one. It seems likely that graphlike structures will turn out to be a good means for describing classes of important game situations in other games as well as in positional ones. It also seems likely that pattern recognition and (perhaps) semantic information-processing techniques will eventually be very valuable to the construction of general game-playing programs.

## NOTES

**4–1.** Whether computers can have "choice" is a debatable question, but for us it is largely irrelevant. One might quibble with the ability of computers to "play" games, on the grounds that their ability to "choose among alternatives" has not been proved, and that this ability is (at least on the surface) required in the von Neumann-Morgenstern formalization of the theory of games. To answer this quibble, we simply note that we are really concerned with the ability of computers to *simulate* playing games, not with whether they "really" make choices, etc. If the reader wishes to pursue the quibble on its own terms, three facts are offered: (1) Computers can reason "causally," that is, take into (perhaps only partial) account the consequences of various actions; (2) a computer's decision can be based (perhaps only partially) on a "random" element; (3) a computer program can be "self-affecting" (see Chapter 8). Each of these facts serves either to diminish our ability to say that the computer's operation is necessarily predetermined or to increase our ability to say that the computer can have a "sense of purpose," that its actions can be "purposeful." When we combine fact (1) with fact (3), we come to the conclusion that a computer can change the way it reasons causally about a problem and, in a sense, display "free will." Whether its "will" is really as "free" as ours may seem to depend upon its abilities to sense and act upon the "real world"; still, we may note that in certain limited realms of commonly shared sensation and action (such as games), the computer's "freedom of will" may roam more widely, and more successfully, than our own. Thus, Koffman's computer program (discussed in the last section of this chapter) could develop its own strategies for the game of 4x4x4 Tic-tac-toe and within 12 games "learn" to start beating its human opponents.

**4–2.** How well can people play games? This is a very devious question: Actually the significant thing seems to be that people can improve their ability to play a game. Are there limits? For example, how close are the current Chess Grandmasters to playing their game with the optimum strategy? We know that optimum strategy must exist, but the game-theoretic procedure for determining what it is lies beyond the bounds of computational ability. Thus, we don't really know what the optimum strategy is unless we can find some better way to compute it. At the moment, all we can do is look at people who play Chess better than average players, and even their performance tells us little about how well the game might be played in theory.

From a theoretical standpoint, there may be limits to how well a game can be played by machines. It may be possible to prove that there are games which *cannot* be played "perfectly," in a practical sense. Although such games would be finite, their optimum strategies would be beyond the bounds of computational procedure (using the game-theoretic, enumerative procedure), and all games that were "strategically isomorphic" to them would be of at least the same size. (See the last section of this chapter.) Assuming they could be shown to exist, we might call these games "grin-and-bear-it" games. Some interesting questions would, of course, be: Are there grin-and-bear-it games that have finite descriptions (rules and payment functions) small enough so that they can be played by humans and computers? What are some grin-and-bear-it games?

**4–3.** Kriegsspiel is played with two players and an "umpire." Each player has his own chessboard, which cannot be seen by the other player, but the umpire can see both chessboards. As in Chess, the players choose oppositely colored pieces and alternate in making moves. Each player's board is empty except for his own pieces, which are initially arranged in the standard formation. Generally, neither player knows exactly what moves the other player has made. Instead, when one player (say, $A$) makes a move, he makes a sequence of choices. After each choice, the umpire informs him whether or not the choice is "legal" (i.e., consistent) according to the rules of ordinary Chess, with the moves so far made by both players. If the choice is not legal, then it has no effect upon the boards of either player. If the choice is legal, then the configuration of pieces on $A$'s board is transformed accordingly and it becomes the other player's turn to move. Neither player hears the choices that are made by the other player.

**4-4.** Such a game might still have aspects of strategy and problem solving: Suppose the payment function specifies that payments shall be received only when the game reaches a terminal node, that is, at the end of a play. Suppose that for different plays of the game the payment function specifies a different "total payment," and suppose that the payment function has a maximum: that is, there is a possible play for which the "total payment" is greater than or equal to that for any other possible play. Finally, suppose that for any play of the game the payment function specifies that the "total payment" is to be divided equally among all players. We then have a "strictly noncompetitive" game in which each player has the problem of cooperating with the other players so as to bring about a play that yields the maximum total payment. One can design strictly noncompetitive games that are very difficult to play.

**4-5.** Alternatively, one can view a game as a problem in which the solution is a tree, rather than a sequence, of operators. Usually, such a representation of the complete strategy for playing a game cannot be stored explicitly in a computer, but must instead be stored implicitly, as a *procedure* for finding the operator to apply in a given situation. The reader who is familiar with the procedural epistemology of Hewitt (1968 et seq.) may anticipate with the present author the desirability of writing some game-playing programs in languages of the PLANNER genus (see Chapters 6 and 7).

**4-6.** Playing "perfectly" in this sense is really playing cautiously, and is equivalent to making the assumption that one's opponent(s) also have infinite time and resources. In fact, if one has extra knowledge about one's opponent, not specified in the rules of the game, it may well be possible to play "better than perfectly." Thus, in reality, a player may intentionally choose an alternative that he knows to have a poor theoretical value—if he thinks that his opponent will not see how to exploit his "mistake" and will instead fall into a trap. As one might expect, neither classical game theory nor the field of game-playing programs currently being developed by AI research has very much to say about "opponent-oriented" strategies. The ability to develop such strategies is clearly possessed by intelligent human game-players, and thus we would expect that AI research might eventually program computers to simulate it. However, it may be a long time before this can happen, since the human development of an opponent-oriented strategy often makes use of knowledge about the opponent which is not limited strictly to his past performance at the game. The development of a good opponent-oriented strategy would require that the computer be able to make a "model" of its opponent's game-playing abilities and goals, but computers currently do not have the ability to gather, represent, or use the information necessary to make models that would be sufficiently accurate. For the reader who is interested in pursuing this subject, Samuel (1967) mentions the desirability of programming game-playing computers to formulate "deep objectives" as part of their strategies, and to hypothesize on their opponent's deep objectives. Colby and Tesler (1969), Colby and Smith (1969), and Abelson and Carrol (1965) discussed the ability of computers to simulate human "belief systems" (though not in the context of game playing); Clarkson (1963) presented an early program that could model human decisions about stock purchasing. (There are probably other relevant papers in the field of "simulation of cognitive processes" of which the present author is not aware.) Also, von Neumann and Morgenstern (1944) treated the subject of "bluffing" in Poker, although not from a "model making" standpoint.

**4–7.** The value of the alpha-beta technique is indicated by the fact that its use in programs which play the game of Kalah has evidently removed this game from the sphere of human dominance; that is, the Kalah-playing programs are probably unbeatable by humans, even though the optimum strategy for the game is beyond the bounds of computational ability (Kalah is, however, less difficult than Checkers). For further information on Kalah, see Russell (1964).

## EXERCISES

*4–1.* Estimate whether the complete generation and minimax evaluation of the game trees for Chess and GO can be performed by (a) a "conventional" machine; (b) an "attainable" machine; (c) a "theoretical serial" machine; (d) a "theoretical parallel" machine (see Chapter 2, "Limits to Computability.") (e) Make the corresponding estimates as to whether these machines could carry out a dynamic search of the complete "reasonable game trees" of these games (see the section "Checkers" in this chapter).

*4–2.* Investigate whether it is epistemologically adequate to describe real-world phenomena as the plays of a partially specified game, for which it is necessary to infer some of the rules. Is such a description metaphysically adequate? (See Chapter 3.)

*4–3.* (a) Show how White can move to gain at least a draw.

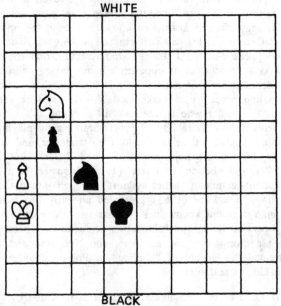

(b) What subproblems did you consider in finding a solution to (a)? (c) Discuss how a computer might be programmed to solve Chess end-game problems.

*4–4.* (*Poker Coins.*) * (a) Find the optimal strategy for the game of Poker Coins, the rules of which are:
   (1) A player throws $N$ coins; he then puts one or more aside and rethrows the rest.
   (2) This throwing is repeated until he no longer has any coins to throw (i.e., all the coins have been put aside).
   (3) Each of the other players takes a turn at throwing $N$ coins, according to rules 1 and 2; the winners are those players with the maximum number of heads.

(b) Analyze Poker Dice, which is played according to the same rules except that $N$ dice are thrown and those players with the highest score are the winners.

*4–5.** (a) Analyze Giveaway Chess, played as follows:
   (1) Captures must be made, although a player may choose which capture to make, if more than one is available.
   (2) Pawns must be promoted to queens if they reach the eighth row.
   (3) The kings obey the same rules of moving and capturing as in ordinary chess, but there is no such thing as "mate," and neither player loses if his king is captured.

---

* From Beeler, Gosper, and Schroeppel (1972). Reprinted with permission.

(4) The first player to lose all of his pieces wins.

(b) Analyze Escalation Chess, where white gets 1 move, black 2, white 3, etc. If a player is in check, he must get out of check on his first move. A player may not move into check or take his opponent's king, but he can place his opponent in a "multiple check," etc. A player is checkmated if he can't get his king out of check on his first move.

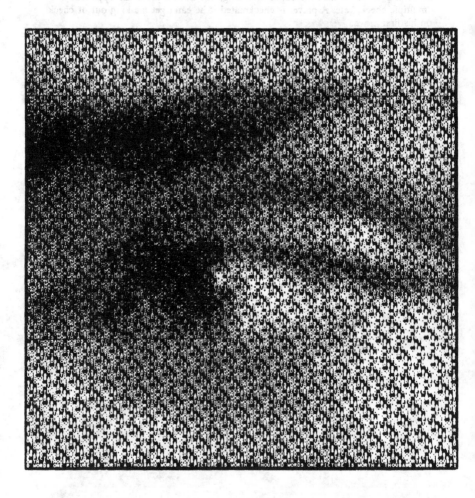

Wordy Eye. (Reprinted with permission from the computer artwork of M. R. Schroeder. Copyright © Bell Laboratories, 1973.) See Example 5–5.

# 5

# PATTERN PERCEPTION

## INTRODUCTION

This chapter discusses ways that machines can simulate "pattern perception." Roughly speaking, pattern perception is the ability to find a simple, useful description for something, given an initial description that is very complex, or of low utility. In order to find the simple description, one might make use of some property ("form," "design," or "regularity") that is possessed by the more complex description. If there is such a property, then the complex description is said to be an example of a "pattern." Pattern perception may operate on descriptions of either physical or abstract things. Thus, it is common to talk of "visual patterns," "sound patterns," "symbol patterns," and, even, "reasoning patterns." Not all of these have been explicitly investigated by AI research. However, it should be clear that a machine which can solve problems in a real-world environment must be able to make and use descriptions of that environment. Machines can make some descriptions rather easily (e.g., photographs), but they have difficulty in using them to "understand" (recognize and solve problems involving) what is being described. From a practical standpoint, the extent to which machines are able to perceive patterns is a limit to the extent that they can solve real-world problems.

This chapter concentrates on the use of machines to do "visual" pattern perception, or *scene analysis,* both because this is the area in which the largest amount of work has been done to date, and because there are good grounds for believing visual pattern perception to be one of the current, major problems confronting AI research. However, other types of pattern perception will be discussed in the next two sections and in the last section of this chapter. A wide variety of approaches have been followed toward visual pattern perception by machines. An attempt will be made to summarize some of the most important approaches and indicate the ways in which each approach is related to the others. However, there is not space in this chapter for a complete survey of the subject. For a more complete summary of vision systems, refer to the book by Duda and Hart (1973), and the survey papers by Rosenfeld (1972) and Turner (1971).

# SOME BASIC DEFINITIONS AND EXAMPLES

AI researchers have adopted a set of basic definitions for the word "pattern" which are fairly consistent with the definitions used by researchers in other fields (e.g., "numerical taxonomy," "behavioristic psychology," "theoretical linguistics"). The definitions are not very hard to understand. However, since the word "pattern" is usually not defined in everyday conversation, this section is devoted to an explication of its use in AI research and a discussion of some general problems involving "patterns" that have been considered by AI researchers.

A pattern is a collection of objects, each of which has the property that it satisfies a certain criterion, known as the pattern rule for the pattern. The objects in a pattern are said to be pattern examples. (Research papers sometimes confuse these ideas, using the word "pattern" to denote what we have chosen to call pattern rules and pattern examples.) Artificial intelligence research has been concerned with several basic problems involving patterns, pattern rules, and pattern examples.

1. (*Classification*) Given an object and a collection of pattern rules, determine which pattern rules are satisfied by the object.

2. (*Matching*) Given a pattern rule and a collection of objects, find those objects which satisfy the pattern rule.

3. (*Description, or Articulation*) Given an object, find a description for it in terms of pattern rules that are satisfied by the parts of the object, or by the object itself.

4. (*Learning*) Given a collection of objects, some of which do

and some of which do not belong to a given pattern, determine a pattern rule for those that do belong to the given
pattern.

Each of these problems may occur in a way which involves the
others as subproblems. In addition, there are important problems of
*representation,* which involve finding languages with which to state pattern rules.

EXAMPLE 5–1. "SUNFLOWER" PATTERNS. This example was used
in Chapter 2 for a brief discussion on the nature of mathematical
descriptions. Figure 5–1 shows an example of a sunflower pattern. This pattern example of a sunflower pattern is a collection
of dots in the plane. For simplicity's sake, each dot is considered to be simply a "point." A dot can be described by giving its position relative to some pair of fixed reference points in
the plane, one to serve as the origin and the other to establish
a scale and a baseline for angular measurements. Thus,
$r = 11.1, \theta = 2$ is a (polar coordinate) description of a dot. We
say that a dot belongs to a sunflower pattern example if and
only if it satisfies the pattern rule for the sunflower pattern. This
pattern may be described either by presenting some of its pattern examples (we presented one in Fig. 5–1) or by stating a
pattern rule for it. An English statement of a pattern rule for the
sunflower pattern example shown in Fig. 5–1 is: A collection
of dots is an example of the sunflower pattern if and only if
each dot is the intersection of 2 of the 24 Archimedean spirals
that have equations obtained by substituting for $k$ any value
between 1 and 12, inclusive, and by substituting for $i$ either
$+1$ or $-1$, in the expression

$$r = i\left(\theta + \frac{k\pi}{6}\right),$$

when a suitable pair of reference points is chosen. An infinite
number of dots can belong to such a collection.

EXAMPLE 5–2. RECOGNIZING PRINTED CHARACTERS. Much early
research in pattern recognition was motivated by a desire to
build machines (known as *optical-character recognizers,* or
OCR's) that would be capable of reading alphabet and number
characters, either written or printed on paper. OCR's currently
are very good at reading certain special types of machine-
printed characters, rather good (about 80% accurate) at recog-

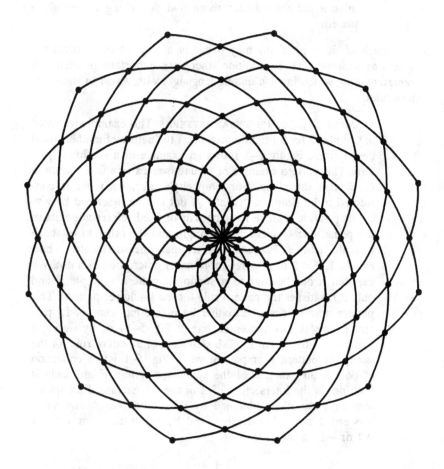

Figure 5–1. The Archimedean sunflower pattern.

nizing typed and hand-printed characters, and very poor at recognizing handwritten or script letters (e.g., "*recognize*"). When we say a machine can "read" or "recognize" certain characters, we are essentially talking about a problem of pattern classification. For example, there are several possible ways of writing or printing the letter A. Most ways produce one of several possible distributions of ink on a paper surface that an (English-literate) human will be capable of identifying as an example of the letter A. Thus, the letter A is a pattern; each distribution of ink on paper that is identified by people as being an A is a pattern example of the letter A. A machine *recognizes* the letter A if, whenever it is presented with a pattern example of A, it outputs some signal corresponding to A (it may, for example, print its own version of A), and if it never outputs that signal when it is not presented with a pattern example of A.[1]

Similarly, we can define what it means for a machine to recognize other characters (b, n, f, l, 2, etc.). If a machine is to recognize a character or pattern, it must have a corresponding pattern rule that can be applied to anything which is presented to it, to test whether or not the thing presented is a pattern example. Some OCR's are given the pattern rules they use to recognize characters and others are designed to develop their own pattern rules (see the next section). In each case, when presented with a distribution of ink on paper, the OCR is required, in effect, to *classify* that distribution of ink as being a pattern example of some pattern. (It may classify it as being ambiguous.) For further information on OCR's, see Holt (1968), Munson (1968), and Duda and Hart (1968).

EXAMPLE 5-3. SEQUENCE PREDICTION. Our definition of "pattern" makes no reference to time or sequentiality. However, it is possible in our formalism to talk about perception of sequential patterns. For example, consider the problem of "sequence prediction." Initially, one is presented with some finite sequence of objects; say, numbers. Thus, one might be shown the sequence $\sigma = 0,1,1,2,3,5,8,13$. The assumption is that the sequence will continue; one's problem is to "predict" how it will continue. In other words, we assume that $\sigma$ is an *initial portion* of some unknown, infinite sequence of numbers. Given an initial portion of the infinite sequence, we attempt to predict the remainder of the sequence. We may make our prediction either by presenting some finite sequence $\sigma'$ as an *immediate* continuation of $\sigma$ or by presenting some Turing machine $T$ so that $T(\sigma)$ will effectively print out the *complete* continuation of $\sigma$. Thus, for the sequence $\sigma$, we might predict an immediate continuation of $\sigma$ to be the sequence $\sigma' = 21,34$. Or we might predict a com-

---

[1] Presenting a pattern example of A to the machine usually means placing the piece of paper with its distribution of ink in an appropriate position before a television camera or equivalent scanner. The camera will make an "initial description" of the piece of paper; this description will be a collection of electric signals that can be processed by the machine.

plete continuation of the sequence by presenting a Turing machine that would implement the rule: "Given an initial portion $\sigma$, generate the number that follows the last element of $\sigma$ by adding the last *two* elements of $\sigma$ to each other. Reset $\sigma$ to be the old initial portion followed by the number generated, and begin again." (Thus, $T$ would generate 21 by adding 8 and 13, $T$ would generate 34 by adding 13 and 21, etc.) In effect, each initial portion of the sequence to be predicted can be considered as a *pattern example* of that sequence.

It is rather easy to see how a Turing machine that predicts the complete continuation of a sequence can be used to construct a *pattern rule* (Turing machine) that will tell us what sequences are pattern examples (initial portions) of the infinite sequence.

However, the "problem of sequence prediction" is complicated by two facts:

1. There are infinite sequences of numbers that cannot be effectively enumerated by any Turing machine (see Chaitin, 1966, 1969).
2. Given any two sequences of numbers, say $\sigma$ and $\sigma'$, it is possible to find a Turing machine $T$ which predicts that $\sigma'$ will be the immediate continuation of $\sigma$.

In other words, there exist sequences that cannot be predicted with complete accuracy by any Turing machine, and it is theoretically possible to justify any finite prediction of the continuation of a given sequence by reference to some Turing machine.

Consequently, the problem of sequence prediction may be restated as: "Find a simple Turing machine that can, given a blank tape, enumerate the sequence $\sigma$ and its complete continuation within a given, required 'accuracy'." The concepts of "simple" and "accuracy" can be given mathematical definitions (e.g., see Arbib, 1969, p. 229). We may therefore suppose that we have chosen some definitions. Let us hold the accuracy required of our prediction at a constant level and imagine looking at all the Turing machines (Tm's) that, with this accuracy, predict (enumerate) $\sigma$ and its continuation. Some Tm's will be simpler than others, but it is possible that more than one Tm will have the greatest value of simplicity. Thus, there may be many predictions for the sequence $\sigma$, all of which are equally valid. We should therefore generalize the problem of sequence prediction and state: "Given $\sigma$, find the set of most simple Turing machines that, within a given required accuracy, predict the continuation of $\sigma$."

Most real-world problems of sequence prediction cannot be solved very easily by using a Turing machine formalization. In fact, no very good formalization (language) for sequence prediction in real-world problems has yet been developed. Aside from its metaphorical, theoretical relationship to subjects like the theory of scientific inquiry (see Chapter 2), there has been some question as to the relevance of the problem of sequence prediction to practical robotics and artificial intelligence. To quote McCarthy and Hayes (1969):

> Imagine a person who is correctly predicting the course of a football game he is watching; he is not predicting each visual sensation (the play of light and shadow, the exact movements of the players and the crowd). Instead his prediction is on the level of: team A is getting tired; they should start to fumble or have their passes intercepted.

Similarly, attempts to use numerical sequence prediction techniques to forecast the stockmarket are shortsighted unless they also process information about the multitude of events in the real world which can affect the market. From the standpoint of AI research, a more relevant kind of sequence prediction to investigate would be the prediction of sequences of relational structures. The problem of sequence prediction also occurs in AI research into language understanding, where it may be necessary to predict the next word or phrase in a sentence, given the preceding words. Here the prediction must be made relative to a grammar for the language and to some model for the possible meanings of the sentence. Finally, a paper by Slagle and Lee (1971) shows how game-tree searching techniques can be applied to sequential pattern recognition.

EXAMPLE 5–4. RELATIVELY PRIME NUMBERS. This example is similar to the sunflower pattern discussed in Example 5–1. Two integers are said to be *relatively prime* if and only if they have no common divisor other than unity. Thus, 4 and 9 are relatively prime because the divisors of 4 are 1 and 2 and the divisors of 9 are 1 and 3. Similarly, 12 and 21 are not relatively prime because both can be divided by 3. Figure 5–2 shows part of a pattern $\mathcal{P}$ of dots in the plane (here the dots are colored white and the plane is colored black), which has the following pattern rule: "A dot is a pattern example of the pattern $\mathcal{P}$ if and only if its $x$ and $y$ coordinates are relatively prime integers." Figure 5–2 shows all those dots (pattern examples) of $\mathcal{P}$ whose integer coordinates are each greater than or equal to zero and less than

Figure 5–2. The relatively prime integers from 0 to 256.

or equal to 256. The figure,[2] to quote Reichardt (1971), "shows the intriguing combination of regularity and randomness which characterizes the distribution of prime numbers and the property of joint divisibility."

EXAMPLE 5–5. WORDY EYE. The Frontispiece to this chapter is a picture that contains pattern examples of at least five patterns: the letters of the English alphabet; the words of the English language; the sentences of the English language; the sequence

---

[2] The Frontispiece to this chapter and Fig. 5–2 are reprinted with permission from the computer artwork of M. R. Schroeder; copyright © Bell Laboratories, 1973.

formed by repeating *ONE PICTURE IS WORTH A THOU-SAND WORDS;* and the set of pictures that depict a human eye. This picture nicely illustrates the hierarchical, structural nature of many patterns. A system for understanding patterns in the real world must be capable of dealing with the ways in which patterns can be made up of patterns. Thus, we may choose to state a pattern rule for the letter A as follows: An object is a pattern example of A if it is made up of an object that is a pattern example of the "upward-angle" pattern and an object that is a pattern example of the "horizontal-line" pattern, and these two objects are related to each other in a certain way. Our discussion of vision systems will trace a hierarchy of patterns (point, line, curve, region, texture, . . . , object, scene) which should be recognized by machines that can see. Especially relevant in this regard is the explication of "hierarchical synthesis" given by Barrow et al. (1972).

EXAMPLE 5–6. SALT AND PEPPER SHAKERS. Mr. and Mrs. Jones of A.D. 2100 are eating a quiet dinner at home. Mrs. Jones decides her fried seaweed is not salty enough and reaches for the saltshaker, only to discover that the table has been inadequately set, and there is no saltshaker on it. "Robbie," she calls, "would you bring us the saltshaker?" Robbie the Robot floats into the kitchen and proceeds to look for a saltshaker. It finds two objects, each of which might be a saltshaker (they are the right shape and size), but they are each opaque—the robot can't see their contents. Looking more closely at the objects, Robbie notices that there are holes in the top of one of the objects and that these holes are placed so as to form a pattern example of the letter S. Robbie therefore takes this object to the dinner table. By this time Mrs. Jones also wants the peppershaker and Robbie, having been too literal-minded (but next year's models will be better . . .), must go back to the kitchen. However, it has successfully recognized a pattern example of the pattern "saltshaker."

EXAMPLE 5–7. EXTRATERRESTRIAL PLANETARY EXPLORATION. Let us suppose that a team of robots is conducting a slow, but patient, geological exploration of the Moon. Because of the time lag in communications from Earth, the robots form a largely self-directing collection of machines. In fact, each robot is somewhat independent of the others because they are too thinly distributed about the Moon's surface to be in frequent

contact with each other. One of the robots, M65, is safely
navigating a narrow path between two craters when a moonquake sends it sliding out of control over the edge of the path
and down the slope of one of the craters. M65 arrives intact
but disoriented at the crater bottom. It doesn't know precisely
where it is or where to go next. The caterpillar-treaded robot
crawls back up the slope of the crater. Reaching the edge, M65
takes a panoramic picture of its surroundings, and generates a
description of the scene's major details (shape and placement
of pattern examples of the patterns "mountain," "large boulder,"
"crater," etc.) It compares this description to another description that it had generated of its surroundings shortly before
the moonquake occurred. Noting some similarities, it attempts
to reestablish its old position and orientation and to proceed
with its business.

Examples 5–6 and 5–7 are, of course, entirely fanciful and beyond
the current state-of-the-art in AI research. Indeed, for Robbie the Robot
to behave as it did in Example 5–6, it would have to be able to solve
the problems of recognizing and understanding human speech, which
are at least as difficult as simply recognizing and distinguishing salt
and pepper shakers. Similarly, the techniques necessary for robot M65
to "reestablish its orientation" and "navigate successfully" over long
distances of lunar terrain (without human assistance) may not be
available for a few decades. However, it should be noted that AI researchers have made serious proposals that artificial intelligence techniques be used to construct machines that could carry out less pre-
tentious, but still somewhat self-directing, explorations on Mars (see
McCarthy, 1964a; Glaser, McCarthy, and Minsky, 1964).

EXAMPLE 5–8. RECOGNIZING A CUBE. Succeeding sections will
discuss techniques for using a television picture of a scene to
produce a line drawing of the scene. Suppose the scene contains only a cube resting on an unknown surface. We might then
obtain a line drawing something like

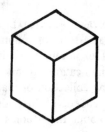

Now this line drawing is a simple description of the original television picture of the scene. The line drawing itself may be considered as an object, however, and techniques that recognize some line drawings as being examples of the pattern "descriptions of a cube" can be considered. Thus, depending on the orientation of the camera with respect to the cube and the surface, any one of an infinite number of line drawings might be obtained that would describe a pattern example of the pattern "cube." We recognize each of these as belonging to a pattern different from that to which the line-drawing below belongs.

As is illustrated by the last example, most pattern-recognition programs really work with descriptions of things rather than with the actual things themselves. Thus, to find pattern examples of various patterns ("cubes," "boxes," etc.) in a real-world environment, the computer will typically make use of a television camera picture of that environment. This picture constitutes its initial description of the environment. The initial description may be processed to yield other descriptions of the environment, or of parts of the environment, and these descriptions may be recognized as "descriptions of a cube," "descriptions of a box," etc. The computer may then print out that it has found a pattern example of the pattern "cube" in the environment; if necessary, it may use its descriptions to help guide a mechanical arm that would attempt to pick up the pattern example of "cube" that was found. Of course, when the computer does so, it may find that its descriptions are incorrect.

It is usually possible to describe a given object in many different ways. The kind of description one uses will in general depend upon the problem at hand. The major kinds of descriptions that are currently used by pattern processing systems all have the structural characteristics of *vectors, matrices, strings, lists,* and *graphs.*

EXAMPLE 5–9. PATTERN MATCHING AND TEMPLATES. One representation that has been developed for stating pattern rules having each of these five kinds of structure is the use of templates in pattern matching. An early example of this technique was presented by Uhr and Vossler (1963), who described a program that successfully generated its own set of template matrices, which it used to recognize handprinted characters. Similarly, in Chaper 4 we discussed the work of Koffman (1967) and Citrenbaum (1972), who presented programs that could develop and use templates with the structure of graphs to play positional games. Most of the recent programming languages for AI research make extensive use of templates with the structure of lists for *pattern matching:* pattern matching in this case means locating subexpressions in a larger expression or data base (set of expressions), and perhaps naming the located subexpressions by assigning them as values to variables. As an example, we shall briefly describe the pattern matching language used in the AI programming language QA4 (Rulifson, Derksen, and Waldinger, 1972).

In this language, a pattern rule can be any list expression that is correctly made up of atoms, variables, and certain "pattern operators" defined for QA4. Intuitively, two expressions match if their elements have the same values, at all levels. Thus, an atom (essentially, an alphanumeric string) is treated as a constant, and normally will only match another instance of itself; if an atom is to be treated as a variable it must have one of six possible *variable prefixes:* ←, ?, $, ←←, ??, and $$. The first three prefixes restrict the variable to match only individual terms (expressions), while the second three allow variables to match "fragments," or segments of lists. Thus, $X, A1263$, and $ATOM$ are constant atoms (when they occur in a pattern rule); $?Y, $Z$, and $←W$ are variables restricted to individual terms; $←←P, ??A1$, and $$$C$ are fragment variables. The prefix ← permits a variable to match any individual term, regardless of the variable's previous value, and specifies that after the match the variable will have as its value the term it is matched against. The prefix ? allows a variable to match only its previous value, if any (QA4 allows variables to not have values); otherwise it is allowed to match any individual term, and acquire that term as its value. Finally, the prefix $ allows a variable to match only its previous value; if the variable does not have a value initially, then it is not allowed to match anything. The three double-character prefixes have analogous meanings to those of their single-character counterparts, except that they restrict variables to match only fragments of lists.

Thus, the expression $(\leftarrow X \; (?Y \; \$Z) \; 2 \; \leftarrow\leftarrow W)$ is a pattern rule in the QA4 language; a wide variety of expressions will satisfy, or match this pattern rule (template), given that the variables $X, Y, Z,$ and $W$ have the proper initial values, where required. Some expressions which might match this pattern rule are *(PLUS (SIN A) 2 4.5 PI)* and *((THE AMERICAN CONGRESS) (HAS EXACTLY) 2 HOUSES)*. Other expressions cannot match this pattern rule, regardless of the initial values of its variables: examples of such expressions are *(TIMES 2 3)* and *((AN OUT) REQUIRES 3 STRIKES)*. It should be clear how this language allows a pattern rule to specify the structural nature of the pattern examples which satisfy it.

Among the special operators which further extend this capability are . . , *PAND*, and *POR*. If the subexpression . . *pat* occurs in a pattern rule (where *pat* is itself a pattern rule), then this subexpression matches an argument expression if *pat* matches some subexpression of that argument, perhaps the entire argument itself. Thus, if the initial values of the variables $X$ and $Y$ are $C$ and $D,$ respectively, then the pattern rule $(.. \; \$X \; ..\$Y)$ matches the expression $((A \; B \; C) \; D)$. The operators *PAND* and *POR* allow pattern rules to make use of logical combinations of pattern rules. A pattern rule of the form *(PAND patl . . . patn)* matches an expression if and only if that expression matches all the pattern rules from *patl* through *patn*. Similarly, a pattern rule of the form *(POR patl . . . patn)* is satisfied by an expression if and only if that expression matches at least one of the pattern rules from *patl* through *patn*. Thus, the pattern rule $(PAND \; \leftarrow X \; (TUPLE \; 1 \; \leftarrow Y))$ matches the expression *(TUPLE 1 2)*, assigning $X$ the expression *(TUPLE 1 2)* as its value, and making 2 the value of $Y$.

This kind of pattern matching language has been useful in many ways, perhaps most notably as an *interlingua* (intermediate language) for question-answering systems. For example, Winograd's English understanding program (see Chapter 7) demonstrated how a wide variety of English questions can be translated into PLANNER theorems (see Chapter 6) that can use such pattern rules to represent the "essential unknowns" of their respective questions. Similarly, the English question "Why did the chicken cross the road?" might be translated into a QA4 expression like:

```
(AND (EXISTS (CHICKEN ?Y))
     (EXISTS (ROAD ?Z))
     (EXISTS (CROSS ?Y ?Z ?EVENT))
     (EXISTS (CAUSE ?X ?EVENT))
```

Evaluation of this expression will cause a search of the current *data base* (set of expressions that a QA4 program may treat as assertions about the world) for expressions matching, successively, the pattern rules (CHICKEN ?Y), (ROAD ?Z), (CROSS ?Y ?Z ?EVENT), and (CAUSE ?X ?EVENT). The pattern matching facilities within such programming languages as PLANNER, QA4, and CONNIVER (see citations in the Bibliography under Hewitt, Rulifson, and Sussman) provide one of the most general formalizations for pattern processing yet developed by AI researchers. This generality derives from the utility of storing symbolic data in list structures, the expressiveness of the pattern rule notation for describing list structures, and the fact that the formalization of these systems does not require the use of any specific terminology or facts associated with particular real-world pattern-perception problems.

In closing this section, reference should also be made to an earlier, but still very general group of formalizations for pattern processing systems, which includes *perceptrons* and *statistical decision theoretic pattern recognition* (note 5–1). There is not space here to discuss these topics but, fortunately, excellent summaries of them are given in the books by Minsky and Papert (1969), Duda and Hart (1973), and Mendel and Fu (1970). In Chapter 7, the topic of statistical decision theoretic pattern recognition is briefly discussed in comparison with the *grammatical inference* approach to pattern recognition.

# EYE SYSTEMS FOR COMPUTERS

The most basic part of a computer system that performs visual pattern perception is the *eye system,* which is simply the collection of *computer eyes* that it can control, and from which it can receive information. A computer eye is a device for producing descriptions of the electromagnetic radiation in space. In general, such an eye consists of a sensor, optics, and usually an illuminator (Earnest, 1967). The purpose of the illuminator is to direct electromagetic radiation into the environ-

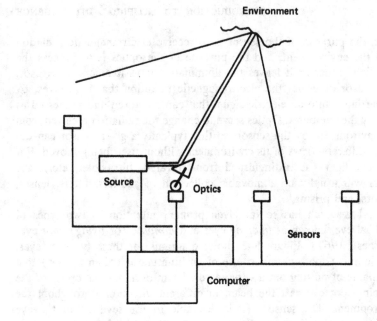

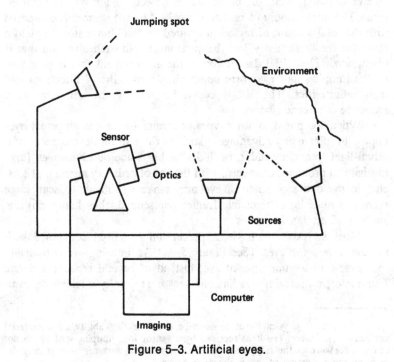

Figure 5–3. Artificial eyes.

ment, the purpose of the sensor is to receive electromagnetic radiation from the environment, and the purpose of the optics is to process the radiation, either as it leaves the illuminator or as it enters the sensor. The sensor describes the electromagnetic radiation that it receives, by converting it into an electric signal that can be stored and processed as data by the computer. Optics serve to change the radiation received from the environment by the sensor, so that typically a given sensor can describe different views of its environment without itself being moved. For ordinary light (as distinguished from infrared, ultraviolet, etc.), the optics will usually be a movable collection of shutters, filters, lenses, mirrors, and prisms.

AI research has so far given primary attention to two types of artificial eye, known as *imaging* eyes and *jumping* (or *flying*) *spot* eyes (Earnest, 1967). Figure 5–3 shows diagrams for these types of eyes. The jumping-spot eye makes use of an illuminator (often a laser) that is capable of putting out a very narrow beam of light. The optics of the jumping-spot eye cast the beam in different directions throughout the environment. The sensors (it is desirable to use several) of the eye receive radiation from the beam that is reflected back by the environment. The total amount of radiation received by the sensors is compared with the total amount of radiation emitted by the illuminator, to yield a score for the "reflectivity"[3] of the environment in each direction that is illuminated. The initial description of the environment that is produced by the jumping-spot eye corresponds simply to a list of directions and their reflectivities. This list is coded for use by the computer as a sequence of electric signals.

When compared to other types of artificial eyes, jumping-spot eyes appear to offer many advantages (such as the natural development of a visual-light frequency radar, or "lidar"), but some disadvantages (e.g., mechanical problems connected with the use of ordinary mirrors, prisms, etc., in the optics of such an eye may make it difficult to scan large scenes at rates faster than five frames per second, thus hampering the analysis of motion in scenes).

Most AI research on visual perception has been concerned with the use of *imaging* eyes. (See Figure 5–4.) An imaging eye is basically the reverse of a jumping-spot eye; instead of several sensors and one illuminator, an imaging eye has one sensor (typically a television cam-

---

[3] The proper physical term to describe the reflecting ability of a material surface is *reflectance*. The light received by a sensor in a jumping-spot eye is not really a measure of the reflectance of any one material surface, since it may depend on the placement of many objects in space.

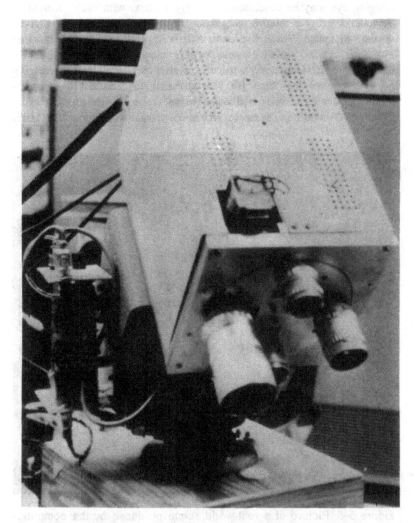

Figure 5–4. A computer-controlled television camera. (Courtesy of Karl Pingle and Lynn Quam, Stanford AI Project.)

era) and often has several illuminators. Thus, the sensor used in an imaging eye is usually more complex than those in a jumping-spot eye. The optics in an imaging eye generally control the way light is directed into the sensor rather than out of the illuminators. With proper use of the optics in an imaging system, the pictures produced by the eye can be "focused," "magnified," "zoomed," etc. A picture produced by an imaging eye may be described as a large matrix, with each element of the matrix being a number measuring the intensity of light in a given volume of space. When a picture matrix is produced by an imaging eye, it is, again, coded for use by the computer as a sequence of electric signals. Generally, a picture matrix produced by an imaging eye will contain less than 100,000 elements (in contrast to approximately 300 million rods and cones in the retina of the human eye). Figure 5–5 shows an example picture of a real-world scene of fairly

Figure 5–5. Picture of a real-world scene produced by the computer-controlled television camera shown in Figure 5–4. (Courtesy of Karl Pingle, Stanford AI Project.)

simple objects, produced by an imaging eye at the Stanford Artificial Intelligence Project.

   Imaging eyes have the advantages that their illumination requirements are roughly compatible with those necessary for humans and that their optical systems have been already extensively developed for use in ordinary photography. Furthermore, there is no difficulty in using imaging systems to make motion pictures of scenes.

# SCENE ANALYSIS

## Picture Enhancement and Line Detection

This section discusses techniques that can be used by computers for the analysis of pictures. As in the preceding section, a picture is considered to be a large matrix of numbers, each number representing the intensity of light in a portion of space. The total portion of space described by a picture will be referred to as a *scene*. Our primary concern is to show how a computer can analyze a single picture of a given scene. Techniques for analyzing and comparing several pictures of the same scene are described in Quam (1971) and Duda and Hart (1973). The techniques we discuss can be grouped into three classes: "picture enhancement and line detection," "perception of regions," and "perception of objects."

*Picture-enhancement techniques* are methods for using one picture to produce another. When used correctly, they can be of help in discovering significant details in a picture.[4] However, because the picture that results from the use of such a technique usually has less information content than the original picture, picture enhancement techniques currently seem to be of more value to human photographers than they are to computer vision systems. Some relatively simple picture-enhancement techniques will be presented here, and the reader is referred to Duda and Hart (1973) for a discussion of other, more complex methods.

One of the simplest picture-enhancement techniques is that of noise *"removal,"* or *smoothing.* Usually, in developing a picture-matrix description of a scene, some noise will be picked up, causing various elements of the matrix to deviate from their correct value. If the noise is random, such that noise in adjacent elements of the picture matrix is uncorrelated, then a spatial averaging or smoothing technique may be applied to reduce it. This technique consists simply of resetting the value of each element of the picture matrix to be the average of the old values of the picture elements in a "window" surrounding it. To illustrate, suppose we smooth the picture matrix

$$
\begin{array}{ccccc}
1 & 0 & 2 & 5 & 7 \\
0 & 2 & 4 & 5 & 6 \\
9 & 2 & 8 & 4 & 7 \\
9 & 2 & 7 & 5 & 7 \\
6 & 1 & 5 & 3 & 6 \\
\end{array}
$$

---

[4] Quam (1971, pp. 78, 101) shows how picture-enhancement techniques were used to detect a cloud on Mars, which would probably not have been recognized without the use of these techniques. (Also see Leovy et al., 1971.)

using 3 x 3 windows. Then the picture element that has value 8 will be reset to have the value 4.3. Smoothing a picture usually introduces some "blurring" in the picture matrix that is produced.

Another technique for noise removal consists of finding each picture element that differs greatly from a surrounding set of approximately equivalent picture elements, and then replacing it with their average value. In the example displayed above, this technique might give a new value of 6 to the element that has value (intensity) 3. This technique is often referred to as *salt-and-pepper removal,* and has the advantage that it will usually reduce most of the random shot noise in a picture without causing the "blurring" created by smoothing.[5]

*Contouring,* or *isodifference detection,* is often used in terrestrial map making to emphasize lines of constant altitude. The technique consists of establishing a sequence of brightness levels,

$$\theta_0 < \theta_1 < \theta_2 < \cdots < \theta_n$$

for which each picture-element $P_{i,j}$ in a given matrix has an intensity $I_{i,j}$ such that $\theta_k \leqq I_{ij} < \theta_{k+1}$ for some $k$. Each picture element is then given a new intensity value corresponding to the appropriate $\theta_k$.

*Edge enhancement,* or *sharpening,* of a picture will produce a new picture similar to that obtained by contouring. In the edge enhancement of a picture only those picture elements that separate elements of greatly varying intensity are shown. For each picture-element $P_{i,j}$ with intensity value $I_{i,j}$ of the matrix, we compute the "cross operator"

$$R_{i,j} = \left( (I_{i,j} - I_{i+1,j+1})^2 + (I_{i,j+1} - I_{i+1,j})^2 \right)^{\frac{1}{2}}$$

We then form the new picture matrix with elements $P_{i,j}$ that have intensity $I'_{i,j} = 1$ if $R_{i,j} \geqq \theta$, where $\theta$ is some *threshhold value* and $I_{i,j} = 0$ otherwise. The threshhold value $\theta$ determines how greatly the intensity must vary in order to show a given picture element. (See Roberts, 1963.)

Other techniques developed for picture enhancement make use of spatial frequency analysis and Fourier transforms. These are well explained by Duda and Hart (1973). Figures 5–6 and 5–7 illustrate the power of these techniques applied to a picture of the Martian moon Phobos, taken by Mariner 9.

*Line-detection techniques* are methods for finding significant curves in a picture matrix that can be used to produce a *line drawing.* The problems of making a good program for line detection in pictures are significant and still largely unsolved. The value of having such a program is great, however, as the reader will see from the discussions in

---

[5] Quam (1971) referred to this technique as "Custering," after General Custer (U.S. Army) who was defeated when surrounded by Indians.

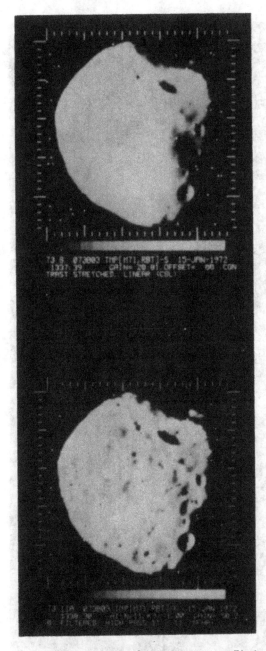

Figure 5–6. (Top) Original picture of Martian moon Phobos, taken by Mariner 9. (Bottom) High-pass spatial frequency filtering of the original. (Courtesy of Lynn Quam and Robert Tucker, Stanford AI Project and Jet Propulsion Laboratory.)

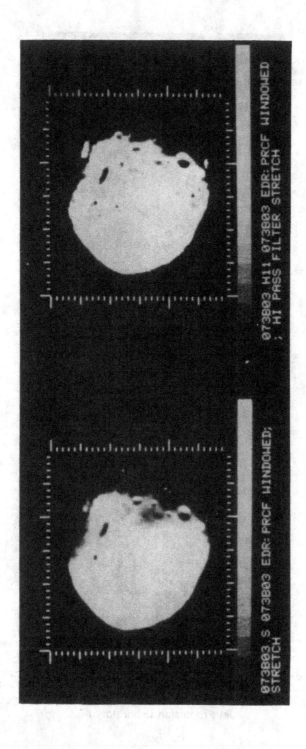

Figure 5–7. (Left) Custering of the original picture of Phobos (see Fig. 5–6). (Right) High-pass filtering of the custered picture. (Courtesy of Lynn Quam and Robert Tucker, Stanford AI Project and Jet Propulsion Laboratory.)

this chapter on "identification of objects" and "learning to recognize structures of simple objects." The edge-enhancement technique described above is one simple type of line-detection program. Another simple method for detection of lines is based on the use of *coincidence predicates*. A simple version of this method is the following: For a given picture matrix with elements $P_{i,j}$, having intensity-values $I_{i,j}$, form a new picture matrix with elements $P'_{i,j}$ having intensity-values $I'_{i,j}$, where $I'_{i,j} = 1$ if $(I_{i,j} - I_{i+1,j})$ and $(I_{i,j+1} - I_{i+1,j+1})$ are both large and of the same sign, or if $(I_{i,j} - I_{i,j+1})$ and $(I_{i+1,j} - I_{i+1,j+1})$ are both large and of the same sign, where "large" is determined by the specification of some threshhold value. Other methods for line detection have been investigated by Heuckel (1969), Herskovits (1970), Griffith (1970), Kelly (1970a,b), Montanari (1971), Hayes and Rosenfeld (1972), and many others. Figure 5–8 shows a computer-produced line drawing of a real-world scene like that shown in Figure 5–5. This figure illus-

Figure 5–8. Line drawing of a real-world scene produced from a television picture like the one in Figure 5–5. (Courtesy of Karl Pingle, Stanford AI Project.)

trates some of the problems that currently plague attempts to develop good line-detection programs. It is difficult to develop programs that can overlook "meaningless" variations in light intensity (e.g., shadows) and still detect "meaningful" ones (e.g., the actual boundaries and edges of objects).

## Perception of Regions

Given a line drawing, a *vertex* can be defined as a point where two or more lines meet, and a *region* as an area of the picture that is entirely enclosed by lines (and usually contains no lines). The problems involved in finding and identifying "meaningful" regions in a picture are similar to those for identifying lines, and are still somewhat unsolved.

As might be expected, several researchers have investigated the use of local operators to detect regions in a picture, similar to, but not requiring, the use of local operators to detect lines, as discussed above. Brice and Fennema (1970) present a good description of a vision system following this approach. The study of such operators has led to many abstract results in *digital topology,* that may be of interest to the reader (e.g., see Rosenfeld, 1973).

Perhaps the most intuitive method for recognizing and describing regions is to make use of line detection programs to find lines in the picture, use one of many possible algorithms for locating vertices, and then trace along the lines and vertices searching for closed curves. (A closed curve is a sequence of connected lines leading back to the first line in the sequence.) Each closed curve is part of the boundary of a region, and the shape of the region can be described in terms of the lines and vertices that enclose it. This technique is suggested by Winston (1970) for recognizing regions with geometric shapes in a line drawing; Winston's general approach is described in the next section.

An interesting technique for describing the shape of a region is *medial axis transformation,* or *prairie fire analysis.* Given a region such as shown here, we may imagine that the interior of the region is covered with highly flammable grass and the exterior of the region is empty (presumably covered with asphalt). Suppose we simultaneously light

a fire all along the boundary of the region. The fire will then spread inward and be quenched where it meets itself. Each point where two or more fire fronts meet and quench each other is known as a *quench point*. The collection of quench points for our example will look something like the next drawing. This collection of quench points, or *skele-*

*ton,* may be taken as a description of the shape of all regions that will produce it. (The precise initial region may be constructed if some additional information is given.) Duda and Hart (1973) discuss this and other methods of region recognition and description in greater detail.

## Perception of Objects

Historically, the first program to successfully use vision to recognize objects in an environment was written by Roberts (1963). This program used local operators to transform a digitized picture into a line drawing, which was then searched for vertices and regions. Relevant information about each line, vertex, and region would be computed and stored in a list structure; e.g., each vertex would have associated with it a description of the regions surrounding it. The program was given a set of similar list structures that presented the same kind of information about each of the edges, vertices, and surfaces of the three basic objects it could recognize (cubes, wedges, and hexagonal prisms). The program would attempt to make a preliminary, consistent matching of each vertex, line and region of the line drawing against a corresponding element in one of these three objects. Given this matching, the program would compute the projective geometry transformation that would yield the best fit between each portion of the line drawing and the object to which it had been corresponded—with a good enough fit, the object would be "recognized" as having produced that portion of the line drawing. Roberts' program was able to recognize *compound objects,* made by piecing together transformations of cubes, wedges, and hexagonal prisms.

Guzman (1968a,b) made the next significant advance in visual perception by machines. He wrote the first program which did not re-

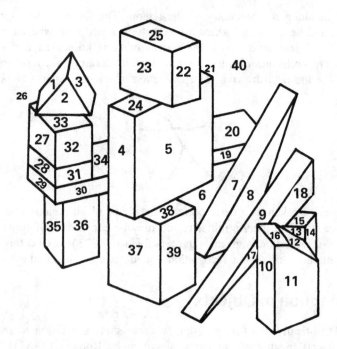

Figure 5–9. (Guzman, 1968, reprinted with permission.)

quire stored descriptions of the objects it could recognize, and did not
proceed by trying to match such descriptions against line drawings of
the scene. Given a picture such as that in Fig. 5–9, in which the lines
and vertices have been detected and correctly labeled and in which the
regions of the picture have been numbered as indicated, Guzman's program (called SEE) will identify 12 objects, as indicated in Table 5–1.

TABLE 5–1.   Identification of Objects by SEE

| Object | Regions | Object | Regions |
|---|---|---|---|
| 1. | 3,2,1 | 7. | 25,23,22 |
| 2. | 32,33,27,26 | 8. | 14,13,15 |
| 3. | 28,31 | 9. | 10,16,11,12 |
| 4. | 19,20,34,30,29 | 10. | 18,9,17 |
| 5. | 36,35 | 11. | 7,8 |
| 6. | 24,5,21,4 | 12. | 38,37,39 |

What is most impressive about SEE is that it can make this identification without knowing anything in detail about specific polyhedra or
about what to expect in Fig. 5–9. The operation of SEE is based only
on the use of information collected locally at each vertex in the picture.

SEE begins operation when it is presented with a special description of a picture. The description contains information about the *regions* in the picture, the *vertices* in the picture, and the *background* of the picture. For the simple picture ONE shown in Fig. 5–10, SEE would be given the following information:

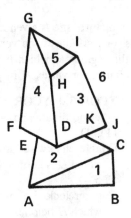

Figure 5–10. ONE, a picture of a simple scene. (Guzman, 1968, reprinted with permission.)

Regions:     (1 2 3 4 5 6)
              A list (not necessarily ordered) of the regions composing scene ONE

Vertices:    (A B C D E F G H I J K)
              Unordered list of vertices contained in scene ONE

Background: (6)
              Unordered list of regions composing the background of scene ONE

In addition, SEE is given information about each of the regions and vertices named in this description. For regions, this information describes the regions that are *neighbors* to each region; the *kvertices* of each region; and the FOOP property of the region. For region 2 in picture ONE, this information is as follows:

NEIGHBORS:  (3 4 6 1 6)
              Counterclockwise ordered list of all regions that neighbor region 2

KVERTICES:  (D E A C K)
              Counterclockwise ordered list of all vertices that belong to region 2

FOOP:            (3 D 4 E 6 A 1 C 6 K)
                 Counterclockwise ordered list of alternating
                 neighbors and kvertices of region 2

Each of these properties of a given region could be determined rather simply by a program that would scan along the lines in a good line drawing.

For each vertex in a picture, SEE is given information that describes the x and y coordinates, or *position* of the vertex; the other vertices to which it is connected; the regions to which it belongs; and the *type* of the vertex (Fig. 5–11). Thus, for vertex H in picture ONE, SEE is given

POSITION:        XCOR 3.0, YCOR 15.0
                 $x$-coordinate and $y$-coordinate of H
NVERTICES:       (I G D)
                 Counterclockwise ordered list of vertices to
                 which H is connected
NREGIONS:        (3 5 4)
                 Counterclockwise ordered list of regions to
                 which H belongs
TYPE:            FORK
                 Type name of the vertex (see Fig. 5–11)

The type name of a vertex is the name of one of eight possible classes to which it may belong, depending on the number of lines and the size of the angles that form the vertex (see Fig. 5–11). These classes are exhaustive and mutually exclusive in that any vertex must belong to one and only one of them. In addition, for each vertex SEE is given a counterclockwise-ordered list of alternating regions and vertices to which it belongs or is connected, and SEE is given other information about the size of the angles belonging to the vertex, etc. Again, all this information could be determined by a program that would scan along the lines in a good line drawing, such as that for scene ONE.

Given this information, SEE proceeds in a heuristic manner to find evidence (Fig. 5–12) that regions in the picture should be grouped together and considered as surfaces of a three-dimensional object. Initially, SEE considers each region in the picture to be within an individual *nucleus;* no two regions share the same initial nucleus. However, if SEE decides that two regions in separate nuclei should be grouped together (considered part of the same object), it will *merge* their nuclei, placing all regions in both nuclei within the same, new nucleus. Thus, SEE will eventually build up nuclei containing many regions, depending

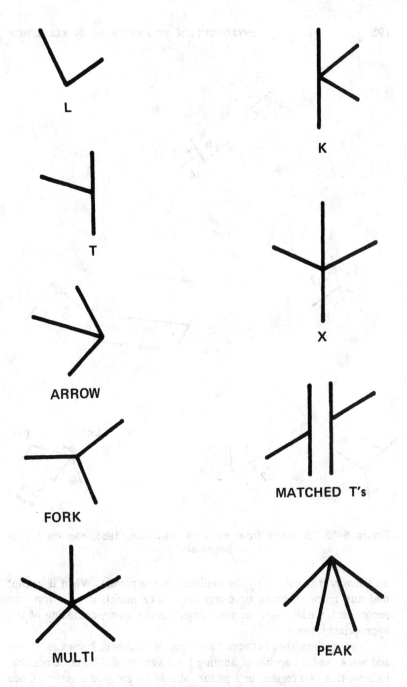

Figure 5–11. Types of vertices. (Winston, 1970, reprinted with permission.)

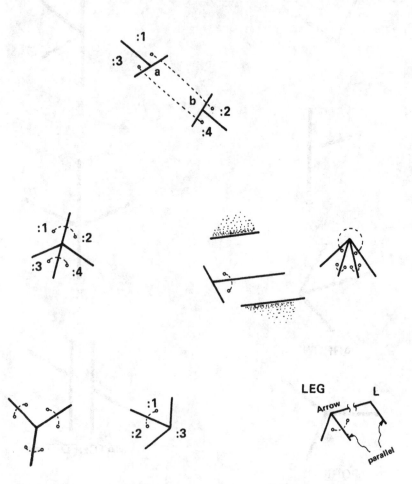

Figure 5–12. Evidence from vertices. (Guzman, 1968, reprinted with permission.)

on the way it is guided by the evidence in the picture. When it cannot find any more evidence or merge any more nuclei, it will stop and report each nucleus as a separate object in the scene, consisting of the appropriate regions.

SEE distinguishes between two types of evidence, known as *strong* and *weak,* and is capable of hunting for a variety of different clues that indicate that two regions in a picture should be grouped together. Once it has found such a clue, it decides whether the clue is strong or weak evidence, and it notes that the clue was found by placing either a strong or weak *link* between the two regions and the nuclei to which they belong.[6] Its decision to merge two nuclei is based solely on the number of strong and weak links between them, not on the clues that caused those links to be formed.

Some of the clues that SEE uses are listed below (see Fig. 5–12).

*Fork.* If three regions meet at a vertex of the FORK type, and none is in the background, strong links between them will be formed (with some exceptions: see Guzman, 1968a,b).

*Arrow.* If three nonbackground regions meet at a vertex of the ARROW type, a strong link will be formed between the two regions that have the small (less than 180°) angles of the vertex.

*X.* If four nonbackground regions meet at a vertex of the $X$ type, and if the vertex is not formed by the intersection of two straight lines, then two strong links are established, as in Fig. 5–12.

*Peak.* If several nonbackground regions meet at a vertex of the PEAK type, all regions except the one containing the obtuse angle (greater than 180°) are given strong links to each other.

*T's.* SEE attempts to find vertices of type $T$ that *match* each other. Two vertices of type $T$ match each other if their central segments are colinear and if they are "facing each other." SEE establishes strong links between regions of matching $T$'s, as in Fig. 5–12, providing these links do not cause a background region to be linked to a nonback-ground region.

*Leg.* An ARROW type vertex is a LEG if one of its small angles leads (if necessary, through a chain of matched T's) to an angle which has one side parallel to the central segment of the arrow. A weak link is formed between the two non-background regions of a LEG type vertex that have the small angles of the vertex.

In addition to these rules, SEE makes use of other clues in its search for strong and weak evidence. For a complete description, the reader is encouraged to see Guzman (1968a,b). However, it should be noted that vertices of types $L$, $K$, and MULTI are not used by SEE to establish links.

When SEE has established as many strong and weak links as pos-

---

[6] The operations of forming, merging, and linking nuclei are all conducted by SEE on a data structure separate from that for the original picture. In essence, SEE builds up a new description of the picture, using nuclei and links, and modifies this description by reference back to the original picture and description of its regions and vertices.

sible between the regions in a picture, it makes use of three rules for the merging of the nuclei that contain the regions:

1. If two nuclei are connected by two or more strong links, they are merged into a single nucleus.
2. If the first rule cannot be applied to any of the nuclei, then if two nuclei are connected by a strong link and a weak link, they are merged; having made this merge, go back to the first rule and see if it can be applied.
3. If neither the first nor second rule can be applied, and there is a nucleus containing a single region that is joined by a strong link to another nucleus (and has no other links leaving it), then the two nuclei are merged.

These heuristic rules are sufficient to enable SEE to identify objects in many rather complex scenes, even when SEE's "view" of an object may be partially occluded by other objects. In general, when SEE makes mistakes, it errs conservatively by not grouping together regions that humans would think plausibly belong to the same object. Thus, for the scene in Fig. 5–13, SEE groups all the regions together in the same plausible manner that humans would, except for regions 41 and 42; it leaves these regions in their initial nuclei and reports them as belonging to separate objects.

It should be pointed out that, given a single picture of a scene, it is impossible to *prove* that any of the regions in the picture actually belong to the same object. Each region in the scene could be the base of a pyramid such that all other faces of that pyramid are hidden from view; thus, no two visible regions would belong to the same object. In effect, any program that identifies objects from a single picture of a scene must be based on notions of plausibility for real-world environments; i.e., it must be a heuristic program.

Guzman discussed a number of extensions that could be made to his program, and the reader is encouraged to investigate his work further. Recently, Huffman (1971), Clowes (1971), and Waltz (1972) have written programs for object recognition which are similar to Guzman's but have a more algorithmic design. Like SEE, these programs rely on local information about the vertices in a line drawing. Huffman's work also discusses the recognition of smooth, curved, nonpolyhedral objects, and the recognition of "impossible" objects. Waltz devotes special attention to the recognition of shadows and the detection of missing lines in a line drawing—this is especially important because the performance of these programs is highly dependent on the quality of the line drawings available to them.

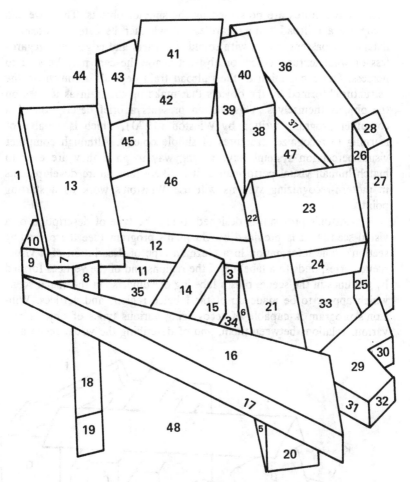

Figure 5–13. Grouping of regions as surfaces of three-dimensional objects. (Guzman, 1968, reprinted with permission.)

# LEARNING TO RECOGNIZE STRUCTURES OF SIMPLE OBJECTS

The problem of object identification by visual perception systems would be intractable if all objects in the real world were to be identified visually using only such features as their texture, color, abstract shape, and the angles formed by their edges. Many objects in the real world are composed of other, simpler objects, somewhat independently of how these features are possessed by the simpler objects. Thus, we can recognize a railroad train regardless of whether its cars are boxcars, flatcars, or passenger cars with rounded corners and edges, and regardless of what texture, color, or abstract shape the cars may be said to possess. Our recognition of the railroad train depends as much on the "structure" formed by the objects that make up the train as it does on the objects themselves. This section presents a brief description of a computer program, written by Winston (1970), which is capable of learning to recognize structures of simple objects. Although computer visual-perception systems have a long way to go if they are ever to match human visual performance, it is likely that future developments in pattern-recognizing systems will use Winston's work as a starting point.

Winston's program is designed to use the type of description of a visual scene that is provided by Guzman's program (see the preceding section). The information in a Guzman type of description of a visual scene corresponds to a labeling of the regions and of the vertices formed by the lines in the scene, plus a labeling of "objects" in the visual scene which appear to be made up of the labeled regions and vertices. Winston's program is capable of recognizing various types of objects and various relations between them, and of describing the visual scene as a

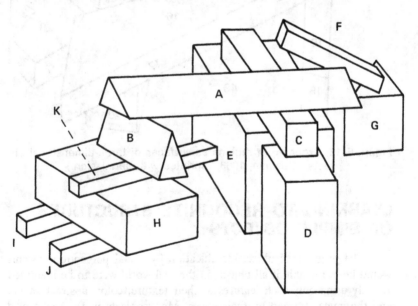

Figure 5–14. Blocks and wedges. (Winston, 1970, reprinted with permission.)

*structure* made up of certain objects and relations. The major types of objects and relations recognized in Winston's (1970) program are *bricks, wedges,* and *above, supports, in-front-of, right-of, left-of,* and *marrys.*[6] Winston showed that his program could be modified to recognize other objects and relations. When shown the scene in Fig. 5–14, Winston's program will recognize the objects and relations listed in Table 5–2. The program will generate a description of the scene that

TABLE 5–2.  Objects and Relations for Fig. 5–14

| | supported-by | | | in-front-of | | |
|---|---|---|---|---|---|---|
| A | | B | C | | F | G |
| B | | | K | | | — |
| C | | D | E | | | — |
| D | | | — | | | E |
| E | | | — | | | — |
| F | | | E | | | — |
| G | | | — | | | — |
| H | | I | J | | | — |
| I | | | — | | | — |
| J | | | — | | | — |
| K | | | H | | | E |

corresponds to the graphlike structure shown in Fig. 5–15. Such a description will be called a *description graph*. The greater part of Winston's program is concerned with comparing description graphs of visual scenes to each other, and with developing general description graphs that can represent *sets* of visual scenes. To do this, Winston allows his description graphs to contain nodes that may represent *groups* of objects and to contain arcs that may represent relations between groups of objects and objects. For example, one such relation is *one-part-is,* which holds for nodes *A* and *B* if *A* represents a group of objects and *B* represents an object "in" *A*. Furthermore, Winston allows relations themselves to be described by description graphs in which nodes may represent relations, and arcs may represent relations between relations (illustrated below). Winston's use of description graphs is sufficiently general that not only objects and structures of objects (i.e., scenes), but also relations, sets of scenes, relations between scenes, comparisons of scenes, relations between description graphs, and comparisons of description graphs may all be described by description graphs.

As an example, we may define an *arch* to be a group of objects (*A, B,* and *C*) such that *B* and *C* are each "a-kind-of" brick and *A* is "a-kind-of" object; *B* and *C* are "standing" and *A* is "lying"; *A* "must-be-supported-by" *B* and *A* "must-be-supported-by" *C; B* and *C* "must-

---

[6] These objects and relations all have approximately the meanings that humans normally give them, except for *marrys.* Two objects are said to "marry" each other if the objects have faces that touch each other and have at least one common edge.

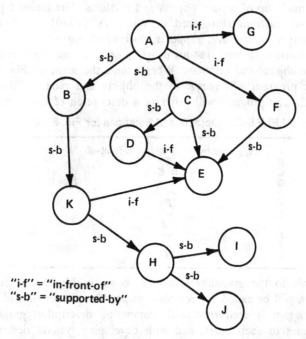

Figure 5–15. A description graph of Figure 5–14. See Table 5–2.

not-abut" (i.e., have faces that touch each other). Figure 5–16 shows some simple examples of groups of objects, some of which are arches and some of which are not. Figure 5–17 shows part of a description graph for the set of scenes that should be recognized as "arches," according to the definition. Note that this description graph includes nodes that represent relations (*must-be-supported-by, supported-by,* etc.) and arcs that represent relations between relations (*modification-of, must-be-satellite,* etc.). Winston's computer program can use this description graph to identify correctly those groups of objects shown in Fig. 5–17 which are arches and those groups of objects shown in Fig. 5–17 which are not arches. Moreover, Winston's program can use this description graph to recognize that the entire group of objects shown in Fig. 5–18 is "a-kind-of" arch. In fact, the computer program will find and identify five groups of objects in Fig. 5–18 that are each "a-kind-of" arch.

What is most impressive about Winston's program is the fact that it is a "learning" program. By this we mean that Winston's program is capable of *developing its own description graph* for a set of scenes that it is told are examples of some pattern. Thus, the program is capable of developing the description graph for "arch" shown in Fig. 5–17, if

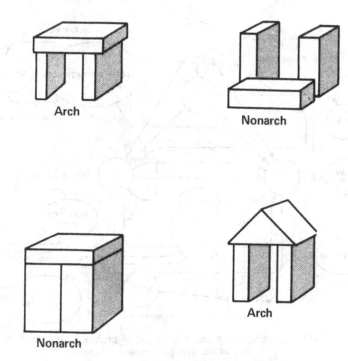

Figure 5–16. Arches and nonarches.

it is shown only the groups of objects (scenes) in Fig. 5–16, and if it is told whether each group of objects in Fig. 5–16 is or is not an arch. It can then use this description graph to identify other, previously unpresented groups of objects (such as that in Fig. 5–18) as being arches, without being told that they are arches. Thus, we can reasonably say that the program "learns" to recognize the pattern "arch." Similarly, the program can learn to recognize "columns," "houses," "pedestals," "tents," "tables," and "arcades" (Fig. 5–19).

Although Winston's program is a "learning" program, it does require a "teacher" to tell it what patterns to recognize (e.g., "arch" and "house") and to give it pattern examples (scenes) for each pattern. Winston's thesis had a great deal to say about the subject of "teaching." In particular, he emphasized the value of presenting to the computer scenes that are "near misses." A *near-miss* is a scene that is not an example of the pattern being taught because it fails to satisfy only one condition of the pattern rule for the pattern. Each of the nonarches in Fig. 5–16 is a near-miss to the pattern "arch."

Because comparisons of description graphs are themselves repre-

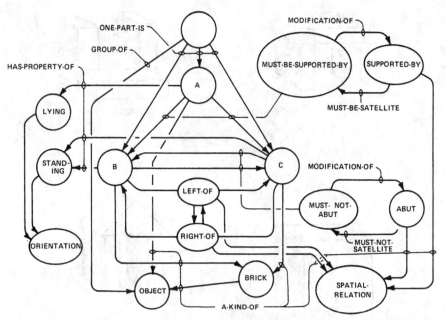

Figure 5–17. A description graph for the set of arches. (Winston, 1970, reprinted with permission.)

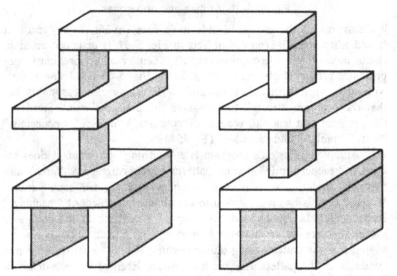

Figure 5–18. "A kind of" arch. (Winston, 1970, reprinted with permission.)

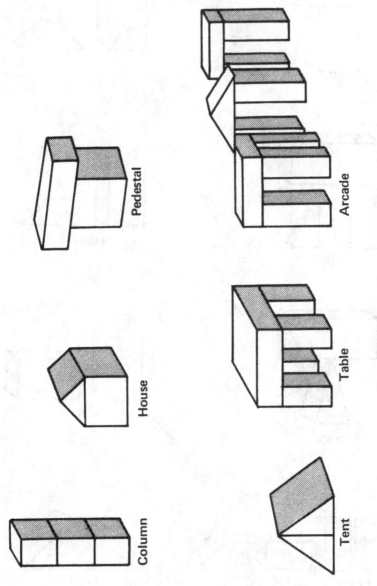

Figure 5–19. Examples of shapes.

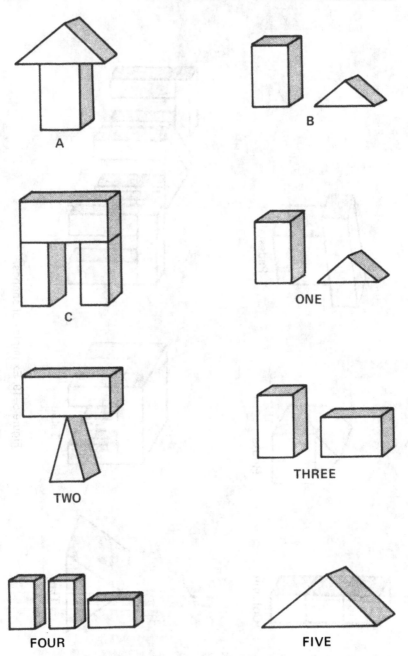

Figure 5–20. A simple analogy problem: Find X such that A:B :: C:X.
(Winston, 1970, reprinted with permission.)

sented as description graphs, Winston's program can be used to solve three-dimensional analogy problems similar to those solved by Evans' ANALOGY program (see Exercise 5–4). Thus, if presented with scenes that are labeled as in Fig. 5–20, and if asked to find a value for $X$ such that "$A$ is to $B$ as $C$ is to $X$" will be true, Winston's program will choose "$X$ = FOUR."

Winston presented a number of suggestions for further work on pattern perception systems. One of the most desirable extensions he suggested is the design of programs that can learn to recognize patterns with pattern rules that are partly "functional." A pattern is said to have a functional pattern rule if its pattern examples are each required to be capable of "performing a function." Thus, the pattern "table" has a functional pattern rule if we require that each of its pattern examples be capable of supporting a plate with food, and silverware, glasses, etc. The subject of functional pattern rules is still an open problem of AI research.

The use of graphlike structures as descriptions for pattern examples and rules has been considered by other researchers, including Shaw (1968), Clark and Miller (1966), Pratt and Friedman (1971), and Barrow, Ambler, and Burstall (1972). This approach is related to attempts (viz., Miller and Shaw, 1968; Banerji, 1968; Narasimhan, 1964; Tachibana, 1972) to develop linguistic methods for visual pattern perception. The discussion of this subject is resumed in Chapter 7.

# SOME PROBLEMS FOR PATTERN PERCEPTION SYSTEMS

At the moment, computers are effectively blind. Pattern (especially visual pattern) perception is one of two[7] major areas of investigation for which AI research has not yet been able to give computers a "level of competence" approaching that of people. This is true despite the impressive results of Guzman, Winston, and others. Also, visual pattern perception is a major bottleneck to the development of many useful mechanical intelligences: It is a necessity, for example, for machines that would work intelligently in a factory or could navigate independently on another planet. This section summarizes some of the current, major problems confronting AI research on pattern perception systems.

---

[7] The other area is "semantic information processing." Many of the problems in these areas appear to be strongly related.

First and foremost, it is desirable to put together the hierarchy we have described. Both Guzman's and Winston's programs require perfect line drawings, which currently are supplied by people. There does not yet exist a line detection program that can consistently supply good line drawings of real-world scenes, because the ability to find a line often requires global information that local information about the picture should be ignored or given special attention. Thus, either program should be able to cause the eye system or the line finder to search particular areas of the scene, change focusing or threshhold settings for those areas, and perform other functions. Either program should be able to check new lines that are produced in this manner, and use those lines that will make their own tasks of object and structure perception easier.

Besides integrating the hierarchy, a number of extensions can be easily suggested. Programs should be able to detect and make use of curved lines, color, and texture. Programs should be able to recognize structures that are pattern examples of patterns with functional pattern rules. Programs should be able to generate descriptions of motion occurring in scenes and (ultimately) make real-time use of such descriptions. Programs should be capable of detecting optical illusions, and compensating for them. Programs should be able to accept, and describe in visual terms, information provided by other perceptual systems (e.g., auditory or tactile information).

Although current work is being done on these matters (viz., Bajcsy, 1972; Shirai, 1972) it is likely that computers will not approach human visual competence for some time, depending upon the rate at which the processes of visual perception can be understood, implemented and tested in high-level programming languages (such as LISP, PLANNER, CONNIVER, and QA4) and, ultimately, implemented in hardware. Even so, substantial progress has been made in the study of pattern perception, if only because the statement of these goals is more meaningful now than it would have been ten years ago.

## NOTES

**5-1.**  At the root of most pattern perception models is the "Pandemonium" paradigm devised by Selfridge (1958). The Pandemonium machine is composed of decision makers, or demons (physicists may recall Maxwell's demon, an imaginary being capable of acting intelligently on a microscopic level and thus controverting the law of entropy) arranged in a latticelike structure such as shown here. At the bottom of this lattice is the real world.

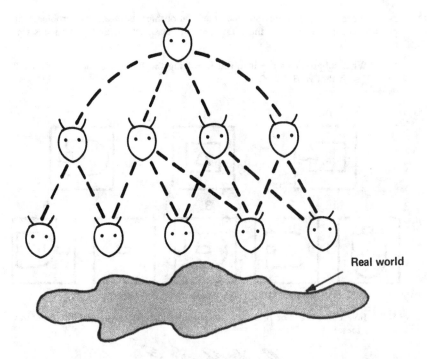

Real world

Each of the demons immediately above the real world scans it and makes a decision concerning the existence of some feature (i.e., the extent to which the real world satisfies the pattern rule for a pattern); the demons at higher levels scan their predecessors and make decisions concerning them. The topmost demon makes the final decision as to whether a pattern example is present. (Essentially this much had been extensively developed by von Neumann, 1951, in a general model for experiments or observations on physical systems, especially quantum mechanical ones.) In addition, Selfridge suggested the use of feedback to alter the nature of the lattice, and proposed an evolutionary scheme for "demon selection." (See Chapter 8.) In their description of *hierarchical synthesis,* Barrow, Ambler, and Burstall (1972) provide an elegant and efficient extension of this idea.

## EXERCISES

*5-1.* Design a computer program that, given the line drawing of the Maze of Dedalus (Exercise 3-1), can find a path out.

*5-2.* What subproblems might a computer need to solve in order to put together jigsaw puzzles?

*5–3.* Write a paper discussing the interrelationships between the problems of pattern recognition, pattern matching, pattern classification, and pattern description.

*5–4.* What subproblems are involved in solving the following analogy problem? Find $X$ such that $A:B::C:X$.

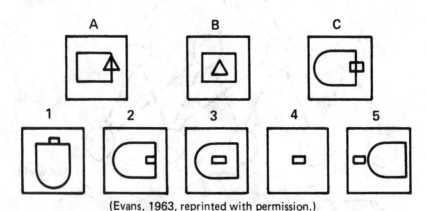

(Evans, 1963, reprinted with permission.)

*5–5.* Investigate ways of describing and generating potentially infinite structures such as these:

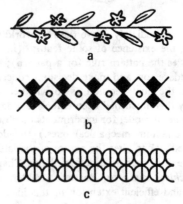

(Watanabe, 1971, reprinted with permission.)

*5–6.* Describe how a computer might be programmed to recognize human faces.

*5–7.* What are the visual subproblems to be solved by a computer program for tying and untying knots?

# A Syllogism worked out.

That story of yours, about your once meeting the sea-serpent, always sets me off yawning;
I never yawn, unless when I'm listening to something totally devoid of interest.

### The Premisses, separately.

### The Premisses, combined.

### The Conclusion.

That story of yours, about your once meeting the sea-serpent, is totally devoid of interest.

# 6

# THEOREM PROVING

## INTRODUCTION

The ability to prove theorems in mathematics is a good example of an intellectual faculty and one that is relevant to the construction of reasoning programs. This chapter is an introduction to the study of computer programs that are capable of finding proofs for theorems within mathematical theories. Such programs are called *theorem provers*. The first part of this chapter introduces the reader to the *predicate calculus*, which is essentially a mathematical framework for the statement of mathematical theories. Later sections discuss the *binary resolution procedure*, a relatively simple procedure that has been the basis for many theorem provers. Alternate means of theorem proving are also discussed. This chapter concludes by showing how theorem provers can be used as problem solvers for problems stated in the state-space paradigm, how they can be used to construct other computer programs and prove the correctness of them, and how analogies can be used to improve the effectiveness of theorem provers. The use of theorem provers is one way to solve the "inference problem" in language-understanding programs.

## FIRST-ORDER PREDICATE CALCULUS

The invention of predicate calculus was one of the major advances in the nineteenth-century development of mathematics. Although Chapter 2 stressed the fact that any mathematical description is essentially a

215

finite description, this does not imply that all mathematical theories can be described within the same finite framework, that there is a mathematical theory of mathematical theories, or *meta-mathematics*.

Predicate calculus is part of the notation for current attempts to develop a theory of metamathematics (note 6–1). Predicate calculus is a language for the expression of mathematical theories. When a mathematical theory is expressed in this language, it becomes a set of *statements* (or *sentences,* or *formulas*), each of which says something about the thing described by the theory.[1] Predicate calculus provides a set of inference rules (for deriving new statements from the ones that are given) and a set of symbols (to be used in making statements) that seem to be adequate for most mathematical theories. Thus, to insure generality, almost all AI work on theorem provers has been concerned with developing machines that handle sets of statements in predicate calculus (note 6–2).

In fact, almost all work in the subject of theorem proving has concerned itself with theorems stated in first-order predicate calculus, which is discussed in this section. Ultimately, it is desirable to extend theorem-proving methods to higher-order logics, because they are more natural for the statement of most mathematical theories. (The difference between first- and higher-order logics is defined below.) Work in this direction has been undertaken (e.g., Robinson, 1969; Hewitt, 1968 et seq.; Pietrzykowski and Jensen, 1972). The first-order predicate calculus is general enough, though, so that if Church's thesis is correct, then all mathematical theories can be expressed using it. In principle, the AI research that has been done in first-order predicate calculus is no less general than any work that may be done in higher-order predicate calculus. However, it is stressed again that, in practice, first-order predicate calculus is not adequate for the statement of mathematical theories about most real-world environments and problems. The first-order expressions of such theories would be extremely long, complicated, and inefficient (if they were at all obtainable), just as it would be extremely complicated and inefficient to try to describe a real-world, problem-solving procedure (e.g., SIN, DENDRAL) as a Turing machine. The AI research on first-order predicate calculus has been valuable as a relatively simple demonstration that computers can be made to "reason" in a general way about logical problems. (This discussion is continued in the section of this chapter entitled "Applications to Real-World Problems.")

---

[1] Some mathematical theories are not descriptions of "real" things. For example, group theory is a description of a class of mathematical theories that are often used to describe many different things.

Briefly, then, predicate calculus provides a framework for making and deriving statements that belong to mathematical theories. Our conception is that statements express "logical thoughts" about things, that statements should be made up of symbols, and that it should sometimes be possible to prove or disprove the truth of a statement with respect to a set of statements that are known to be true.

The *symbols* of first-order predicate calculus are:

1. The *variables* $x, y, z, \ldots$, which will be called *individual symbols*. A variable is a symbol that may represent any object about which we can make a logical statement. The set of things that a variable may represent in a mathematical theory is known as the *universe*, or *domain of discourse*, of that theory.

2. For each $n \geq 0$, the *n-ary function symbols* $f, g, h, \ldots$, and the *n-ary predicate symbols* $P, Q, R, \ldots$. For any given $n$, the number of such symbols may be zero or nonzero, finite or infinite.[2]

3. The *logic symbols* $\forall$, $\exists$, $\neg$, $\wedge$, $\vee$ which stand for "for all," "there exists," "not," "and," and "or," respectively.[3]

4. *The punctuation symbols* "," and "(", and ")".

To define statements, or *formulas,* we must also define *terms* and *atomic formulas*. We define terms as follows:

A variable is a term.

If $T$ is a sequence of $n$ terms ($n$ greater than or equal to 0) and $f$ is an $n$-ary function symbol, then $fT$ is a term.

No other expressions are terms.

We define atomic formulas as follows:

If $T$ is a sequence of $n$ terms ($n \geq 0$) and $P$ is $n$-ary predicate symbol, then $PT$ is an atomic formula.

No other expressions are atomic formulas.

Finally, we define formulas as follows:

An atomic formula is a formula.

If $U$ is a formula, then so is $\neg U$.

---

[2] A complete formalization of first-order predicate calculus provides an infinite number of variable, function, and predicate symbols. These are generally written $x_i$, $f_j^n$, $p_k^n$, respectively, with the subscripts $i, j, k$ being allowed to take numerical values. However, the examples we present require only a few symbols.

[3] We actually provide a formalization only for first-order predicate calculus without equality; predicate calculus with equality contains an extra logic symbol "=" which stands for "equals." Theorem provers that work in theories with equality have encountered difficulties; for a discussion, see Robinson (1970).

If $U_1, \ldots, U_n$ are formulas, then $\wedge (U_1, \ldots, U_n)$ is a formula.

If $U_1, \ldots, U_n$ are formulas, then $\vee (U_1, \ldots, U_n)$ is a formula.

If $U$ is a formula, then, for any variable $x$, $\forall x U$ is a formula.

If $U$ is a formula, then, for any variable $x$, $\exists x U$ is a formula.

These definitions make possible some strings .of symbols as statements in the first-order predicate calculus and rule out others. Thus,

$$\exists x (\forall y ( \wedge (P(x,y), Q(z))))$$

is a formula, but

$$)xQR(\exists)^{\neg}yz$$

is not.

The first-order predicate calculus expression of a mathematical theory consists of a set $S$ of sentences, each of which is a formula according to the rules given above. Such a set $S$ is called a *system*. It is possible for a system to correspond to many different mathematical theories: A system can be taken as a description for many different things, depending on how one "interprets" its formulas.

For example, in predicate calculus the most basic sort of statement one can make is an atomic formula: $R(x)$, $P(x,y,z)$, and $G(f(x,y)$ are all examples of atomic formulas. Each of these could "mean" anything, depending on how one interprets the symbols involved. Thus, a convenient interpretation of $P(x,y,z)$ might be "the number $x$ plus the number $y$ is the number $z$"; or, $P(x,y,z)$ might mean "$x$ and $y$ are the parents of $z$." An interpretation of a set of atomic formulas is given when we specify interpretations for the variable, function, and predicate symbols used in that set of formulas.

An interpretation for a set of variable-symbols is given by specifying the universe of discourse; that is, the set of values they can assume. The universe of discourse for a set of variable symbols is denoted by the letter $D$. For example, $D$ might be the set of numbers $\{-1, 0, +1\}$. If $D$ denotes a set, then $D^n$ denotes the set of all $n$-tuples of $D$. Thus,

$$\{(-,0,+1)\}^2 = \{(-1,-1),(-1,0),(0,-1),(-1,+1),$$
$$(+1,-1),(0,0),(0,+1),(+1,0),(+1,+1)\}$$

(If $D$ contains $m$ elements, then $D^n$ contains $m^n$ elements.) An interpretation of an $n$-ary predicate symbol $P$ associates each element of $D^n$ with exactly one element of the set {true, false}. Thus, if an interpretation of $P$ gives $(-1,0)$ the *truth-value* "false," we say "$P(-1,0)$ is false." An interpretation of an $n$-ary function $f$ associates each element of $D^n$ with exactly one element of $D$. Thus, if an interpretation of $f$

gives to $(+1,+1)$ the value $+1$, we say "$f(+1,+1)$ is $+1$." For $n = 0$ we define $D^n$ to be the set containing $()$, the zero-tuple. By our definition, the interpretation of a zero-ary function must be a *constant;* rather than write expressions like "$f()$," we can usually denote constants by the letters $a,b,c$.[4]

From our definition of a formula it is clear that there are many types of formulas more complicated than atomic formulas, and that these sentences can be constructed using the logical symbols; that is, the *operators* $\neg, \vee, \wedge$, and the *quantifiers* $\exists, \forall$. Explaining the meaning of such a formula is rather straightforward: The operator $\neg$ produces the *negation* of the statement it is applied to: thus, if $P(-1,0)$ is false, then $\neg P(-1,0)$ is true, and vice versa. If the operator $\vee$ is applied to a sequence of statements, it produces their *disjunction;* that is, it produces the statement that is true if and only if at least one of the statements in the sequence is true. Applying the operator $\wedge$ to a sequence of statements produces their *conjunction,* that statement which is true if and only if all the statements in the sequence are true. Thus, if $P(-1,0)$ is false, $P(1,1)$ is true, and $P(0,-1)$ is true, then

$\vee(P(-1,0),P(1,1))$ is true.
$\wedge(P(-1,0),P(1,1))$ is false.
$\vee(P(1,1),P(0,-1))$ is true.
$\wedge(P(1,1),P(0,-1))$ is true.
$\vee(P(-1,0),P(1,1),P(0,-1))$ is true.
$\wedge(P(-1,0),P(1,1),P(0,-1))$ is false.

It is common to introduce a fourth logic symbol "$\rightarrow$", to be read "implies," and to rewrite any formula of the form $\vee(\neg U, V)$ in the form $U \rightarrow V$. Of course such a formula may be true or false, regardless of the form in which one writes it. The truth or falsity of $U \rightarrow V$ depends only on the truth or falsity of $U$ and $V$. Thus, according to our example,

$P(1,1) \rightarrow P(0,-1)$ is true.
$P(1,1) \rightarrow P(-1,0)$ is false.
$P(-1,0) \rightarrow P(1,1)$ is true.

In fact, if $U$ is false, then $U \rightarrow V$ is true, regardless of the value of $V$.

If the *existential* operator $\exists$ is used to quantify a variable in a formula, it produces the statement that there is some value of the variable in the universe of discourse for which the formula is true. Thus $\exists x P(-1,x)$ means "there is a value of $x$ such that $P(-1,x)$ is true."

---

[4] Similarly, a zero-ary predicate is the same thing as a *proposition*. Propositional (zero-order) predicate calculus is not considered in this book. See Suppes (1957) for a good introduction.

When the *universal* operator $\forall$ is applied to a variable in a formula, it produces the statement that "for all values of the variable in the universe of discourse the formula is true." $\forall x P(-1,x)$ would mean "for all values of $x$, $P(-1,x)$ is true." To find the truth value of a very complex formula, with many logical operators and quantifiers, we start with its simplest components and work outward. For example, suppose $\exists x P(-1,x)$ is true, $\forall x P(-1,x)$ is false, and $P(1,1)$ is true: then

$$\wedge\,(\ \vee\,(\ \underbrace{\exists x P(-1,x)}_{\text{true}},\ \underbrace{\forall x P(-1,x)}_{\text{false}}\,)\,,\ \underbrace{P(1,1)}_{\text{true}}\,)$$

with the intermediate evaluations shown: the inner $\vee$ is true, and the outer $\wedge$ is true.

is true, as the preceding diagram of its evaluation shows. Variables that are quantified in a formula are said to be *bound*, while those which are not quantified are said to be *free*. First-order predicate calculus does not permit predicate or function symbols to be quantified or to be used within predicate arguments in formulas; both of these things are natural for human mathematicians and may happen in higher-order predicate calculus. Henceforth, the expression "predicate calculus" will be used to refer only to first-order predicate calculus, unless otherwise specified. However, the remarks in the remainder of this section are valid for predicate calculus in general.

The interpretation of logic symbols is standard throughout predicate calculus, so the interpretation of the formulas of a system really depends only on the interpretation of the atomic formulas of that system. If the domain of discourse of the variable symbols in a system has been specified, and interpretations for all functions and predicates involved in the system have been given, then we have an interpretation for the system itself. If each formula in a system turns out to have the value "true" with respect to an interpretation, then the interpretation is said to be a *model* for the system. A system may have zero, one, or many models. In the first case it is said to be unsatisfiable; in the other cases it is said to be satisfiable.

So far nothing has been said about *rules of inference,* that is, ways of deriving one formula from other formulas. Suppose $S$ is a system (set of formulas) and $U$ is a formula. Then we say that $S$ *logically implies* $U$ if and only if $U$ has the value "true" with respect to every model for $S$. Trivially, every formula in $S$ is logically implied by $S$. A *rule of inference* is a procedure that, given a set $S$ of formulas, may produce only formulas that are logically implied by $S$. Different formalizations of predicate calculus make use of different inference rules. Five of these are:

EXPANSION RULE. If $U$ and $V$ are formulas and $U$ is logically implied by $S$, then $\vee(U,V)$ is logically implied by $S$.

CONTRACTION RULE. If $\vee(U,U)$ is logically implied by $S$, then $U$ is logically implied by $S$.

ASSOCIATIVE RULE. If $\vee(U, \vee(V,W))$ is logically implied by $S$, then $\vee(\vee(U,V),W)$ and $\vee(U,V,W)$ are logically implied by $S$, and vice versa.

CUT RULE. If $\vee(U,V)$ and $U{\to}W$ are logically implied by $S$, then $\vee(V,W)$ is logically implied by $S$.

∃-INTRODUCTION RULE. If $x$ is not free in $V$ and $U{\to}V$ is logically implied by $S$, then $\exists\times(U{\to}V)$ is logically implied by $S$.

The next section presents a special inference rule, the *resolution procedure*, which can be used in place of all of the preceding five rules. If a logical implication of $S$ can be derived using them, then it can be derived using the resolution principle.

In general, any formula that is logically implied by a system $S$ is referred to as being a *theorem* of $S$. Many systems will contain and logically imply an infinite number of formulas, and will be called *infinite systems*. An attempt can be made to describe an infinite system $S$ by presenting some finite set $S_a$ of formulas and comparing the set $\text{Imp}(S_a)$, of all formulas that are logically implied by $S_a$, with the set $\text{Imp}(S)$. If $\text{Imp}(S_a) = \text{Imp}(S)$, then we say $S_a$ is an *axiomatization* for $S$, and we call the formulas in $S_a$ *axioms* for $S$. Those formulas that are theorems of $S_a$, but not axioms of $S_a$, are called *consequences* of $S$, with respect to the axiomatization of $S_a$.

Two things remain to be pointed out: First, using a given inference rule (or set of inference rules) will not necessarily enable one to produce all the theorems of a given system $S$. An inference rule is an "if ... then ... " statement that enables one to establish the logical implication of some formulas, given the logical implication of other formulas. Given an initial set $S_a$, one can establish formulas not in $S_a$ as being logically implied by $S_a$. Given these formulas plus $S_a$, one can establish more formulas as being logically implied by $S_a$, etc. However, one cannot necessarily establish *every* formula that is logically implied by $S_a$, using a given inference rule. In fact, for some systems, one can show that there is no set of inference rules that will enable one to establish in a finite number of steps each formula that is logically implied by the system. Such a system is said to be *undecidable*.[5] Artificial

---

[5] For example, number theory is undecidable. The proof of the existence of undecidable systems is Godel's famous result (Godel, 1931). The existence of undecidable systems is equivalent to the unsolvability of the Halting Problem (see the section entitled "Limits to Computational Ability" in Chapter 2).

intelligence research cannot produce a consistent theorem prover (a set of inference rules) that is capable of proving (establishing the logical implication of) every theorem of an undecidable system.

Second, if a formula is not logically implied by a given system, it may still be true for some models of the system. Formulas that are true for some models of a system and false for others are said to be *contingent* (Kleene, 1967, p. 29).

# THEOREM-PROVING TECHNIQUES
## Resolution

### Groundwork

If $U$ is a formula and $S$ is a set of formulas, and $S$ logically implies $U$, then $U$ has the value "true" with respect to every model for $S$. Thus, $\neg U$ has the value "false" with respect to every model for $S$. Let us consider the set $S'$, which contains all the formulas in $S$ and also contains the formula $\neg U$. Does $S'$ have a model?

The answer is no, for the following reason: If an interpretation is a model for $S'$, then every formula in $S'$ must have the value "true" with respect to that interpretation. Thus, the interpretation must be a model for any subset of $S'$ (for any collection of formulas that belong to $S'$). In particular, the interpretation must be a model for $S$. However, the formula $\neg U$ must have the value "false" with respect to any model for $S$. Thus, the formula $\neg U$ must have the value "false" with respect to any model for $S'$. However, $\neg U$ is one of the formulas that belongs to $S'$, and so by definition it must have the value "true" with respect to any model for $S'$. If $S'$ had a model, $\neg U$ would therefore have both the value "true" and the value "false." In predicate calculus an interpretation can specify at most one value for any given formula. Consequently, $S'$ does not have a model. Thus, by definition, $S'$ is said to be *unsatisfiable*.

Similarly, we can show that if $S'$ is unsatisfiable, and yet $S$ is satisfiable, then $S$ logically implies $U$. Thus, if we want to show that a satisfiable set of formulas $S$ logically implies a formula $U$, it is sufficient to show that the $S'$ set ($=S \cup \{\neg U\}$) is unsatisfiable.

This is a technique that is often used by the theorem provers developed in AI research. The theorem prover is given a set $S_a$ of formulas, which is called its *data base*. It is also given a formula $U$, called the *conjecture*. The problem for the theorem prover is to prove that $U$ follows from $S_a$; that is, that $U$ is logically implied by $S_a$. The procedure followed by the theorem prover is to construct the formula $\neg U$, called the *negated conjecture,* and to attempt to show that the set $S'_a$, which contains the formulas in $S$, and the negated conjecture, is unsatisfiable. One way in which many theorem provers attempt to show the unsatisfiability of a set of formulas is through the use of the *resolution procedure.* This procedure was originally developed by J. A. Robinson (1965 et seq.). Extensions and other theorem-proving techniques have been developed by Wos, P. B. Andrews, G. A. Robinson, Slagle, Sibert, Luckham, Nilsson, Prawitz, Loveland, Hayes, Kowalski, Meltzer, Darlington, Guard, Gilmore, Gelernter, Reiter, Pietrzykowski, Coles, Green, Kling, Hewitt, and others (see the Bibliography). Some of the early work which led to the development of these techniques was done by Davis, Quine, Dreben, Newell et al., and Wang. This section describes the steps involved in the application of the binary resolution principle. A more detailed presentation is given in Nilsson (1971).

## Clause-Form Equivalents

The first step in the application of the resolution principle to a set of formulas $S'_a$ is to replace each formula in $S'_a$ by an expression known as its *clause-form equivalent.* Every formula in first-order predicate calculus has a clause-form equivalent, which may be obtained by applying the following sequence of operations: First, *eliminate implication signs.* Wherever an expression of the form $A{\rightarrow}B$ occurs in a formula, we replace it by $\lor(\neg A,B)$. For example, if we are finding the clause-form equivalent of the formula

$$\forall x \forall y\,((A(x){\rightarrow}\,\neg C(x,y)){\rightarrow}\,\neg\forall x \exists z \land (P(x,z),R(z)))$$

then this first step produces the formula

$$(\forall x \forall y)\lor(\neg\lor(\neg A(x),\neg C(x,y)),\neg\forall x\exists z\land(P(x,z),R(z)))$$

Our next step is to *reduce the scope of all negation signs,* making each negation sign apply to at most one predicate, using these substitutions:

| | | |
|---|---|---|
| Replace | $\neg\lor(A,\ldots,B)$ | by $\land(\neg A,\ldots,\neg B)$ |
| Replace | $\neg\land(A,\ldots,B)$ | by $\lor(A,\ldots,\neg B)$ |
| Replace | $\neg\neg A$ | by $A$ |
| Replace | $\neg(\forall x A)$ | by $\exists x(\neg A)$ |
| Replace | $\neg(\exists x A)$ | by $\forall x(\neg A)$ |

The application of this step to our example yields

$$\forall x \forall y \lor(\land(A(x),C(x,y)),\exists x \forall z \lor(\neg P(x,z),\neg R(z)))$$

Our third step is to *standardize variables;* that is, rename the variables in our formula so that each quantifier binds a unique variable symbol. Within the scope of a given quantifier the variable that is bound by that quantifier is really a dummy variable, and it doesn't matter what letter we use to represent it. If we standardize the variables in our example, we obtain

$$\forall x \forall y \lor (\land (A(x), C(x,y)), \exists u \forall z \lor (\neg P(u,z), \neg R(z)))$$

Next, we *eliminate the existential quantifiers* from our formula. To see how this may be done, consider the expression

$$\forall x \forall y \exists z P(x,y,z)$$

It is clear that the value of $z$ which will satisfy $P(x,y,z)$ may depend on the values of $x$ and $y$. We can indicate this possible dependence by an undefined function, known as a *Skolem function,* and writing our expression

$$\forall x \forall y P(x,y, f(x,y))$$

We may interpret the Skolem function $f(x,y)$ as specifying for any given values of $x$ and $y$ a value for $z$ that "exists" and is such that $P(x,y,z)$. In general, we obtain the *Skolem transform* of a formula by replacing each existentially quantified variable by a Skolem function of those universally quantified variables that are bound by universal quantifiers whose scopes include the existential quantifier being eliminated. The function letter used to replace a given existentially quantified variable must be different from those function letters (for either ordinary or Skolem functions) that already occur in the formula. Eliminating existential quantifiers, our original example now becomes

$$\forall x \lor y \lor (\land (A(x), C(x,y)), \; x \lor (\neg P(g(x,y),z), \neg R(z)))$$

where $g(x,y)$ is the Skolem function introduced. Since all the variables that occur in the formula are unique, we may move the universal quantifiers to the leftmost part of the formula. This action is known as converting the formula to *prenex form.* The formula now consists of a *quantifier string* (or *prefix*) followed by a *matrix.* Our example becomes

$$\forall x \forall y \lor z \lor (\land (A(x), C(x,y)), \lor (\neg P(g(x,y), z), \neg R(z)))$$

Our next step is to put the matrix in *conjunctive normal form.* Any matrix can be written as the conjunction of a finite set of disjunctions, atomic formulas, and negatives of atomic formulas. This may be done by repeated application of the rule

$$\text{Replace} \lor (A, \land (B, \ldots, C)) \text{ by } \land (\lor (A,B), \ldots, \lor (A,C))$$

Thus, our example becomes

$$\forall x \forall y \forall z \wedge (\vee (A(x), \neg P(g(x,y),z), \neg R(z)), \vee (C(x,y), \neg P(g(x,y),z), \neg R(z)))$$

Finally, since all the variables in our formula are now universally quantified, we may *eliminate the universal quantifiers,* and simply write our example formula as

$$\wedge (\vee (A(x), \neg P(g(x,y), z), \neg R(z)), \vee (C(x,y), \neg P(g(x,y), z), \neg R(z)))$$

These remarks indicate that we can make the following definitions: A *literal* is either an *atom* (atomic formula) or the negation of an atom; a *clause* is a disjunction of literals; a formula is a conjunction of clauses. Disjunctions and conjunctions can be identified simply by their sets of disjuncts and conjuncts, and we can speak of a literal $L$ as being an element of a clause $C$. The *null disjunct nil,* which is the disjunction of the set containing no literals, always has the truth-value "false."

Thus, a formula can be expressed as a *set* of clauses. The "clause-form equivalent" of our example is

$$\{\{A(x), \neg P(g(x,y), z), \neg R(z)\}, \{C(x,y), \neg P(g(x,y), z), \neg R(z)\}\}$$

The clause-form equivalent of a set of formulas is the union of the sets of clauses representing each formula (provided the variables used in each formula are made distinct from those used in the other formulas). As a final example, a clause-form equivalent for the set of formulas

$$S = \{\forall x \forall y P(x) \rightarrow N(y), \forall x \exists z \wedge (Q(x,z), \neg P(z)), \forall y \exists x R(x,f(y,a))\}$$

is the set of clauses

$$\{\{\neg P(x),N(y)\}, \{Q(u,g(u))\}, \{\neg P(g(u))\}, \{R(h(w),f(w,a))\}\}$$

(The Skolem functions are $g(u)$ and $h(w)$.)

Given a set of formulas $S_a$ and a formula $U$, the theorem provers we describe will attempt to show that $S_a$ logically implies $U$ by forming the set $S'_a$, which contains the formulas in $S_a$ and the negated-conjecture $\neg U$, and then attempting to show that $S'_a$ is unsatisfiable. The first step in showing the unsatisfiability of $S'_a$ is to find the clause-form equivalent for $S'_a$. Having found the clause-form equivalent of $S'_a$, the theorem prover will attempt to find new clauses that are logically implied by the clauses in $S'_a$. If it can show that the empty clause, *nil,* is logically implied by $S'_a$, then it will have shown that $S'_a$ is unsatisfiable, since *nil* is false for any interpretation. The theorem provers discussed here use an inference rule known as the *binary resolution principle* to find clauses that are logically implied by other clauses. The basic process used in the binary resolution principle is known as the *unification procedure*.

## The Unification Procedure

To describe this procedure, some terminology must be introduced.

A *substitution* $\theta = \{(t_1,v_1),\ (t_2,v_2),\ \ldots,\ (t_n,v_n)\}$ is an operation that, when applied to a clause $C$, yields another clause $C\theta$, obtained by replacing each occurrence in $C$ of the variables $v_i$ by the corresponding terms $t_i$ (we require for any given substitution $\theta$ that $i \neq j$ implies $v_i \neq v_j$). For example, application of the substitution

$$\theta = \{(g(z),x),(a,y)\}$$

to the clause

$$C = \{ \neg P(x,y),Q(b,y)\}$$

yields the clause

$$C\theta = \{ \neg P(g(z),a),Q(b,a)\}$$

Although it is required in a substitution that all the individual symbols $v_i$ be distinct, it is not required that all the terms $t_i$ be distinct. The *empty substitution* $\epsilon = \{\ \}$ consists of not replacing anything so that for all $C$, $C\epsilon = C$. If for two clauses $C$ and $C'$, there is some substitution $\theta$ such that $C\theta = C'$, then $C'$ is said to be an *instance* of $C$. If $C'$ contains no variables, then $C'$ is said to be a *ground clause* and to be a *ground instance* of $C$. Thus, for the two clauses

$$C = \{R(x,y,z),S(u,f(x))\} \quad \text{and} \quad C' = \{R(c,a,b),S(c,f(c))\}$$

the substitution

$$\lambda = \{(c,x),(a,y),(b,z),(c,u))\}$$

makes $C\lambda = C'$, so $C'$ is a ground instance of $C$. We say a clause $C$ is *unifiable* if there is a substitution $\theta$ such that $C\theta$ contains only one literal. Such a substitution is said to be a *unifier* for the clause $C$. Thus, the substitution $\alpha = \{(a,x),(b,y)\}$ unifies the clause

$$C = \{P(x,f(y),b),P(x,f(b),b)\}$$

producing the clause

$$C\alpha = \{P(a,f(b),b)\}$$

If $\alpha$ is a unifier for a clause $C$, then the clause $C\alpha$ is known as a *unification* of $C$, and both $C$ and the literals in $C$ are said to be *unified* by $\alpha$.

A given clause may have several unifiers. Thus, the clause $C$ considered above has, besides the unifier $\alpha$, the unifier $\beta = \{(b,y)\}$, which produces the unification

$$C\beta = \{P(x,f(b),b)\}$$

The unifier $\beta$ is, in a sense, "more general" than the unifier $\alpha$ because $\beta$ does not specify a substitution for the variable symbol $x$. The resolution procedure described in the next section works most successfully if we are able to find very general unifiers for clauses, and it is the purpose of the unification procedure to find the *most general unifier* (*mgu*) for any given clause, provided the clause can be unified. To state the unification procedure, we need to define the notion of a "composition" of substitutions.

The *composition* of two substitutions $\alpha$ and $\beta$ is denoted by the expression $\alpha\beta$ and is that substitution obtained by applying $\beta$ to the *terms* of $\alpha$ and then adding to $\alpha$ any $(t_i,v_i)$ pairs in $\beta$ that have variable symbols $v_i$ not occurring among the variables of $\alpha$. Thus, if

$$\alpha = \{(g(x,y),z),(f(a,w),w)\}$$

and

$$\beta = \{(a,x),(b,w),(c,z)\}$$

then

$$\alpha\beta = \{(g(a,y),z),(f(a,b),w),(a,x)\}$$

It can be shown that applying $\alpha$ and $\beta$ successively to any clause $C$ yields the same result as applying $\alpha\beta$ to $C$. Thus, $(C\alpha)\beta = C\alpha\beta$. Similarly, one can show that composition is associative; that is, that for any clause $C$ and substitutions $\alpha$, $\beta$, and $\gamma$, we have $(C\alpha\beta)\gamma = (C\alpha)\beta\gamma$.

A substitution $\lambda$ that unifies a clause $C$ is said to be *most general* if, given any other unifier $\theta$ of $C$, one can always find a substitution $\gamma$ such that $\lambda\gamma = \theta$; that is, such that $C\lambda\gamma = C\theta$. The unifications of a clause $C$ produced by its most general unifiers are all *alphabetic variants* of each other; that is, each of them may be obtained from any of the others by a substitution of variable symbols for variable symbols. Thus $C = \{P(x,f(y),b),P(b,f(w),b)\}$ has the most general unifications $\{P(b,f(y),b)\}$ and $\{P(b,f(w),b)\}$ and the second unification may be obtained from the first by application of the substitution $\{(w,y)\}$. The most general unifiers for these unifications are $\{(b,x),(y,w)\}$, and $\{(b,x),(w,y)\}$, respectively.

We can now state a unification procedure that finds the most general unifier for any given clause $C$ if $C$ is unifiable, and reports failure if $C$ is not unifiable. The unification procedure makes use of two "program variables," $\lambda_k$ and $k$, which are initially set to $\epsilon$ and 0; throughout its operation, unification alters their values. Thus, $k = 0$ and $\lambda_0 = \epsilon$. The eventual value of $\lambda_k$ is the most general unifier of the given clause $C$ (subject to our comments about alphabetic variants) if $C$ is unifiable, and is E if $C$ is not unifiable. The steps of the procedure are as follows:

1. If $C\lambda_k$ contains only one literal, then return $\lambda_k$ as the mgu for $C$ and *stop*.
2. If $C\lambda_k$ contains more than one literal, find the first symbol position for each literal in which not all literals have the same symbol. For example, if

$$C\lambda_k = \{P(g(x),a,f(u,v)),P(u,a,z)\}$$

then the first symbol positions are as marked by the arrows.
3. Construct the *disagreement set* for $C\lambda_k$, which contains the well-formed expressions (terms or literals) from each literal in $C\lambda_k$ that begins at the marked positions. Thus, the disagreement set for the example is $\{g(x),u\}$.
4. If there exist two terms $s_k$ and $t_k$ in the disagreement set such that $s_k$ is a variable symbol and $t_k$ does not contain $s_k$, then take any two such terms $s_k$ and $t_k$, replace $\lambda_k$ by $\lambda_{k+1} = \lambda_k\{(t_k,s_k)\}$ and replace $k$ by $k + 1$, and *go to* step 1. For our example, $s_k$ may be taken to be $u$, and $t_k$ may be taken to be $g(x)$. Thus,

$$\lambda_{k+1} = \lambda_k\{(g(x),u)\}$$

and

$$C\lambda_{k+1} = \{P(g(x),a,f(g(x),v)),P(g(x),a,z)\}.$$

5. If there do not exist two such terms $s_k$ and $t_k$ in the disagreement set, then report that $C$ cannot be unified and *stop*.

No proof will be offered that the unification procedure does in fact find the most general unifier (see J. A. Robinson, 1965, or Luckham, 1967). For the example shown in the explication of the procedure, if $C$ were initially the clause $\{P(g(x),a,f(u,v)), P(u,a,z)\}$, then the procedure would return the mgu $\lambda_2 = \{(g(x),u),(f(g(x),v),z)\}$, and the most general unification of $C$ would be

$$C\lambda_2 = \{P(g(x),a,f(g(x),v))\}$$

Examples of some clauses and their most general unifications conclude this discussion.

| *Clauses* | *Most General Unifications* |
|---|---|
| $\{P(x),P(a)\}$ | $\{P(a)\}$ |
| $\{Q(x,y,a),Q(x,y,b)\}$ | Not unifiable |
| $\{R(z,f(x),y),R(a,y,f(x))\}$ | $\{R(a,f(x),f(x))\}$ |
| $\{P(x,z,y),P(u,i,a),P(w,u,a)\}$ | $\{P(a,a,a)\}$ |
| $\{P(f(x)),P(x)\}$ | Not unifiable |
| $\{P(f(x),y,g(y)),P(f(x),z,g(x))\}$ | $\{P(f(x),x,g(x))\}$ |

## The Binary Resolution Procedure

An inference procedure used by many theorem-proving programs may now be stated. This procedure is known as the *binary resolution procedure,* and it constitutes an inference rule that enables us to construct some of the clauses that are logically implied by any given set of clauses. The *resolution process* will be explained first, followed by a description of its use in a procedure to prove that a given set of clauses is unsatisfiable, when the set is in fact unsatisfiable.

Suppose we wish to find clauses that are logically implied by two given clauses, say, $C_1$ and $C_2$. Let us denote the literals belonging to $C_1$ by $L_i$ and those belonging to $C_2$ by $M_j$. Thus, $C_1 = \{L_i\}$ and $C_2 = \{M_j\}$. Let us suppose that $C_1$ and $C_2$ have no variables in common (we can always rename the variables in one or the other clause to accomplish this). Let $\{l_i\} \subseteq \{L_i\}$ and $\{m_j\} \subseteq \{M_j\}$ be two subsets of $\{L_i\}$ and $\{M_j\}$ such that a most general unifier $\lambda$ exists for the set of literals $\{l_i\} \cup \{\bar{\phantom{m}}m_j\}$. Then the clause

$$C_3 = \{\{L_i\} - \{l_i\}\}\lambda \cup \{\{M_j\} - \{m_j\}\}\lambda$$

is logically implied by $C_1$ and $C_2$. Depending on how we choose $\{l_i\}$ and $\{m_j\}$ we may obtain other clauses that are logically implied by $C_1$ and $C_2$. It may not be possible to choose $\{l_i\}$ and $\{m_j\}$ such that $\{l_i\} \cup \{\bar{\phantom{m}}m_j\}$ can be unified. However, if $\{l_i\}$ and $\{m_j\}$ can be so chosen, then the two clauses $C_1$ and $C_2$ are said to *resolve* and to be *parent clauses* for the *resolvent(s)* $C_3$. The process of choosing $\{l_i\}$ and $\{m_j\}$ sets to find resolvents for two given parent clauses is called the *resolution process*. Since the clauses $C_1$ and $C_2$ we consider are always finite, there are only a finite number of ways we can choose $\{l_i\}$ and $\{m_j\}$. Thus, $C_1$ and $C_2$ can have only a finite number of resolvents.

As an example, consider the two clauses

$$C_1 = \{P(f(x),y),P(z,f(a)),Q(u)\}$$

and

$$C_2 = \{\neg P(y,z),\neg Q(f(x))\}$$

If we choose $\{l_i\} = \{P(f(x),y)\}$ and $\{m_j\} = \{\neg P(y,z)\}$, then $\{l_i\}\cup\{\neg m_j\} = \{P(f(x),y),P(y,z)\}$, which has the mgu

$$\lambda = \{(f(x),y),(f(x),z)\}$$

The corresponding resolvent for $C_1$ and $C_2$ is

$$C_3 = \{P(z,f(a)),Q(u),\neg Q(f(x))\}\lambda$$
$$\quad = \{P(f(x),f(a)),Q(u),\neg Q(f(x))\}$$

Similarly, if we choose $\{l_i\} = \{P(f(x),y),P(z,f(a))\}$ and $\{m_j\} = \{\neg P(y,z)\}$, then we find that $\{l_i\}\cup\{\neg m_j\} = \{P(f(x),y),P(z,f(a)), P(y,z)\}$ has the mgu $\lambda' = \{(f(a),y),(f(a),z),(a,x)\}$, and we obtain the corresponding resolvent

$$C_3 = \{Q(u),\neg Q(f(x))\}\lambda' = \{Q(u),\neg Q(f(a))\}$$

Altogether, $C_1$ and $C_2$ have four different resolvents, of which three may be obtained by *resolving on P* and one may be obtained by resolving on $Q$.

Again, let $C_1 = \{\neg P(x),R(x)\}$ and $C_2 = \{\neg R(x),Q(x)\}$. If we choose $\{l_i\} = \{R(x)\}$ and $\{m_j\} = \{\neg R(x)\}$, we obtain the resolvent $C_3 = \{\neg P(x),Q(x)\}$. These three clauses correspond respectively to the predicate calculus formulas

$$\forall x(P(x)\rightarrow R(x))$$
$$\forall x(R(x)\rightarrow Q(x))$$
$$\forall x(P(x)\rightarrow Q(x))$$

The English meanings of these formulas are

"Everything with property $P$ has property $R$."
"Everything with property $R$ has property $Q$."
"Everything with property $P$ has property $Q$."

In this case it should be intuitively clear that the third statement is logically implied by the first two.

Given a pair of clauses $C_1$ and $C_2$, we obtain their resolvents by attempting to apply the *resolution process* to $C_1$ and $C_2$ with respect to each possible combination of their subsets $\{l_i\}$ and $\{m_j\}$. (A computer program can be designed not to investigate some combinations that obviously will not work, such as those combinations that use more than one predicate symbol.) If one of the resolvents of the clauses $C_1$

and $C_2$ is the empty clause *nil,* then we know that $C_1$ and $C_2$ cannot both be satisfied (either one of them might be satisfiable, but there is no model which will make them both true). For example, the resolvent of $C_1 = \{P(x)\}$ and $C_2 = \{\neg P(y)\}$ is the empty clause.

The *binary resolution procedure* for showing that a set $S$ of clauses is unsatisfiable can now be very simply stated. Let $S$ be a set of clauses $\{C_1, C_2, \ldots, C_n\}$. We apply the resolution process successively to each pair of clauses $C_i$, $C_j (i \neq j)$, and place any resolvents obtained in a new set $R(S)$. When we have gotten all possible resolvents from $S$ (for any finite set of finite clauses there are only finitely many possible resolvents, and we can tell when all the possibilities have been tried), we apply the binary resolution procedure to the set $R_1(S) = S \cup R(S)$, which contains all the clauses in $S$ plus all their resolvents. This yields the set of all resolvents of $R_1(S)$, which is denoted by $R(R_1(S))$. Next we form the set $R_2(S) = R_1(S) \cup R(R_1(S))$, which contains all the clauses in $R_1(S)$ plus all their resolvents. We apply the resolution procedure to $R_2(S)$ to obtain the set $R(R_2(S))$, and we form the set $R_3(S) = R_2(S) \cup R(R_2(S))$. In general,

$$R_{i+1}(S) = R_i(S) \cup R(R_i(S))$$

where $S$ is our initial set of clauses; if $X$ is a set of clauses, then $R(X)$ denotes the set of all resolvents of the clauses in $X$. The set $R_i(S)$ is called the *i*th *level* of clauses that are logically implied by $S$. The resolution procedure consists of developing in succession the levels of clauses that are implied by $S$ until either we run out of computation time (in which case the answer is "no proof found") or the empty clause *nil* is produced as a resolvent in some level. If *nil* is ever produced, then we know that $S$ is unsatisfiable. The resolution procedure corresponds to a *breadth-first* development of the clauses that are logically implied by $S$.

A graph that (1) associates the empty clause with one of its nodes, (2) associates only the ancestors (parents, parents of parents, etc.) of the empty clause with the rest of its nodes, and (3) connects each node only to those clauses that are its parents or of which it is a parent (as determined by the resolution procedure) is called a *refutation graph* of $S$ and constitutes a simple proof that $S$ is unsatisfiable. Figure 6–1 shows a refutation graph for the unsatisfiable set of formulas.

$$S = \{\forall x P(x), \forall x P(x) \rightarrow Q(x), \forall x Q(a) \rightarrow E(x), \neg E(d)\}$$

which is equivalent to the set of clauses

$$S = \{\{P(x)\}, \{\neg P(y), Q(y)\}, \{\neg Q(a), E(z)\}, \{\neg E(d)\}\}$$

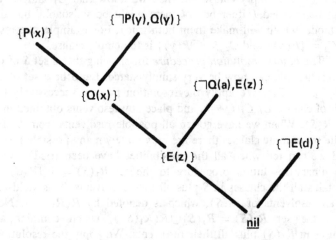

Figure 6–1. A refutation graph.

Of course not all the clauses that are logically implied by $S$ are shown in a refutation graph.

It is possible to prove that the resolution procedure is a valid way of showing that a set of clauses is unsatisfiable (i.e., if *nil* is produced by the procedure, then the set must be unsatisfiable; if the corresponding set of formulas can be shown unsatisfiable, using the five inference rules presented in the first section of this chapter, then *nil* will be produced), but space does not permit the presentation of such a proof. The reader is referred to (Nilsson, 1971, pp. 181–183).

## Summary

To summarize the theorem-proving technique described above, the theorem prover

1. is given a set $S_a$ of axioms and a conjecture $U$.
2. forms the negated-conjecture $\neg U$.
3. forms the set of formulas $S'_a$, consisting of $S_a$ and $\neg U$.
4. produces the clause-form equivalent of $S'_a$.
5. applies the resolution procedure to this set of clauses until it either runs out of time or produces the empty clause *nil*.
6. constructs a refutation graph if *nil* is produced and announces that it has found a proof for the conjecture.

A theorem prover that uses the resolution procedure is said to be *resolution-based*.

# Heuristic Search Strategies

## Extensions

Methods of theorem proving such as the resolution procedure are not practically applicable to systems with more than a few axioms. They correspond to a simple "breadth-first" development of the consequences of the system. It is necessary to use resolution in a *selective* manner, if one wishes to develop a theorem prover that can operate with systems of more than about ten axioms. Various selective techniques for resolution-based theorem provers have been developed. These techniques are generally known as *heuristic search strategies* because of the way in which they alter a theorem prover's development of the consequences of a given system. Using such strategies, it is possible for a theorem prover to prove fairly difficult theorems in systems having up to two dozen axioms (the proof of such a theorem might be 50 steps long). This section reviews some of the currently used search strategies for theorem provers. These strategies fall into three basic categories: *refinement* strategies, *simplification* strategies, and *ordering* strategies (see Nilsson, 1971).

## Simplification Strategies

Often it is possible to eliminate literals or clauses from a set of clauses, in a manner that does not affect the unsatisfiability of the set of clauses. (That is, if the set is unsatisfiable before simplification, it will be unsatisfiable afterward. Conversely, if a set is satisfiable before simplification, it will be satisfiable afterward.) When this can be done, it will reduce the rate at which irrelevant clauses are generated. Three ways of simplifying a set of clauses are to eliminate tautologies, evaluate predicates where possible, and to eliminate clauses that are *subsumed* by other clauses.

A *tautology* is a statement of the form "*A* or not *A*." In predicate calculus every tautology is trivially true. The clause-form equivalent of a tautology is a clause that contains both a literal and the complement of that literal. If such a clause belongs to a set of clauses, then it may be eliminated from the set without affecting the unsatisfiability of the set. Thus, clauses like

$$\{Q(x,y), \neg R(z), R(y)\} \qquad \text{and} \qquad \{P(f(x)), \neg P(f(x))\}$$

may be eliminated.

Sometimes it is possible to evaluate the truth value of a literal immediately after the clause containing it is generated. In such a case, if

234                INTRODUCTION TO ARTIFICIAL INTELLIGENCE

the literal has the value "true," then the entire clause may be eliminated without affecting the unsatisfiability of the set $S$ of clauses. If the literal has the value "false," then it may be eliminated from the clause in which it occurs. Generally, it is possible to evaluate a literal only if one has some specific information about the nature of its predicate. Thus, one might have a predicate $P(x,y,z)$ equivalent to "the sum of number $x$ and number $y$ is number $z$." In such a case, literals using this predicate can be immediately evaluated by machine.

A clause $C_1 = \{L_i\}$ is said to *subsume* a clause $C_2 = \{M_j\}$ if there is a substitution $\theta$ such that $\{L_i\}\theta$ is a subset of $\{M_j\}$. For example, $\{P(x),Q(a)\}$ subsumes $\{P(f(a)),Q(a),R(y)\}$. If $C_1$ and $C_2$ are clauses in $S$ and $C_1$ subsumes $C_2$, then $C_2$ may be eliminated from $S$ without affecting the unsatisfiability of $S$. Intuitively, $C_1$ is "more general" than $C_2$.

Usually, it is wise to eliminate tautologies and, where possible, evaluate predicates before eliminating by subsumption. Subsumed clauses should be eliminated only after each level $R'(S_a)$ of $S$ has been completely developed (see Kowalski, 1970a,b).

## Refinement Strategies

As we have indicated, the resolution principle presented in the preceding section can be generalized. It can also be modified to produce new inference rules that restrict the possible clauses in $S$ which may be resolved, beyond the simple requirement that they be resolvable. Such a modification is known as a *refinement strategy,* and is equivalent to a new inference rule $R_c$ that permits resolutions only between clauses that satisfy a *refinement criterion* $C$. A refinement strategy $R_c$ is said to use resolution *relative* to $C$. Many different refinement strategies for resolution have been developed. One of these, the *ancestry-filter* (AF) strategy, will now be presented. For a discussion of other strategies, see Nilsson (1971).

If clauses $C_1$ and $C_2$ can be resolved to form clause $C_3$, we say $C_1$ and $C_2$ are *parents* of $C_3$. Given a sequence $C_1, C_2, \ldots, C_n$ such that, for $1 \leq i \leq n$, $C_i$ is a parent of $C_{i+1}$, we say $C_1$ is an ancestor of $C_n$ and $C_n$ is a *descendant* of $C_1$ (refer to the terminology for graphs in Chapter 3).

The refinement criterion for ancestry-filtered resolution can now be stated. Two clauses will be resolved if and only if either

    *i*) Both belong to $S_a$.
    *ii*) One belongs to $S_a$ and the other is a descendant of a clause in $S_a$.
    *iii*) One is an ancestor of the other.

The use of this criterion in effect gives us a new inference rule, which is denoted as $R_{AF}$. If $S$ is a set of clauses, then $R_{AF}(S)$ denotes the set of all resolvents between pairs of clauses that belong to $S$ and satisfy the ancestry-filter criterion. Thus, defining,

$$R_{AF}^1(S) = S \cup R_{AF}(s)$$

and

$$R_{AF}^{n+1}(S) = R_{AF}^1(R_{AF}^n(S)),$$

it can be proved that if $S_a$ is unsatisfiable, then there is some $n$ such that $R_{AF}^n(S_a)$ contains the empty clause. Conversely, if $S_a$ is satisfiable, then there is no $n$ such that $R^n(S_a)$ contains the empty clause. Thus, resolution relative to ancestry filtering can be used in place of ordinary resolution. Figure 6–2 shows a refutation of a simple set of clauses, produced according to the ancestry-filter refinement of resolution.

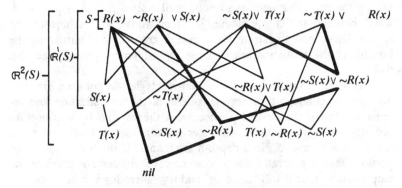

Figure 6–2. A search for refutation using the AF strategy.
(Nilsson, 1971, reprinted with permission.)

In practice it is possible to add further restrictions to the AF strategy. Any such restriction will, in effect, add to the refinement criterion used by the theorem prover. Before using a refinement criterion it is important to prove the *completeness* of resolution with respect to that criterion; that is, one must show that if $S$ is unsatisfiable, then there is some $n$ such that $R_c^n(S)$ contains the empty clause; if $S$ is satisfiable, then there is no $n$ such that $R_c^n(S)$ contains the empty clause. It is also important to show that the total cost of using the criterion (in terms of computation time and memory space used by the computer) is less than the cost of generating, storing, and resolving the clauses eliminated by the criterion. The AF refinement strategy for resolution generally tends to produce much deeper but less broad searches than would be produced by unrefined resolution. Other refinement strategies include the "set-of-support" strategy, and "model" strategies.

## Ordering Strategies

Ordering strategies are the most "heuristic" of the three types of search strategies considered in this section. Ordering strategies correspond to the use of evaluation functions for searching state-space graphs (discussed in Chapter 3). A given ordering strategy does not necessarily prohibit resolution between certain pairs of clauses (as do refinement strategies). Rather, an ordering strategy provides that resolution between certain pairs of clauses shall be performed before resolution between other pairs.

Suppose we have an inference rule $R$ (possibly a refinement of resolution). The search for a refutation of a set of clauses $S_a$ corresponds to the development of successive *levels* $R^i(S_a)$, each level containing the preceding ones as subsets, until a level $R^n(S_a)$ is produced, which contains the empty clause. In other words, it is a breadth-first search. Up to now the strategies discussed are means of narrowing the breadth of the refutation search done by a theorem prover (see Fig. 6–2).

Theorem provers that use ordering strategies do not do a breadth-first search, although they may make use of the simplification and refinement strategies discussed previously. A theorem prover that uses an ordering strategy selectively generates *portions* of the levels below an initial set of clauses $S_a$, in a depth-first manner. If the first sequence of portions that it generates down to some level $n$ does not produce the empty clause, then it will "back up" and try generating another sequence of portions (see Fig. 6–3). Perhaps the two most common ordering strategies are the *unit-preference* strategy and the *fewest-components* strategy.

A *unit* is a clause that contains only one literal. Similarly, a *doubleton* contains two literals, etc. In the unit-preference strategy the theorem prover first attempts to resolve units against units (if this succeeds, then it has produced the empty clause), then units against doubletons, then units against tripletons, etc. Thus, given a set $S$ of clauses, the theorem prover generates the "unit preference" portion of $R^1(S)$. It then generates the unit-preference portion of $R^2(S)$, etc., and continues until it either produces a null clause or reaches its *level bound* $n$; that is, until it generates the unit-preference portion of $R^{n-1}(S)$. If it reaches its level bound without producing the empty clause, it might back up to $S$ and begins generating "doubleton-preference" portions of the levels $R^i(S)$

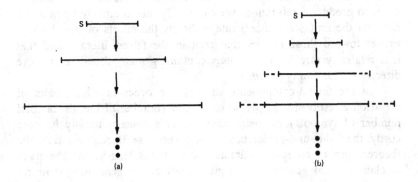

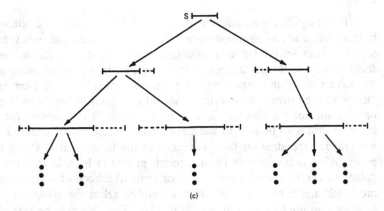

Figure 6–3. A schematic indication of the refutation searches produced by (a) resolution, (b) AF resolution, (c) AF resolution with an ordering strategy.

below $S$, but within the same level bound. Similarly, it could continue to "tripleton-preference" portions, etc. If, in the course of generating a "$q$-tupleton-preference" portion of some level, the theorem prover produces a clause containing $p$ literals, where $p$ is less than $q$, the theorem prover reverts to the generation of $p$-tupleton-preference portions.

    If the empty clause can be produced at all within the level-bound $n$, it will be produced by a theorem prover using the unit-preference ordering strategy. Quite often the unit-preference strategy will enable a theorem prover to greatly reduce the amount of resolution it does in order to produce a refutation. We can justify this intuitively by pointing out that the unit-preference strategy directs the efforts of the theorem prover toward those clauses that contain the fewest literals, and that it is relatively rare for two clauses containing many literals to resolve directly to the empty clause.

In the fewest-components strategy the order in which pairs of clauses are resolved depends on the length (number of literals, or total number of symbols) of their resolvent. This strategy usually is more costly than the unit-preference strategy because it requires that the theorem prover compute estimates of resolvent lengths for the pairs of clauses it has generated at a given level before generating their resolvents.

### Review

The simplification, refinement, and ordering strategies discussed in this section are all *syntax-oriented:* A theorem prover that uses them searches more selectively for a refutation than it would if it did not use these strategies, though it does so in a way that is dependent more on the structures of the expressions it generates rather than on their relations to each other. Its selectivity has little relation to the "meaning" or *semantics* of the theorem it is trying to prove. It is desirable for a theorem prover to be able to select the clauses it resolves in some manner that is dependent on their relevance to the theorem it is trying to prove. Also, it is desirable for a theorem prover to be able to form a "plan" or description of a proof that has some likelihood of corresponding to the actual proof for the theorem, and to select the clauses it resolves according to how well they fit its plan. That this may be feasible will be evident from the following sections.

## Reasoning by Analogy

Chapter 3 discussed various types of analogies and the value of "reasoning by analogy" as an ability of a general problem solver. Kling (1971a,b) developed a method whereby theorem provers can select the clauses they resolve in a manner that corresponds to one type of reasoning by analogy. The presentation of his method in this section is highly schematic and is intended merely to give the reader a good idea of Kling's approach. For more detailed information the reader should see Kling's own explications.

Let us consider the case of a theorem prover being used to prove theorems about abstract algebra.[6] Such a theorem prover might have a standard set $S_a$ of clauses that would constitute its basic knowledge, or *axiomatization,* of abstract algebra. The user of the theorem prover would supply it with a theorem $T$ to be proved, stated using the predicates and functions that occur in $S_a$. The theorem prover would form the negation $\neg T$ of the theorem, reduce $\neg T$ to clause form, add it to $S_a$ to form a set of clauses $S$, and attempt to prove the unsatisfiability of $S$. This procedure could be followed for virtually any theorem $T$ about abstract algebra. In principle, the generality of the theorem provers as problem solvers depends only on the extent to which problems can be described by sets of statements in the predicate calculus. As later sections will show, it is certainly possible to express some aspects of real-world problems within predicate calculus formalizations.

In fact, however, the generality of theorem provers as problem solvers is limited by considerations of computation time and memory space. A difficult theorem $T$ might require 50 steps in the proof generated for it by a good theorem prover, using an axiomatization $S_a$ that contained a dozen clauses. For such an $S_a$ and $T$ the theorem prover might generate 200 clauses altogether before finding the proof. If the theorem prover were given more axioms than necessary (say, $S_a$ containing 30 clauses), it might generate 600 clauses altogether before finding a proof, and run out of space. That is, it would generate about 400 irrelevant clauses. Even with optimal use of the heuristic search strategies discussed in the preceding section, current theorem provers usually are unable to prove nontrivial theorems when $S_a$ contains more than about 30 clauses. And a good axiomatization $S_a$ for a subject like abstract algebra requires about 250 clauses.[7]

Thus, theorem provers as we have so far described them cannot be general problem solvers for nontrivial subjects such as abstract algebra. The axiomatizations (or *data bases*) for such subjects are simply too large for a program (possessing current limitations in time and memory space) to solve problems in them without some way of estimating which clauses are "relevant" to the problem at hand and should be resolved or generated first. We can expect the situation to be much worse for real-world problems, where the number of clauses necessary

---

[6] For the purposes of this book it is not necessary to know abstract algebra. It is chosen simply for the sake of exposition and because Kling chose it for his work.

[7] An axiomatization for a theory must contain not only the clauses that interrelate the basic undefined words ("point," "line," "between," etc.) of the theory, but also those clauses that define the nonbasic words ("circle," "triangle," "congruent")—predicates or functions—used within the theory.

for an adequate axiomatization of reasoning-program knowledge about the world might be very large indeed.[8]

The situation is amply illustrated with the use of "Venn diagrams" (see Fig. 6–4). In proving a theorem $T$ from a data base $S$, a theorem

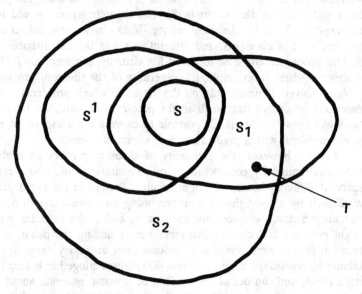

Figure 6–4. Venn diagram for the data-base problem.

prover might generate the set of clauses indicated by the area labeled $S_1$. Given the larger data base $S'$ (which includes $S$ as a subset), the theorem prover will usually generate the much larger set of clauses $S_2$ before it obtains a proof for $T$. In general, we want the theorem prover to have the data base $S'$ available, since there is no a priori information as to which theorems it will be required to prove. But, we would like to have some program that could often select, for any given theorem $T$, a data base $S$ from which $T$ could be easily derived. We could attempt either to modify the theorem prover itself or to write a new program that would select data bases for the theorem prover.

Because of the undecidability of the predicate calculus, this problem cannot be completely solved. However, Kling provided a partial solution to it in the form of an *analogy generator* that, given some help,

---

[8] Minsky (1968a, p. 26) makes a very rough estimate that would correspond by the present author's interpretation to $10^5$ or $10^6$ clauses, belonging to a high-order predicate calculus, for a reasoning program at the level of human intelligence. But, of course, it is only a guess.

is often capable of making a good selection for the data base $S$. Kling's analogy-generating program is called ZORBA-1.[9] It is designed to operate in conjunction with a theorem prover (see Green, 1969), of the type we have described.[10]

ZORBA-1 is given the following information as input:

1. A theorem $T$, which is to be proved by QA3
2. A theorem $T'$, which has already been proved
3. The proof of $T'$; that is, an ordered sequence of clauses $C_1, \ldots, C_n$ such that each clause $C_i$ is either an element of $S'$ (the large data base) or an element of $\neg T'$, or the resolvent of two clauses, say, $C_j$ and $C_k$ which occur prior to it in the sequence (i.e., such that $j$ and $k$ are both less than $i$)
4. The large data-base $S'$
5. A "semantic template" for $S'$

The first three items of this list are problem-dependent; they vary with the theorem $T$ which is to be proved by QA3. The fourth and fifth items do not depend upon $T$, but do depend on $S'$. To the extent that ZORBA-QA3 is being used as a general problem solver relative to $S'$, they can be considered problem-independent.

Given this information, ZORBA-1 produces an *analogy A* consisting of:

1. $A^p$, a one-to-one association (or *map*) between the predicates used in the proof of $T'$ and predicates that may occur in the proof of $T$. That is, each predicate used in the proof of $T'$ is associated with exactly one predicate that occurs in $S'$ and might be used in the proof of T.
2. $A^c$, which associates each clause used in the proof of $T'$ with a set of clauses that each occur in $S'$ and might be used in the proof of $T$.
3. $A^v$, which associates sets of variables used in the proof of $T'$ with sets of variables that might be used in the proof of $T$.

These associations are represented in the computer by appropriate data structures, and are referred to as predicate analogies, clause analogies,

---

[9] ZORBA is an acronym for (ZO) Reasoning By Analogy. Zorba was a passionate, intuitive Greek, and many contemporary thinkers consider analogy an intuitive process outside the realm of reason (Kling, 1971a, p. 4).

[10] QA3 was developed at Stanford Research Institute, principally by Green and Raphael. It is resolution-based and uses the unit-preference heuristic. However, it does not use ancestry-filter refinement, but instead uses the set-of-support refinement (which has not been discussed; see Nilsson, 1971, pp. 223–224).

and variable analogies, respectively. $T'$ will be called an *analog* of $T$, and vice versa.

Thus, an analogy developed by ZORBA consists of a predicate analogy, a clause analogy, and a variable analogy. Predicate and variable analogies are used within ZORBA-1. Its output to QA3 is a clause analogy $A^c$. QA3 uses $A^c$, together with $\neg T$, as the data base which it attempts to prove unsatisfiable. If QA3 succeeds, then we are justified in saying that the two programs, ZORBA and QA3, have together "reasoned by analogy" from the proof of $T'$ to obtain a proof of $T$. Of course there may be many types of analogical reasoning that are not described by this particular paradigm. The importance of this paradigm to AI research depends only on how well it works; that is, whether ZORBA and QA3 are in fact able to prove theorems that could not be proved by QA3 alone, relative to the same large data base $S'$.

In fact, the ZORBA-QA3 program pair is rather successful, at least with respect to the data base for abstract algebra developed by Kling (1971a). Kling's abstract algebra data base $S'$ contains 239 clauses. Given two analogous theorems $T$ and $T'$ and a proof for $T'$ requiring 20 clauses, ZORBA-1 could usually select a clause analogy $A^c$ containing less than two dozen clauses; that was sufficient for QA3 to use in proving $T$. For the reader who is acquainted with abstract algebra, the following example is quoted:

> $T'$: "The intersection of two normal groups is a normal group."
> $T$: "The intersection of two ideals is an ideal."

Either of these theorems would have been unprovable for QA3, given the data base $S'$. However, ZORBA-1 and QA3 together are able to develop a proof for $T$, "reasoning by analogy" from a proof for $T'$.

In practice, ZORBA-1 must select its clause analogy heuristically by searching through a space of *partial analogies* (see Fig. 6–5). For an $S'$ containing 239 clauses, the number of possible clause analogies, each containing 24 clauses, is extremely large (about $10^{60}$; see Kling, 1971a, p.110). ZORBA-1 first develops a partial analogy $A_1$, which is relatively small and contains less than a dozen clauses. For each partial analogy $A_i$ that it develops, it either adds or deletes a few clauses in order to create $A_{i+1}$. ZORBA-1 is guided in its development of partial analogies by the "semantic template" for $S'$, which is provided to it. Usually ZORBA-1 needs to generate less than ten partial analogies. The semantic template is a small table of descriptions for the predicates occurring in $S'$. For example, the predicate "group" is given the description

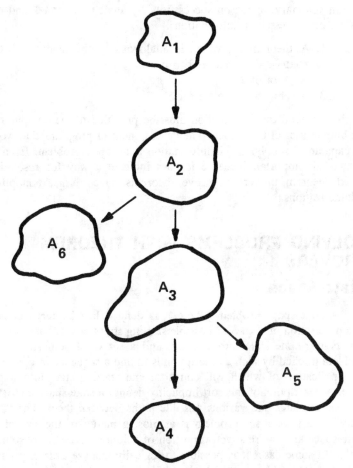

Figure 6–5. Heuristic search through an analogy-space.

STRUCTURE (SET;OPERATOR)

Thus, ZORBA-1 knows automatically whenever it sees "group $(A;*)$" that $A$ is a set and * is an operator. ZORBA-1 uses the semantic template to generate descriptions of those clauses occurring in $S'$ and in the proof of $T'$. The clauses it chooses from $S'$ for its partial analogies are those that have descriptions similar to the descriptions of the clauses in the proof of $T'$. ZORBA-1 terminates its search when it has found analogs for each of the clauses in the proof of $T'$. It then submits the resulting clause analogy to QA3.

In summary, to reason successfully by analogy, ZORBA-1 and QA3 require certain essential information:

1. A theorem $T'$, which is analogous to the theorem $T$ that ZORBA-QA3 is required to prove
2. A proof of $T'$
3. A semantic template for $S'$

At the moment, this information must be provided to the computer by the human user of the system. The development of programs that would be capable of supplying this information is an open problem. Even so, ZORBA-1 is important because it does indicate a way for resolution-based theorem provers to prove theorems, given large, nonoptimal axiomatizations.

# SOLVING PROBLEMS WITH THEOREM PROVERS
## State-Space

A state-space problem $(S,F,G)$ as defined in Chapter 3 consists of a description of a set $S$ of possible starting states, a set $F$ of operators that convert one state into another, and a set $G$ of goal states. The problem implied by such a description is to find a sequence of operators, the application of which will convert some starting state into a goal state. A description of this sort implicitly defines a state-space consisting of a set of states and various possible paths between them. The "difficulty" of a state-space problem is at most a matter of the size of the state-space and the relative proportion of solution paths to nonsolution paths. In some cases it is possible to logically analyze a description of a state-space problem and show that its solutions are equivalent to those for a problem with a simpler description, a smaller state-space or set of operators. With this possibility in mind, the "problem-reduction" problems, the "problem of problem-representation," and "global" analysis of games were discussed in Chapter 3.

This section describes how resolution-based theorem provers can be used as general problem solvers for state-space problems. We consider three questions:

1. How can predicate-calculus theories be used to describe state-space problems?
2. How can theorem provers be used to construct paths through state-spaces?

3. To what extent can the techniques we describe be applied to real-world problems?

Our discussion of these questions is based on McCarthy (1959, 1963a, 1964, 1968), McCarthy and Hayes (1968), Black (1964), Green (1969), Waldinger and Lee (1969), Amarel (1970), Nilsson (1971), and Fikes and Nilsson (1971). Other relevant papers include Amarel (1967, 1968), Slagle (1965, 1970), and Slagle and Bursky (1968).

## Predicate-Calculus Descriptions of State-Space Problems

A given state-space problem can usually be described in many ways, depending on how one chooses to describe its states and operators. The state-space problems discussed previously have the following features:

1. Each state can be described as a collection of objects, each having certain relations to the others. For example, in the 15-Puzzle, the "objects" might be blocks, positions, and the null-block or empty position.
2. Each operator can be described as a procedure for changing one state into another. For a given state, zero, one, or many operators might be applicable.

Thus, our terminology for state-space problems includes "states," "objects," "relations," and "operators." A predicate-calculus description of a state-space problem may reflect these concepts in the following ways:

1. "States" and "objects" can be represented by variable symbols called *state* (or *situation*) *variables* and *object variables*. In this discussion, $s, s', s'', \ldots$ represent state variables and $o, o', o'', \ldots$ represent object variables. Particular constant states and objects will be denoted by $s_0, s_1, s_2, \ldots$ and $a, b, c, \ldots$, box, monkey, $\ldots$, respectively.
2. "Relations" between objects, and properties of states and actions can be indicated by *fluent symbols*, which are either predicate or function symbols. We are primarily concerned with *situational fluents* (McCarthy and Hayes, 1968), which are functions or predicates that include states among their arguments. For example, "ON(monkey, box, $s_0$)" might be a situational-fluent predicate with the value "true" if the ob-

ject "monkey" is *on* the object "box" in state $s_0$. Similarly, "MOVE(monkey, box, $a$, $b$, $s_0$)" might be a situational-fluent function which has as its value the state $s$ produced when the object "monkey" *moves* the object "box" from position $a$ to position $b$ in state $s_0$.

First-order predicate calculus formulas that use the ordinary logic symbols ($\wedge, \vee, \neg, \rightarrow, \exists, \forall$) and use state variables, object variables, and fluent symbols as their nonlogical symbols, can be used to express facts about state spaces. A collection of such formulas can be considered a description of a state-space, provided the collection is satisfiable and contains formulas that use situational-fluent functions which represent the operators associated with the state-space. Problems associated with the state-space can be represented by formulas that are to be proved, using the formulas that describe the state-space.

EXAMPLE 6–1. THE MONKEY-AND-BANANAS PROBLEM. This problem (McCarthy, 1963a) was given as an exercise in Chapter 3. It is one of the classic "toy problems" considered by AI researchers as an example of an extremely simple problem that involves common-sense reasoning about situations, actions, tools, etc. The problem is repeated here along with a predicate-calculus formalization for it:

A monkey is in a room where a bunch of bananas is hanging from the ceiling, too high to reach. In the corner of the room is a box, which is not under the bananas. How can the monkey get the bananas? The solution to the monkey's problem is to move the box under the bananas and climb onto the box, from which the bananas can be reached.

The objects used in our description of this state-space problem are *monkey, box, bananas, place1, place2, place3*. The operators used are *goto, move, climb*, and *reachfor*, each of which will be a situational-fluent function. The relations used in the description are *under, on, at*, and *has-bananas*, each of which will be a situational-fluent predicate. Table 6–1 gives the first-order predicate calculus formulas that correspond to a description of the monkey-and-bananas state-space, using these objects, operators, and relations. The monkey's problem is represented by the formula

$$(\exists s)(\text{has-bananas}(s)) \tag{6-1}$$

which is to be proved, using the formulas in Table 6–1. The formulas in Table 6–1 do not say explicitly that it is possible for the monkey to get the bananas. However, if we can prove formula 6–1 from the formulas in Table 6–1, then we may conclude that there is some sequence of applications of the operators that will convert state $s_0$, in which the monkey does not have the bananas, into a state $s$, in which he does. We can conclude this because the only state that is explicitly mentioned in Table 6–1 is the state $s_0$. If formula 6–1 holds for all models of the formulas in Table 6–1, then it must hold for the model in which the only states that exist are $s_0$ and those which can be obtained from $s_0$ by the application of some sequence of operators to it. Thus, if the formulas of Table 6–1 logically imply formula 6–1, then there is some way for the monkey to convert $s_0$ into a state $s$ for which "has-bananas(s)" is true.

Figure 6–6 shows the proof of formula 6–1, using the formulas in Table 6–1 and the resolution procedure. The negation of formula 6–1 is added to the set of formulas in Table 6–1, and the resulting set of formulas is shown to be unsatisfiable. Thus, the set of formulas in Table 6–1 logically imply that the monkey can get the bananas.

Simply proving that the monkey *can* get the bananas is not, of course, the same thing as showing a way for him to do it. We would like our proof of the existentially quantified formula 6–1 to be *constructive;* that is, we would like it to produce an actual sequence of operations which, when applied to $s_0$, will produce a state $s$ for which "has-ba-

TABLE 6–1A.  The Monkey and Bananas Problem (Predicate-calculus Axioms)*

---

A1.  $\forall p \forall p' \forall s (at(box,p,s) \rightarrow at(box,p,goto(p',s)))$

A2.  $\forall p \forall p' \forall s (at(bananas,p,s) \rightarrow at(bananas,p,goto(p',s)))$

A3.  $\forall p \forall s (at(monkey,p,goto(p,s)))$

A4.  $\forall p \forall p' \forall s (\wedge (at(box,p,s),at(mon,p,s)) \rightarrow at(box,p',move(mon,box,p,p',s)))$

A5.  $\forall p \forall p' \forall p'' \forall s (at(ban,p,s) \rightarrow at(ban,p,move(mon,box,p',p'',s)))$

A6.  $\forall p \forall p' \forall s (at(mon,p,s) \rightarrow at(mon,p',move(mon,box,p,p',s)))$

A7.  $\forall s (under(box,ban,s) \rightarrow under(box,ban,climb(mon,box,s)))$

A8.  $\forall p \forall s (\wedge (at(mon,p,s),at(box,p,s)) \rightarrow on(mon,box,climb(mon,box,s)))$

A9.  $\forall s (\wedge (under(box,ban,s),on(mon,box,s)) \rightarrow$ has-bananas(reachfor,mon, ban,s)))

A10. $\forall s (\wedge (at(box,p_3,s),at(ban,p_3,s)) \rightarrow under(box,ban,s))$

A11. $\wedge (at(box,p_2,s_0),at(ban,p_3,s_0))$

---

\* (mon = monkey, ban = bananas)

**TABLE 6–1B.** Monkey and Bananas (Clause Form)

A1. $\{\overline{\text{at}}(\text{box},p,s),\text{at}(\text{box},p,\text{goto}(p',s))\}$
A2. $\{\overline{\text{at}}(\text{bananas},q,s'),\text{at}(\text{bananas},q,\text{goto}(q',s'))\}$
A3. $\{\text{at}(\text{monkey},r,\text{goto}(r,r'))\}$
A4. $\{\overline{\text{at}}(\text{box},u,v),\overline{\text{at}}(\text{mon},u,v),\text{at}(\text{box},u',\text{move}(\text{mon},\text{box},u,u',v))\}$
A5. $\{\overline{\text{at}}(\text{ban},t,t''),\text{at}(\text{ban},t,\text{move}(\text{mon},\text{box},t',\ t''',t''))\}$
A6. $\{\overline{\text{at}}(\text{mon},v',v'''),\text{at}(\text{mon},v'',\text{move}(\text{mon},\text{box},v',v''v'''))\}$
A7. $\{\overline{\text{under}}(\text{box},\text{ban},w),\text{under}(\text{box},\text{ban},\text{climb}(\text{mon},\text{box},w))\}$
A8. $\{\overline{\text{at}}(\text{mon},w',w''),\overline{\text{at}}(\text{box},w',w''),\text{on}(\text{mon},\text{box},\text{climb}(\text{mon},\text{box},w''))\}$
A9. $\{\overline{\text{under}}(\text{box},\text{ban},x),\overline{\text{on}}(\text{mon},\text{box},x),\text{has-bananas}(\text{reachfor}(\text{mon},\text{ban},x))\}$
A10. $\{\overline{\text{at}}(\text{box},p_3,y),\overline{\text{at}}(\text{ban},p_3,y),\text{under}(\text{box},\text{ban},y)\}$
A11. $\{\text{at}(\text{box},p_{2},o)\}$
A12. $\{\text{at}(\text{bananas},p_{23},o)\}$
Negated Conjecture (NC): $\{\overline{\text{has-bananas}}\ (z)\}$

*Consequences of the Axioms* (Fig. 6–6)
C1. $\{\text{at}(\text{box},p_2,\text{goto}(p',s_0))\}$
C2. $\{\overline{\text{at}}(\text{mon},p_2,\text{goto}(p',s_0)),\text{at}(\text{box},u',\text{move}(\text{mon},\text{box},p_2,u',\text{goto}(p',s_0)))\}$
C3. $\{\text{at}(\text{box},u',\text{move}(\text{mon},\text{box},p_2,u',\text{goto}(p_2,s_0)))\}$
C4. $\{\overline{\text{at}}(\text{ban},p_3,\text{move}(\text{mon},\text{box},p_2p_3,\text{goto}(p_2s_0))),\text{under}(\text{box},\text{ban},\text{move}(\text{mon},\text{box},$
$p_2,p_3,\text{goto}(p_2s_0)))\}$
C5. $\{\text{at}(\text{bananas},p_3,\text{goto}(q',s_0))\}$
C6. $\{\text{at}(\text{ban},p_3,\text{move}(\text{mon},\text{box},t',t''',\text{goto}(q',s_0)))\}$
C7. $\{\text{under}(\text{box},\text{ban},\text{move}(\text{mon},\text{box},p_2,p_3,\text{goto}(p_2,s_0)))\}$
C8. $\{\text{under}(\text{box},\text{ban},\text{climb}(\text{mon},\text{box},\text{move}(\text{mon},\text{box},p_2,p_3,\text{goto}(p_2,s_0))))\}$
C9. $\{\text{at}(\text{mon},v'',\text{move}(\text{mon},\text{box},r,v'',\text{goto}(r,r')))\}$
C10. $\{\overline{\text{at}}(\text{box},v'',\text{move}(\text{mon},\text{box},r,v'',\text{goto}(r,r'))),\text{on}(\text{mon},\text{box},\text{climb},(\text{mon},\text{box},$
$\text{move}(\text{mon},\text{box},r,r'',\text{goto}(r,r')))))\}$
C11. $\{\text{on}(\text{mon},\text{box},\text{climb}(\text{mon},\text{box},\text{move}(\text{mon},\text{box},p_2,u',\text{goto}(p_2,s_0)))))\}$
C12. $\{\overline{\text{on}}(\text{mon},\text{box},\text{climb}(\text{mon},\text{box},\text{move}(\text{mon},\text{box},p_2,p_3,\text{goto}(p_2,s_0)))),\text{has-ba-}$
$\text{nanas}(\text{reachfor}(\text{mon},\text{ban},\text{climb}(\text{mon},\text{box},\text{move}(\text{mon},\text{box},p_2,p_3,\text{goto}$
$(p_2,s_0)))))\}$
C13. $\{\text{has-bananas}(\text{reachfor}(\text{mon},\text{ban},\text{climb}(\text{mon},\text{box},\text{move}(\text{mon},\text{box},p_2,p_3,\text{goto}$
$(p_2,s_0))))))\}$

---

nanas(s)" will be true. Green (1969b) was the first to devise a resolution-based theorem prover capable of supplying constructive proofs
for existentially quantified formulas.

# Path Finding, Example Generation, Constructive Proofs, Answer Extraction

This section presents Luckham and Nilsson's (1971) generalization of Green's technique. This technique is illustrated by a simple
problem, and then its application to the monkey-and-bananas problem
is described.

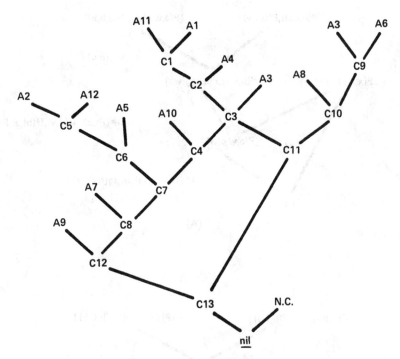

Figure 6–6. A proof that the monkey can get the bananas.
(See Table 6–1.)

Let us suppose we are given a simple set $S_a$, which contains only the axiom

$$\wedge\,(((\exists u P(b,u,u)) \Rightarrow \forall u P(b,u,u)), \vee\,(\forall w \exists r P(a,w,r),\, P(b,b,b)))$$

(6–2)

and let us suppose we are asked to prove the conjecture

$$\exists x \forall y \exists z P(x,y,z)$$

(6–3)

in a constructive fashion; that is, we wish to find values for the variables $x,y,z$ such that formula 6–3 will be true. Our standard procedure is to convert formula 6–2 and the negation of formula 6–3 into clause-form expressions and attempt to show that the set $S$ that contains them is unsatisfiable. Figure 6–7a shows the use of the resolution principle to construct a refutation graph and demonstrate the unsatisfiability of $S$. (Such a refutation graph could be easily obtained by a current resolution-based theorem prover.)

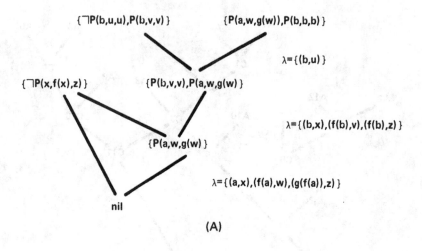

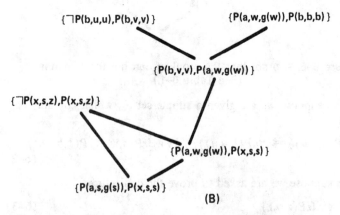

Figure 6–7. A simple refutation graph (A) and its modification (B) to produce an example.

Green's technique provides a way to construct the required ex-ample of formula 6–3, using a refutation graph such as Fig. 6–7a. Once a theorem prover has derived a refutation graph that proves the initial conjecture (disproves the negated conjecture), an "example-construct-ing program" can be used, which does the following things:

1. First, for each application of resolution in the graph, the literals that are unified by that resolution are marked. In Fig. 6–2 these literals are underlined.
2. New variables are substituted for any Skolem functions occurring in the clauses coming from the negation of the conjecture. Thus, the variable $s$ is substituted for the Skolem function $f(x)$.
3. Any clause in the graph which comes directly from the negation of the conjecture is converted into a tautology by adding its own negation to it. Thus, the clause

$$\{\neg P(x,s,z)\}$$

is converted into the clause

$$\{\neg P(x,s,z),P(x,s,z)\}$$

4. Following the structure of the original refutation graph, a modified graph is constructed. Each resolution in the modified graph unifies the same literals as are unified in the original refutation graph (i.e., the marked literals). If a clause being resolved contains "tautology literals" added to it in step 3, the variables in the tautology literals receive the same instantiations as they do elsewhere in that resolution.
5. The clause at the bottom of the modified graph is an example of values for the existentially and universally quantified variables occurring in the conjecture, for which the conjecture will be true.

Figure 6–7b shows the modified graph obtained from Fig. 6–7a. The clause at the bottom node is equivalent to

$$\forall y \vee (P(a,y,g(y)),P(b,y,y))$$

where $g$ is a Skolem function introduced during the construction of the refutation tree. We may interpret this clause as saying, "Either $x = a$, $y = y$ (i.e., $y$ has any value), $z = g(y)$, *or* $x = b$, $y = y$, $z = y$ (i.e., $y$ has any value and $z$ must equal $y$) will make the conjecture true, and at least one of these two cases must be a valid example for any model of the axioms." The presence of a Skolem function indicates that our solution is, to some extent, general; there are many models for formulas 6–2 and 6–3, and each model contains its own set of values for $x,y,z$ which will satisfy the quantifiers in formula 6–3. The Skolem function indicates that the values of certain variables depend on the particular model and on the values for other variables one happens to choose.

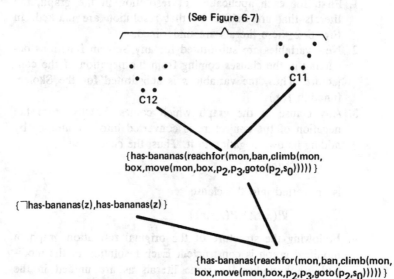

Figure 6–8. Modifying the refutation tree for the monkey-and-bananas state-space. In this case only the bottom part of the tree is affected by the modification process.

The generality of the examples constructed by this technique depends on the refutation graph it is given. Often there will be many ways to prove that a given set of clauses is unsatisfiable, and different ways will yield different examples. Because of the undecidability of the predicate calculus, there usually is no way to guarantee that a given example is the most general.

We can prove the validity of this example-constructing technique by observing that the modified graph represents the inference of the example from the axiom 6–2 and a tautology, consisting of formula 6–3 and its negation. Since the inference itself is valid (i.e., the resolution principle is valid and has been applied correctly), and since a tautology is always true, the example that is constructed must be correct.

Figure 6–8 shows the application of Green's example-constructing technique to the Monkey-and-Bananas Problem, modifying the refutation graph shown in Fig. 6–6. To get his bananas, the monkey should interpret the expression shown at the bottom of the graph, working outward from the innermost subexpression "goto(place2,$s_o$)." Thus, he should perform the following sequence of actions:

> *goto* place2
> *move* monkey, box, from place2 to place3

> *climb* monkey, box
> *reachfor* monkey, bananas

Green (1969a) used his constructive-proof technique to obtain a similar solution for the monkey (note 6–3).

To summarize, the use of resolution-based theorem provers to solve state-space problems involves

1. Describing state-spaces by means of sets of predicate calculus formulas.
2. Expressing problems as conjectures to be proved.
3. Finding proofs for conjectures, using the resolution principle and various heuristic search strategies (refinement methods, etc.).
4. Using Green's technique to convert a resolution proof of the unsatisfiability of the negated conjecture into an example of the conjecture's truth.
5. Interpreting this example as a description of the solution to the state-space problem.

The Exercises at the end of this chapter show how this process can be applied to other state-space problems (besides the Monkey-and-Bananas Problem) that were discussed in Chapter 3.

## Applications to Real-World Problems

The extent to which theorem provers can be used for solving real-world problems depends on several factors, including how well predicate calculus can be used to describe real-world situations and actions, and how efficiently theorem provers can be used to find solutions to problems that are given predicate calculus formalizations. This section concludes with a limited discussion of these factors. The reader is encouraged to see Green (1969a), McCarthy and Hayes (1968), and Hewitt (1969,1970) (discussed in the next section) for more on this subject.

Since, presumably, any mathematical theory can be expressed as a system of predicate calculus formulas, there is no doubt that predicate calculus offers a metaphysically adequate mathematical framework for the description of the real world, if any such framework can be constructed at all. Our major questions must concern its epistemological adequacy (how well it can represent everyday aspects of the real world) and its heuristic adequacy (how well it can be used to express information that is helpful in solving problems).

The epistemological adequacy of predicate calculus is probably satisfactory for the construction of real-world problem solvers. Green, McCarthy, and Hayes have shown that predicate calculus can be used to provide formalizations for such aspects of the real world as time-dependency, causality, and ability. The Monkey-and-Bananas Problem illustrates that predicate calculus can be used to formally express concepts involving objects and spatial relations. Other examples can be provided (see the Exercises and the next section) to show that predicate calculus may be used to represent problems that have solutions which are disjunctive, conditional, and which contain loops or recursive definitions. Perhaps the major questions involve the desirability of using modal and many-valued logics instead of predicate calculus, and the question of whether higher than first-order predicate calculus can be used successfully.

There are strong intuitive reasons for suspecting that many-valued logics are more desirable than predicate calculus. A machine capable of solving problems in a real-world environment must have some way of dealing with ambiguities, inaccuracies, probabilities, multiple interpretations, etc. Chapter 3 presented a list of some aspects of the real world which should be easily representable in the reasoning-language used by such a machine. Again, it is clear that each of these aspects of the real world could be embodied in a predicate calculus machine if they can be embodied in any machine at all. However, any such embodiment in a predicate calculus machine would require a set of axioms to define the functions and predicates that were associated with each of these aspects. The question is whether some other logic, which had these axioms built into its logic symbols and inference rules, would be more efficient. This question must be considered in light of the fact that no completely satisfactory many-valued logic has yet been developed. Perhaps it will be necessary to develop a system with a *variable-valued logic,* one that would be able to learn various functions and predicates and build the most useful ones into its logical apparatus.

As for the use of higher than first-order predicate calculus, essentially the same arguments apply. First-order predicate calculus is epistemologically adequate, but it seems likely that a higher-order system would be much more efficient. Hewitt (1969,1971) developed a theorem-proving language (described in the next section) that in many respects is more powerful than omega-order predicate calculus.

Hewitt's work was also concerned with the *heuristic adequacy* of predicate calculus. He showed that it is possible not only to use predicate calculus formulas as statements of facts, but also to use them as recommendations for how to proceed in solving problems.

The other major question concerns how efficiently theorem provers can be used to find solutions to problems that are given predicate calculus formalizations. In considering this question, it is well to point out that Green's technique and resolution-based theorem provers have so far been applied only *within* given state-space problems. The issue of whether theorem-proving techniques can be used to logically analyze a given state-space problem, and show that its solutions are equivalent to those for a problem with a simpler description (smaller state space or set of operators), remains undecided.

Another problem in the efficient use of theorem provers is the *frame problem,* discussed in McCarthy and Hayes (1968). The frame problem arises from the fact that, in a state-space problem, an application of an operator to a state will usually affect some relations between objects in the state and not affect others. In the predicate calculus formalization for such a problem, there must generally be axioms for each operator to express both the relations that are and are not changed by the application of that operator. For example, in the Monkey-and-Bananas Problem we had to state and use the fact that the application of the operator *climb* would not affect the position of the box (see Table 6–1A). Whenever it is necessary to make use of the fact that a certain relation still holds in a given state, the theorem prover must prove it, using the axioms for each of the operators that have been applied since the relation was last shown to be true. This, of course, greatly increases the work that must be done by the theorem prover.

Various techniques for overcoming the frame problem have been investigated, notably by Fikes and Nilsson (1971) and Hewitt (1969, 1971). Fikes and Nilsson present a GPS-like program that controls the application of a theorem-proving program to various sets $S_i$ of clauses, each set $S_i$ representing a given state in a state space. Each operator has associated with it a collection of "delete" and "add" instructions that identify the relations changed by the application of that operator. The program (called STRIPS) performs a heuristic search in the state space until it finds a sequence of operators that will produce a set $S_g$ of clauses containing the desired relations. Figure 6–8 shows STRIPS solving an expanded problem of the Monkey-and-Bananas type. STRIPS controls a robot (Shakey), which performs tasks in a real-world environment, as indicated by Fig. 6–9. (Also see Chapter 9.)

Hewitt's approach to the frame problem was similar. He defined a general class of procedures for manipulating data bases (sets of expressions) that include the $S_i$ sets of Fikes and Nilsson. This approach is discussed in the next section.

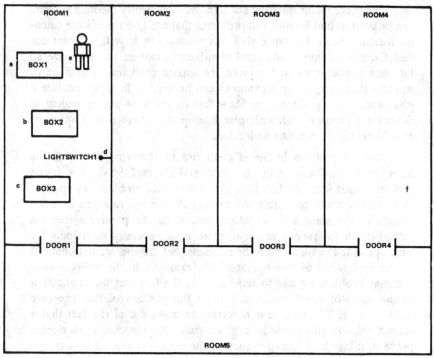

**Tasks**

1. **Turn on the lightswitch**

   Goal wff: STATUS(LIGHTSWITCH1,ON)

   STRIPS solution: {goto2(BOX1),climbonbox(BOX1),climboffbox(BOX1),
   pushto(BOX1,LIGHTSWITCH1),climbonbox(BOX1),turnonlight(LIGHTSWITCH1) }

2. **Push three boxes together**

   Goal wff: NEXTTO(BOX1,BOX2) ∧NEXTTO(BOX2,BOX3)

   STRIPS solution: {goto2(BOX1),pushto(BOX1,BOX2),goto2(BOX3),pushto(BOX3,BOX2) }

3. **Go to a location in another room**

   Goal wff: ATROBOT(f)

   STRIPS solution: {goto2(DOOR1),gothrudoor(DOOR1,ROOM1,ROOM5),
   goto2(DOOR4),gothrudoor(DOOR4,ROOM5,ROOM4),goto1(f) }

**Figure 6–9.** Tasks for STRIPS (initially at position *e*) and its solutions.
(Fikes and Nilsson, 1971, reprinted with permission.)

# THEOREM PROVING IN PLANNING AND AUTOMATIC PROGRAMMING

## Planning

Chapter 3 stressed the value of planning as a process to be performed by a general problem solver; for many real-world problems it is impossible to specify a single sequence of operations that will invariably achieve one's goal. The first and most general solution is a plan. Plans typically describe many alternate sequences of actions and specify conditions according to which different sequences will be followed. "Planning" (i.e., developing a plan) is itself one of the actions that a plan might dictate, and can be considered as an aspect of problem reduction. The feasibility of using program-like structures, such as nondeterministic programs and fuzzy algorithms, to represent plans, has been mentioned.

The use of theorem-proving techniques in planning is still at the stage of preliminary investigation. The few results that have so far been achieved indicate that it may be possible to use theorem provers to construct plans for the solution of real-world problems. Current investigations have followed essentially two approaches (note 6–4) to the development of theorem-proving plan makers: Hewitt (1969,1971) developed the programming language PLANNER, which permits the statement and execution of plans in a theorem-proving format; other researchers (e.g., Green, 1969a; Waldinger and Lee, 1969; McCarthy, 1962; R. W. Floyd, 1967; Manna, 1969,1970) demonstrated that it is possible to use resolution-based theorem provers to develop computer programs, and that it is often possible for people to prove whether or not a given computer program is "correct."

## Planner

A programming language is a way of describing procedures to computers; a description of a procedure, written in a programming language, is a *program*. Computers with a given "language capability" can accept programs written in that language and carry out the procedures they describe. Up to now it has not been necessary to consider any specific programming languages, since these discussions have been more concerned with procedures than programs. Thus, "procedure" and "program" have been used somewhat interchangeably.[11] Theoreti-

---

[11] The text has also been somewhat informal on this point in other respects. Thus, programs have been often said to "do" something or to "perform" some task when in actuality it is the computer that does or performs the procedure described by the program. This informality will be maintained.

cally, any procedure that can be performed by a program written in one programming language can be performed by some program in any other given programming language. The difference between programming languages lies in the simplicity with which various procedures can be stated by their programs. Thus, associative data processing is harder to do in FORTRAN than it is in SAIL (a programming language used at the Stanford Artificial Intelligence Project—see Feldman et al., 1972).

PLANNER is a significant new programming language for artificial intelligence research (see Hewitt, 1968 et seq.). Some of the things that can be done easily with PLANNER will be listed, though space does not allow more than a brief presentation of the language itself. Hewitt's work founded a new genus of programming languages for AI research. Among these are QA4 (Rulifson, 1971), CONNIVER (Sussman and Mc-Dermott, 1972a,b), and SAIL (Feldman et al., 1972). PLANNER is still in the process of being implemented; however, an early version of PLANNER (MICRO-PLANNER—see Sussman and Winograd, 1970; Baumgart, 1972) has been operational for more than a year. LISP is a very desirable background to these languages, and we also suggest reference to McCarthy et al. (1965), Weissman (1967), and Teitelman et al. (1972).

PLANNER is a programming language for the manipulation of data bases. A data base is some set of expressions which a PLANNER program may treat as assertions of knowledge about the world. A program written in PLANNER is a description of a plan for changing the assertions in a data base (or perhaps creating a new data base) depending on the assertions that are already in the data base. The fundamental mechanism that makes PLANNER work is *pattern matching* (see Chapter 5): a PLANNER program (or "theorem") may use pattern matching to search a data base for certain expressions and, if it finds them, add a new expression to the data base. Or a PLANNER program may use pattern matching to search the data base for other programs (theorems) that are designed to add certain expressions to a data base. Thus, the PLANNER "consequent" theorem:

```
(THCONSE  (X)  (FALLIBLE  $?X)
              (THGOAL  (HUMAN  $?X)))
```

is a program specifying a procedure to follow in order to add an assertion of the form (FALLIBLE $?x) to a data base; this procedure consists of attempting to satisfy the statement (THGOAL (HUMAN $?x)), which can be done if the pattern matcher finds an assertion of the form (HU-

MAN $?x) in the data base, or if the pattern matcher finds another con-
sequent theorem in the data base, of the form

(THCONSE (Y) (HUMAN $?Y)...)

and this theorem can be successfully executed with Y bound to the
value of X. If the THGOAL statement can be satisfied in either of these
ways, and the value of X is, say, SOCRATES, then the original THCONSE
theorem will add the assertion (FALLIBLE SOCRATES) to the data base.
In English, this means "something is fallible if it is already known to be
human, or if it can be shown to be human." (The $ and ? are variable
prefixes used for pattern matching—see Chapter 5.)

Similarly, PLANNER makes use of "antecedent" theorems to change
assertions in a data base automatically whenever certain other assertions
are added or erased. Thus,

(THANTE (X Z) (LIKES $?X $?Z)
            (THASSERT (HUMAN $?X)))

is an antecedent theorem (program) which states that whenever an
assertion of the form (LIKES $?x $?z) is added to a data base, the assertion (HUMAN $?x) should also be added.

Thus, a PLANNER theorem is capable of acting as a goal-oriented,
nondeterministic program; it can stipulate various goals for the computer
without stipulating exactly how the computer must try to achieve them.
PLANNER theorems are an example of *pattern-directed* plans.

Furthermore, PLANNER includes the ability to *backup* a plan if a
pattern matching proves to be unsuccessful. Thus, suppose our data
base includes the following simple assertions:

(HUMAN TURING)
(HUMAN SOCRATES)
(GREEK SOCRATES)

and the theorem:

(THCONSE (X) (FALLIBLE $?X)
            (THGOAL (HUMAN $?X)))

We can search the data base to answer the question "Is there a fallible
Greek?" by evaluating the PLANNER program:

(THPROG (X) (THGOAL (FALLIBLE $?X) (THTBF
                                      THTRUE))
            (THGOAL (GREEK $?X)))

This expression will have a successful execution, and return a value
for x, only if both THGOAL's can be satisfied. When PLANNER attempts

to satisfy the first THGOAL, it will make use of the THCONSE theorem to prove the validity of asserting that something (the value matched to x) is fallible; after that, PLANNER will attempt to satisfy the second THGOAL, by trying to prove or find an assertion that the same thing (the same value of x) is Greek. In satisfying the first THGOAL, evaluation of the THCONSE theorem will cause the pattern matcher to attempt to match the pattern rule (HUMAN $?x) against an assertion in the data base. Suppose that the pattern matcher first matches this pattern rule with the assertion (HUMAN TURING), making TURING be the value of x; the THCONSE theorem will then add the assertion (FALLIBLE TURING) to the data base, and control will return to the THPROG which will attempt to satisfy its second THGOAL, by either finding, or proving the validity of, the assertion (GREEK TURING). This attempt will fail, because the assertion (GREEK TURING) does not appear in the data base and there are no theorems in the data base which could be used to add such an assertion to the data base. The failure of this THGOAL will cause the THPROG to *backup*, and attempt to resatisfy its first THGOAL with a different value for x; the THCONSE theorem will be re-executed, and its THGOAL will call upon the pattern matcher to once again match the pattern rule (HUMAN $?x) with an assertion in the data base. However, this time the pattern rule will be matched with the assertion (HUMAN SOCRATES), and the new value for x will be SOCRATES. The THCONSE will succeed, and the assertion (FALLIBLE SOCRATES) will be added to the data base, and so the first THGOAL of the THPROG will succeed again. Lastly, the THPROG will again attempt to satisfy its second THGOAL. This attempt will succeed, because the assertion (GREEK SOCRATES) will be found in the data base. And so, the THPROG itself will terminate execution successfully, and return a value for x that is the answer to our question.

This feature of PLANNER (known as its *hierarchical control structure*) is extremely general. In essence, any decision made during the evaluation of a PLANNER theorem can be undone, if failures "backup" to the point where it was originally made. The generality of this feature has been criticized by some researchers (Sussman and McDermott, 1972), who claim that it is often very inefficient to rely on such a strict depth-first mechanism, and that such a control structure is difficult for human programmers to use without confusion, especially when writing large PLANNER programs. Recently, Sussman and McDermott have implemented a programming language known as CONNIVER, which is similar to PLANNER (in that programs are pattern-directed), but which attempts to provide a more flexible, explicit means of specifying the backup one wants to occur. (Also see Bobrow and Wegbreit, 1972.)

PLANNER has many other important features. The language contains as "primitive functions" many procedures that must be written out in detail within other languages. Knowledge in PLANNER is stored in a generalized "associative memory" (a graphlike structure with labeled nodes and arcs; see Chapter 7). Because PLANNER programs are themselves list structures, it is possible for such programs to be created, changed, or executed by other (and in some cases the same) PLANNER programs. PLANNER theorems are not restricted to first-order predicate calculus, nor do they even necessarily correspond very simply to formulas in higher-order predicate calculus. Thus, a PLANNER theorem might state: "do not use induction on the same variable twice," or "there exist $R$ and $Y$ such that R(Y,TURING) implies Y(TURING)." Predicates may be quantified or included within other predicates. PLANNER is currently being used as the inference mechanism for programs that "understand" natural language (see Chapter 7) and find descriptions of visual scenes. For the interested reader, Figure 6–10 shows a PLANNER program (*Orban's Monkey*) for solving the Monkey-and-Bananas Problem, much of which should be understandable from the discussion thus far. The PLANNER genus is probably the most natural set of programming languages yet developed for the ultimate writing of reasoning programs.

However, barring the aspect of higher-order predicate calculus, there is no direct comparison between theorem-proving systems written in PLANNER and the resolution-based theorem provers. The purposes behind these two approaches to theorem proving appear to be somewhat different. On the one hand, resolution-based theorem provers are designed to be general and complete programs for proving and disproving theorems within mathematical theories. Though we have discussed ways in which they can be designed to take account of the semantic content of mathematical theories (e.g., Kling's analogy generator), the primary accent in their development has been a concentration on their completeness and soundness; that is, on *proving* their applicability to *any* mathematical system and increasing their efficiency as much as possible without relinquishing that applicability (note 6–5).

PLANNER, on the other hand, provides a framework in which it is possible to write very sophisticated programs for special-purpose types of theorem proving. There are many types of information processing and problem solving that involve logical deduction, or theorem proving, without requiring full completeness or generality. When the types of questions, or problems, or theorems to be proved can be anticipated in advance, one can sometimes write a special-purpose program to deal

```
(SETQ MONKEY *(THGOAL (MONKEY GETS BANANAS)(THTBF THTRUE)))
        (THASSERT(CLIMBABLE BOX))
        (THASSERT(BOX AT A))
        (THASSERT(MONKEY AT B))
        (THASSERT(BANANAS AT C))
        (THASSERT(MONKEY OFF BOX))
(DEFPROP REACH (THCONSE (XYZ) (MONKEY GETS BANANAS)
        (THASSERT (MONKEY WANTS BANANAS))
        (PRINT *(THE MONKEY THINKS HE WANTS SOME BANANAS))
        (THOR
        (THGOAL (BANANAS AT (THV XYZ)))
        (THFAIL THEOREM *(YES, HE HAVE NO BANANAS)) )
        (THASSERT (MONKEY AT (THV XYZ))(THPSEUDO)(THTBF THTRUE))
        (THOR
        (THAND
         (THGOAL (MONKEY AT (THV XYZ)))
          (THGOAL (MONKEY ON BOX)) )
         (THFAIL THEOREM *(MONKEY DIDN'T MOVE, MONKEY NOT HELL)))
        (THERASE (MONKEY WANTS BANANAS))
        (PRINT *(MONKEY GETS BANANAS))
        (THSUCCEED THEOREM *SUCCESS)
)THEOREM)
        (THASSERT REACH)
(DEFPROP MOVEBOX (THANTE (X Z Q) (BOX AT (THV X))
        (THGOAL (BOX AT (THV Z)))
        (THOR
            (THAND (EQUAL (THV X)(THV Z))
                    (THSUCCEED THEOREM))
             T)
        (THGOAL (MONKEY AT (THV Q)))
        (THOR   (THOR
                (THGOAL (MONKEY OFF BOX))
                (THAND
                 (THNOT (THGOAL (MONKEY ON BOX)))
                 (THASSERT (MONKEY OFF BOX))) )
                (THAND
                (THERASE (MONKEY ON BOX))
                (THASSERT (MONKEY OFF BOX))
                (PRINT *(MONKEY NOTICES HE IS ON THE BOX))
                (PRINT *(MONKEY GETS OFF THE BOX))) )
        (THOR   (EQUAL (THV Q)(THV Z))
                    (THASSERT(MONKEY AT(THV Z))(THPSEUDO)(THTBF THTRUE)))
        (THERASE (BOX AT (THV Z)))
        (THASSERT (BOX AT (THV X)))
        (THERASE (MONKEY AT (THV Z)))
        (THASSERT (MONKEY AT (THV X)))

        (PRINT (LIST *MONKEY *MOVES *BOX *FROM (THV Z) *TO (THV X)))
        (THSUCCEED THEOREM)
)THEOREM)
        (THASSERT MOVEBOX)
```

Figure 6–10. Orban's Monkey. (Written by Richard Orban; published as an example in Baumgart, 1972. Reprinted with permission.)

```
(DEFPROP CLIMB (THANTE (X Y Z W S Q)    (MONKEY AT (THV X))
          (THGOAL (MONKEY AT (THV Q)))
          (PRINT (LIST *MONKEY *IS *AT (THV Q)))
          (THCOND ((THGOAL (MONKEY WANTS (THV Y))) (THGO B))(T T))
A         (THAND   (THOR
                    (THGOAL (MONKEY OFF BOX))
                    (THAND
                     (THERASE (MONKEY ON BOX))
                     (THASSERT (MONKEY OFF BOX))
                     (PRINT *(MONKEY NOTICES HE IS ON THE BOX))
                     (PRINT *(MONKEY CLIMBS OFF THE BOX))) )
                   (THOR
                    (THGOAL (MONKEY AT (THV X)))
                    (THAND
                     (THERASE (MONKEY AT (THV Q)))
                     (THASSERT (MONKEY AT (THV X)))
                     (PRINT(LIST *MONKEY *GOES *FROM(THV Q)*TO(THV X)))))
                   (THSUCCEED THEOREM *SUCCESS))
          (THFAIL THEOREM (PRINT *(WHAT MONKEY ?)))
B         (PRINT (LIST *THE *MONKEY *WANTS *SOME (THV Y)))
          (THGOAL ((THV Y) AT (THV S)))
          (PRINT (LIST *MONKEY *NOTICES *THAT (THV Y) *ARE *AT (THV S)))
          (THOR    (EQUAL (THV X)(THV S))
                   (THGO A) )
          (THOR
            (THAND
              (THGOAL ((THV W) AT (THV Z)))
              (THGOAL (CLIMBABLE (THV W)))
              (PRINT (LIST *MONKEY *NOTICES *A (THV W) *AT (THV Z))) )
            (THFAIL THEOREM
              (PRINT *(ALONE IN THE WORLD. WITH OUT A FRIEND))) )
          (THOR
            (EQUAL (THV Z) (THV S))
            (THASSERT ((THV W) AT (THV S))(THPSEUDO)((THTBF THTRUE)))
          (THOR
            (THGOAL (MONKEY AT (THV S)))
            (THAND
              (THERASE (MONKEY AT (THV Q)))
              (THASSERT (MONKEY AT (THV S)))
              (PRINT (LIST *MONKEY *GOES *FROM (THV Q) *TO (THV S)))) )
          (THAND
            (THOR
              (THERASE (MONKEY OFF (THV W)))
              T)
            (THOR
              (THAND
                (THASSERT (MONKEY ON (THV W)))
                (PRINT (LIST *MONKEY *CLIMBS *ON (THV W))) )
              (PRINT *(MONKEY ALREADY ON BOX. BUT YOU KNEW THAT))) )
          (THSUCCEED THEOREM)
)THEOREM)
          (THASSERT CLIMB)
```

Figure 6–10. (Continued)

only with these questions, problems, or theorems, according to a pre-determined strategy. PLANNER provides a way of writing special-purpose programs for theorem proving that make use of predetermined strategies. One of the most valuable features of PLANNER is that these predetermined strategies can be, to some extent, self-developing. It is, of course, possible to write complete, general theorem-proving strategies in PLANNER, but so far its most impressive uses have been of a special-purpose nature (e.g., the use of PLANNER and the similar language PROGRAMMAR in Winograd's English-understanding program).

It is possible that the future will see some sort of hybridization between PLANNER-based and resolution-based theorem provers. PLANNER and similar languages may eventually provide the notation for designing reasoning-programs that will process information according to special strategies. These strategies may specify conditions in which general-purpose theorem provers, perhaps resolution-based, will be used.

## Automatic Programming

Green (1969a), and Waldinger and Lee (1969), wrote theorem-proving programs that write simple programs in LISP (see McCarthy et al., 1962, or Weissman, 1967). Both programs are based on the resolution process for theorem proving. A brief description of the nature of these programs is given here. A complete discussion is, of course, given in the papers by the authors. Nilsson (1971, pp. 201–205) has also reviewed Green's results.

First, a few words about LISP are probably necessary. LISP is a programming language for writing programs that manipulate symbolic expressions known as *list structures*. A list structure is a list whose elements may be lists, or lists of lists, etc. Thus, a general definition of list structures is: "X is a list structure if X is an *atom,* or X is an ordered sequence of zero or more list structures." An *atom* is a string of symbols. For example, a, abc, jmc are all atoms. A list structure is usually denoted by a pair of parentheses enclosing the sequence of its elements. Thus, (a (a bc) (abc (jmc))) is a list structure. The empty list, which does not contain any elements, is denoted by ( ) or by the atom NIL. List structures may be *reentrant;* that is, they may contain themselves as elements. Thus, $X = (a\ X) = (a\ (a\ X)) = \cdots$ is a list structure. LISP provides a collection of primitive functions for manipulating list structures. These functions can be used to make more complex programs. Finally, a program written in LISP is itself a list structure. Thus, programs in LISP can be designed to create and manipulate other programs.

However, the ability to write programs that create programs is a solution to only the simplest problem of automatic programming. In general, what we desire are programs that create *correct* programs, where "correctness" is determined by a program's ability to compute a given function. That is, we desire a "program-writing program" *P*, which, when given a description of a function *f*, produces a program *F* that computes the value of *f(x)* for any given value of *x* for which *f(x)* is defined.

There are many ways of describing functions. Programs are themselves descriptions of functions, and it would, of course, be trivial to write a program *P* that could write programs if the descriptions of functions given to it were already in program form. Usually, we must assume that *P* is given some less explicit description of the function *f* than an actual program for *f*.

Probably the least explicit way of describing a function *f* is to specify a *predicate,* say *R(x,y)*, such that *R(x,y)* is true if and only if *f(x)* is defined and equals *y*. It is often possible to specify such a predicate *R* associated with a function *f* without specifying a program that computes *f*. Indeed, it is often much easier to specify predicates than programs. For example, suppose that *x* and *y* are variables that may have as their values any finite sequences of natural numbers. Thus, *x* might equal (4 1 3) and *y* might equal (10 11 91). We say a sequence *x* is *sorted* iff the elements of *x* are arranged in ascending numerical order.[12] Thus, *x* = (4 1 3) is not sorted, whereas *y* = (10 11 91) is sorted. Given this notion of "sorted," we can give the following description of a function *sort:* For any sequence *x*, *sort(x)* is a sequence containing the same elements as *x*, and sort(*x*) is sorted. Thus, sort ((4 1 3)) would be (1 3 4). However, this description of the function "sort" does not specify a procedure for computing the function. It merely specifies a *relation,*[12] holding between *x* and sort(*x*); they must both contain the same elements and sort(*x*) must be sorted—in other words, a *test* we can apply to any proposed procedure to see if it does indeed compute the function "sort" (the test is: "choose any sequence *x;* if the procedure when applied to *x* does not produce a sequence *y* such that *y* contains the same elements as *x* and *y* is sorted, then the procedure fails the test"). The description here gives no explicit in-

---

[12] We can define the property "sorted" by using three relations, "equals," "left of" and "less than": A sequence *x* is sorted if for any elements *e* and *e'* of *x*, *e left of e'* implies *e less than e'* or *e equals e'*.

formation as to how one should go about producing sort$(x)$ if one is given an arbitrary sequence $x$.

Given a predicate $R(x,y)$ which describes a function $f$, Green (1969) identified four basic problems for the field of automatic programming, such that each problem can be stated using a first-order predicate calculus formula and such that each problem can be solved[13] by a theorem prover that has the ability to provide constructive proofs for existential formulas.[14] The four basic problems follow.

*Checking.* This problem is stated to a theorem prover, using the formula $R(a,b)$, where $a$ and $b$ are two specific sequences of numbers. By proving $R(a,b)$ true or false, a theorem prover "checks" whether $b = f(a)$. This problem does not require a theorem prover with the constructive-proof ability.

*Simulation.* This problem is stated to a theorem prover, using an expression of the form $\exists x R(a,x)$, where $a$ is some specific sequence of numbers. By providing a constructive proof of the truth of this formula, a theorem prover "simulates" a program that sorts the sequence $a$; that is, it computes the value of $f(a)$.

*Verifying.* This problem is stated using the formula $\forall x R(x,G(x))$, where $G$ is a program provided to the theorem prover by the person (or machine) who wants the problem solved. By proving the formula true, the theorem prover verifies that $G$ correctly computes the function described by $R$. By constructively proving the formula false, the theorem prover shows that $G$ is not a correct program for the function described by $R$, and the theorem prover provides a value of $x$ for which $G$ needs "debugging."

*Program Writing.* The formula for this problem is $\forall x \exists y R(x,y)$. By constructively proving that this formula is true, a theorem prover can provide a program for the function $f$ described by $R$. By constructively proving the formula is false, a theorem prover would find a value of $x$ for which $f(x)$ would not be defined.

Green considered in detail the use of an example-constructing theorem prover both to construct a program that computes a function described by a relation $R$ and to prove the correctness of the program constructed. The theorem-proving program he used was QA3, and the program he attempted to have it construct was one that sorted an arbitrary finite sequence of numbers. (Green used a different relation

---

[13] . . . if it is decidable, and if the theorem prover has "infinite time and resources."

[14] See the preceding section for a discussion of theorem provers that provide constructive proofs for existential formulas.

*R* to describe the function "sort.") In addition to axiom clauses describing the relation *R*, QA3 must be given clauses that describe the primitive functions of the "target language" (in this case, LISP) in which the program for the function described by *R* is to be written. Waldinger and Lee's (1969) program, known as PROW, has built into it the axioms describing the primitive functions of LISP. PROW contains a special subprogram (which is not a theorem prover) for converting programs from one language into another.

In order to develop programs that have loops or are recursive, both theorem provers must be given axioms for *mathematical induction* because, in general, one cannot specify an upper bound for the number of steps that might be required by an execution of such a program. It is therefore not possible to prove that a given program is correct by tracing through all possible executions of that program. Rather, a theorem prover must show that

1. The program computes the correct value of $f(x)$ for some value of $x$, say, $x = a$.
2. There is a function $s$ such that if the program computes the correct value of $f(x)$ for a given value of $x$, then it also computes the correct value of $f(y)$ for $y = s(x)$.
3. For any possible value of $x$ there is a number $n$ such that $x = s(s(s(\ldots (s(a)) \ldots )))$, where $s$ is applied $n$ times.

The function $s$ is known as the *successor function* utilized by the inductive proof.[15] Proving condition 3 establishes that any possible value of $x$ is, for some $n$ (which may be dependent on $x$), an "$n$th successor" of $a$. Proving conditions 1 and 2 establishes that the program computes the correct value of $a$ and of any $n$th successor of $a$. Thus, the proof of the three conditions establishes that the program will compute the correct value of $f(x)$ for any possible value of $x$.

In all work to date on automatic program writing, both $s$ and the proof of condition 3 are, in effect, given to the theorem prover. The correct choice of a successor function and the proof of its validity are at the moment too difficult for automatic program writers. Currently, automatic program writers are capable of proving conditions 1 and 2 (given $s$) only for programs that are very simple, such as a program that sorts an arbitrary sequence of numbers.

However, the fact that inductive proofs can sometimes be ac-

---

[15] The function $s$ is often generalized to produce a set of possible successors to $x$. For this generalization the identity sign ($=$) in axioms 2 and 3 should be replaced by "is an element of" ($\epsilon$.)

complished by theorem provers is significant, especially when one investigates the extent to which people have been able to use inductive proofs to show properties (such as "correctness") of computer programs. Floyd, McCarthy and Painter, Manna, and others showed that mathematical induction can be used to prove the correctness of a variety of computer programs, including compilers and nondeterministic programs (note 6–6). The discussion of theorem-proving programs will be left at this point, with the observation that the problems of program writing can at least be stated formally and can often be solved in a formal manner by human beings.

## NOTES

**6–1.** Some of those responsible for the development of predicate calculus and early work on meta-mathematics include Boole, Cantor, Russell, Whitehead, Lewis, Dedekind, Peano, Frege, Zermelo, Hilbert, Brouwer, Kronecker, Poincare, Tarski, Skolem, and Godel. The Bibliography contains selected references to current texts on mathematical logic by Kleene, Church, Prior, Quine, Shoenfield, Wang and others. (Also see Benacerraf and Putnam, 1964; van Heijenoort, 1967.)

**6–2.** Many-valued logics, modal logics, and fuzzy logics have often been suggested as the most realistic and desirable frameworks within which to construct theorem provers. These logics differ from predicate calculus mainly in the inference rules they provide; their inference rules do not require that a sentence be completely and exactly true in order for it to be used in deriving other sentences. Rather, sentences are allowed to have many different values besides "true" and "false." Thus, in fuzzy logic, the truth value of a sentence may be any real number between zero and one, inclusive ("false" and "true," respectively). Space does not permit a detailed treatment of these logics; the interested reader is referred to the works of Ackerman (1967), McCarthy and Hayes (1968), Prior (1957), Quine (1961), Feys (1965), Zadeh (1965, 1968), and Tsichritzis (1968) cited in the Bibliography. Recently, R.C.T. Lee (1971) showed that the resolution principle developed in this chapter can be used within a formalization of fuzzy logic. The discussion of "meaning" presented in Chapter 7 is relevant to many-valued logics.

**6–3.** Green called this technique *answer extraction*. The present author prefers to use the phrases "constructive-proof generation" and "example construction," since these do not imply linguistic ability, an aspect of artificial intelligence that we have not yet discussed. However, it should be noted that Green's early papers were largely concerned with question-answering and the ability of machines to use natural languages. Also, the phrase "answer extraction" is in fairly common usage. Chapter 7 discusses the use of theorem provers within language-understanding systems.

**6–4.** A third approach to planning was suggested by Kling, within the context of proving theorems by analogy. Suppose the theorem prover is given a proof for a theorem $T'$ and required to find a proof for an analogous theorem $T$, and suppose that the proof for $T'$ requires the establishment of certain "smaller" theorems or *lemmas*. The proofs of these lemmas are also given to the theorem prover. Kling suggested using an analogy generator to produce analogs for the lemmas associated with $T''$, and having the theorem prover attempt to find proofs for these analogs. If proofs were found, then the clauses associated with the analogs could be used in the data base for $T$. To the present author's knowledge, Kling has not yet implemented this method. Indeed, his analysis (1971a, pp. 145–148) suggested that ZORBA–1 may not be suitable for such an implementation. However, the idea indicates a way in which problem-reduction techniques might be used "by analogy" in theorem proving.

**6–5.** In fact, for reasons that include both theoretical and practical limitations, no theorem prover can be *really* complete. Even though a theorem may be logically implied by a set of axioms, we cannot guarantee that the theorem prover will eventually develop a proof for it, because of (1) the undecidability of the predicate calculus and (2) the limitations of space and time which affect the computational ability of any machine. (However, we should note that our first condition does not hold for the first-order predicate calculus; given an arbitrary sentence and a set of axioms, the *semidecidability* of the first-order predicate calculus guarantees that, if the sentence is logically implied by the axioms, a resolution-based theorem prover—given enough space and time—will eventually find a proof for it; on the other hand, if there is no proof for the sentence—that is, it is not logically implied by the axioms—such a theorem prover may not be able to disprove the sentence, no matter how much space and time we give it.)

**6–6.** A good survey of mathematical induction and the subject of automatic program writing was given by Manna and Waldinger (1970). They suggested partial-function logic (predicate calculus with "undefined" as a truth value; see McCarthy, 1963b) as the most natural language for automatic program synthesis. Other papers on the subject have been written by Balzer (1972), and Feldman (1972). Dijkstra (1965 et seq.) has developed the paradigm of *structured programming* as a framework within which to prove the correctness of programs. Recently, Scott (1971) and Milner (1972) have developed a mathematical logic of computation that is of great relevance to this subject. And, Sussman (1972) describes the general structure of a CONNIVER program (called HACKER) for automatic program writing.

## EXERCISES

*6–1.* Show that $U \to V$ can be rewritten as $\neg \wedge (U, \neg V)$.

*6–2.* Find clause-form equivalents for the following formulas:
(a) $\forall x(P(x) \to \vee(P(x),Q(x)))$.
(b) $(\neg \forall x P(x)) \to (\exists x \neg P(x))$.
(c) $(\forall x \exists y \wedge (P(x),Q(x,y))) \to \exists x \wedge (P(x),Q(x,x))$.

*6–3.* Find most general unifiers for each of the following sets of literals:
(a) $\{Q(x,a,y),Q(a,x,y)\}$.
(b) $\{P(x,f(x)),P(g(x),a))\}$.
(c) $\{R(u,w,f(u)),R(b,x,g(x))\}$.
(d) $\{W(z,c,f(y)),W(a,x,z),W(f(y),u,g(x))\}$.

*6–4.* Use the resolution principle to derive contradictions from the negations of each of the following predicate calculus tautologies:
(a) $\forall x(P(x) \to P(x))$
(b) $(\neg \exists x P(x)) \to (\forall x \neg P(x))$
(c) $(\forall x \vee (P(x),Q(x))) \to (\vee ((\forall x P(x)),(\exists x Q(x))))$.

*6–5.* Construct a predicate calculus formalization for the Missionaries-and-Cannibals Problem (Exercise 3–2); give a resolution-based proof that it is solvable and use the example-construction technique to find a solution.

*6–6.* Present a predicate calculus formalization for the Mutilated Checkerboard Problem (Exercise 3–8), and describe how it might be used to prove the checkerboard cannot be covered by the tiles as required.

*6–7.* (a) Present a predicate calculus formalization for the Confusion-of-Patents Problem (Exercise 3–3) and give a resolution-based proof that it is solvable. (b) Use the technique of example construction to find the solution to the problem.

*6–8.* One nice aspect of the PLANNER "robot calculus" is that it allows a relation or a predicate to have a variable number of arguments. Give some real-world examples illustrating such relations.

*6–9.* In the discussion of PLANNER theorems the following statement was presented:

$$\exists R \exists Y[R(Y,\text{Turing}) \to Y(\text{Turing})]$$

Find two English words that might plausibly be substituted for $R$ and $Y$ to make

$$R(Y,\text{Turing}) \to Y(\text{Turing})$$

a "reasonable" statement.

*6–10.* (*The King-and-the-Wizards Problem.*) (a) Long ago, a wicked king was searching for a new wizard with whom to plot some devious schemes. He summoned to him three wizards who seemed especially promising, and let them into a small room, which was barren except for a lighted candle on a table in the middle of the room. "Listen to me well," he said. "In a few minutes all of you will be blindfolded, and I will paste upon each of your foreheads a uniformly colored spot of black or white paper. At least one spot will be white. The first of you who guesses the color of his own spot will become my new wizard, and ride in his own chariot, with all expenses paid. The other two of you will be sent to a terrible fate that I shall not describe. None of you will be allowed to remove any of the spots, and you will each be allowed only one guess." The king then ordered his guards to blindfold the wizards, proceeded to paste white spots on all the wizards' foreheads, and finally had their blindfolds removed. After a few seconds, one of the wizards correctly identified the color of the spot on his forehead. How did he know it? (b) Present a predicate calculus axiomatization for the wizard's reasoning. (c) What sort of thoughts might the other two wizards have been thinking?

"We could play at questions."
　　　　　—Rosencrantz, in *Rosencrantz and Guildenstern Are Dead.* (Stoppard, 1967)

"Augustine describes the learning of human language as if the child came into a strange country and did not understand the language of the country; that is, as if it already had a language, only not this one."
　　　　　—Wittgenstein, *Philosophical Investigations.*

"I find it difficult to believe that whenever I see a tree I am really seeing a string of symbols."
　　　　　—McCarthy, in a discussion on grammatical inference and pattern recognition.

"As a concluding remark: *could* this art be applied (we put the question in strictest confidence)—*could* it, we ask, be applied to the speeches in Parliament?"
　　　　　—Lewis Carroll, *Photography Extraordinary.*

"There is of course no restriction in the memory format against having concepts without English names, and in fact [its] present memories necessarily include such concepts."
　　　　　—Quillian, describing the structure of the TLC computer program. (Quillian, 1969)

"Danger of tumbling upwards be in deep-sea."
　　　　　—Protosynthex III, a computer program. (Schwarcz, Burger, and Simmons, 1970)

"The challenge of programming a computer to use language is really the challenge of producing intelligence."
　　　　　—Winograd, 1971.

"In any case, these are but steps toward more graphical program-description systems, for we will not forever stay confined to mere strings of symbols."
　　　　　—Minsky, 1970.

"What does meaning mean?"

　　　　　—Anonymous.

"Imagine a people in whose language there is no such form of sentence as 'the book is in the drawer' or 'water is in the glass', but whenever we should use these forms they say, 'The book can be taken out of the drawer', 'The water can be taken out of the glass'.
　　　　　—Wittgenstein, *The Brown Book.*

"I have traveled more than anyone else, and I have noticed that even the angels speak English with an accent."
　　　　　—Mark Twain.

# 7

# SEMANTIC
# INFORMATION
# PROCESSING

## INTRODUCTION

This chapter is concerned with the ability of machines to use languages. We shall first discuss the nature of language, both as it is used by "living creatures" and as it is used by "machines", giving primary attention to two of the most important features that are possessed by human and computer languages: extensibility, and self-reference. A conclusion will be drawn that, of all the machines and animals known to man, computers belong to the handful (also including chimpanzees and dolphins) we might plausibly expect to learn our languages.

Of predominant interest throughout this chapter is the ability of sentences in a language to have "meaning" to those who use the language. A sentence that has meaning is said to contain *semantic information*.[1] The third section of this chapter will describe how machines can "understand" and "create" sentences that convey semantic information, and will discuss computer programs that do this for sentences written in English. A collection of some of the conversations people have had with computers will be presented, primarily

---

[1] The "semantic information" of a sentence should not be confused with the "information" measure described in Chapter 2. (See note 7–7.)

273

oriented toward the ability of computers to solve various kinds of problems stated in English. The final section takes up more general questions, relating the problems of language use and development to those of teaching and learning, pattern perception, and general problem-solving and reasoning programs.

# NATURAL AND ARTIFICIAL LANGUAGES
## Definitions

To facilitate matters, we need some rough definitions for "language," "sentence," and "meaning." These definitions will be refined and amplified throughout the rest of this chapter.

A *language* is a set of *sentences* that may be used as signals to convey semantic information. The existence of a signal naturally implies the existence of an *emitter* and a *receiver* (perhaps more than one) and of some "embodiment," or means of *transmission* for the signal. The *meaning* of a sentence is the semantic information it conveys. For a given sentence (signal), this information may vary with the situation in which it is used; in general, we can think of the meaning of a sentence as being a *description* of three things: (1) whatever causes the sentence to be used; (2) whatever is caused by the use of the sentence; (3) whatever else is described by the sentence. It is the task of those who use a sentence (the emitters and receivers) to "understand" these elements of its meaning—for a computer that uses a sentence, "understanding" may be corresponded to making internal data structures (vectors, lists, graphs, programs, etc.) that *model* these elements of the meaning of the sentence. "Communication" is a word we use to describe processes in which one or more sentences are transmitted and understood.

A few examples will clarify the concept of "meaning" that is advocated. Consider the following sentences:

*A.* I have four aces.
*B.* Our position is 10 miles north of yours.
*C.* Elect me and I will end the war honorably.
*D.* Eat cereal X and grow healthy and strong.
*E.* Why?
*F.* I love you.
*G.* People who apply for marriage licenses wearing shorts or pedal pushers will be denied licenses.[2]

*H.* The sum of 18 and 32 is 50.
*I.* The equation $x^2 = -1$ has a solution.

Surely the "meaning" of these sentences is not something fixed or immutable. The meaning of a sentence generally depends as much on who utters it, and where, when, and to whom it is uttered, as it depends on the sentence itself. In understanding a sentence, one should attempt to model what causes the sentence to be transmitted, what the emitter of the sentences hopes that it will cause, etc., as well as what the sentence itself describes.

Throughout, this chapter stresses the importance of model making in the processes of communication and understanding. However, the student should be warned that, especially for languages such as English and French, there is no current, complete explanation for how computers should go about "understanding" sentences. The problems connected with modeling the semantic information carried by sentences are as deep and complex as the situations these sentences may describe. This chapter can do little more than present some of the requirements that would have to be satisfied by an adequate formalism for "models of meaning," describe how computer programs currently approach the subject, and suggest how research might be continued (see note 7–1).

It will serve us well to distinguish between two types of languages called *natural* and *artificial* languages. The differences between them lie mainly in the uses that are made of them, and in the knowledge we have about them.[3] Although both forms of language are of much interest in themselves, our discussion will center on their relations to each other, and especially on the ability of artificial languages to "simulate" natural languages. By *natural languages* we refer to the languages that living creatures use for communication, whereas by *artificial languages* we mean certain mathematically defined classes of signals that can be used for communication with machines.

## Natural Languages

The natural languages constitute a very broad category, since communication processes are important to virtually every living system in existence. We may group natural languages into two large subcategories, and name them *cell-level* and *organism-level* natural languages.

---

[2] This example sentence is quoted from Kuno (1965).
[3] These are not necessarily differences of substance; probably the distinction between them will become less as our understanding increases.

The cell-level languages are evidently the oldest natural languages. The emitters and receivers that use these languages are living cells: The methods for transmission of sentences (signals) are primarily chemical and electrical. "Sentences" transmitted chemically correspond to molecules (often called "messenger molecules"), which may act by catalysis to affect processes in the receiver. One well-known group of such sentences is that of the RNA molecules, which typically carry information between parts of individual living cells. Very little is known about cell-level "molecular languages" except that there are a huge number of molecular sentences that can have "meaning" to living cells. It may be a long time before scientists can "understand" them. For more information on these languages, see Pribram (1971).

Organism-level natural languages are much more familiar to us. The emitters and receivers that use these languages are living organisms (animals, plants, etc.); the means of transmission include chemical, visual, audial, and tactile techniques. Many species have acquired these languages, primarily to carry information about food, danger, and sex. Typically, the language used by the organisms of a given species will have only a small (say, less than 100) number of sentences or signals, and there will be no provision within the language for extending that number. Usually the organisms which use these languages do so involuntarily, in automatic response to the presence of certain stimuli in their environments.

The only known organism-level natural languages that are not so limited are mankind's spoken and written languages (English, French, Chinese, etc.). In theory, these languages possess an infinite number of possible sentences that can be used as signals by people. However, no one knows how many of the "possible" sentences are "meaningful" in practice. The best we can say is that the number may be "comparable" to that of the meaningful molecular sentences in cell-level languages.

One major difference between human languages and those used by other organisms lies in the *structural nature* of the sentences we use. The sentences of any human language are essentially stringlike structures (sequences) of *words*. Spoken words are themselves essentially stringlike structures of *phonemes* (vocally producible sounds that constitute the "alphabet" of the spoken language), whereas written words are often sequences of *letter-symbols*, which constitute the alphabet of the written language (note 7-2). Various languages, of course, have different spoken and written alphabets. English has a written alphabet of 26 letters and a spoken alphabet of 48 phonemes (J. B. Carroll, 1964).

Currently, about 5,000 languages and dialects are spoken throughout the world. The two most commonly spoken languages are Northern Chinese (or Mandarin) and English, which are used by about 600 million and 350 million people, respectively. English is the language in most widespread use, being spoken by 10% or more of the population in 29 countries; it is also the language with the largest vocabulary, containing about 490,000 words, plus another 300,000 technical terms (McWhirter and McWhirter, 1971). Estimates range on the maximum size of the individual human vocabulary; it is very likely that no individual uses more than 100,000 words (probably the boundary is lower, around 60,000)—normal literature written in English makes use of about 10,000 words, while well-educated conversation uses about 5,000 words. Of course it is possible to converse rather well using much smaller vocabularies. Thus, "Basic English" (C. K. Ogden, 1933) contains only 850 words. The 1971 *Guinness Book of World Records* reports that the language with the smallest vocabulary is Taki taki, a South American language that uses only 340 words.

The sentences in a language are always essentially sequences of words from the vocabulary of that language, but, typically, not every sequence of words constitutes a sentence. A set of rules that allows one to recognize the sequences of words that are sentences in a language is known as a *grammar* for that language.[4] Grammars are said to describe the structural, or *syntactic,* nature of languages.

Of course one wants to do more than simply recognize which sequences of words are sentences in a language; it is of primary importance to be able to "understand" the sequences one recognizes. One of the major problems confronting linguistics today is development of an adequate theory of the relationship between the syntactic nature of a sentence (or set of sentences) and the semantic information it conveys. Two important approaches toward a solution of this problem are the theories of *transformational grammar* (Chomsky, 1959 et seq.) and *systemic grammar* (Halliday, 1961 et seq.). This topic is discussed in the next section, but for now it is important to note two "trivial" things: first, the structural nature (syntax) of a sentence helps one determine its meaning; second, the meaning an emitter wants to convey helps determine the structure of the sentences that convey it.

One of the most valuable aspects of human languages is their *extensibility:* The words and sentences of the English language (for example) are not fixed. Rather, English (like most if not all other

---

[4] If the set of rules also enables one to recognize those sequences of words which are *not* sentences, then it is said to *decide* the language. It is possible for a language to be *undecidable,* that is, such that no grammar can decide it.

human languages) includes provisions for extending its own use. As is evident from the preceding paragraphs, the chief way in which English has been extended has been through the definition of new words. Another way is the introduction of new symbols (e.g., mathematical symbols). It is even possible to extend a language by adding to its syntactic nature. Thus, it is possible to use English sentences to define (at least partially) the words, sentences, and grammar of another language, such as German—this is precisely what an introductory English textbook on German will attempt to do. When English is extended in this way to include German sentences, the German sentences may be said to have been *embedded* in English.

Closely related to the extensibility of human languages is their ability to be *self-referencing*. An English sentence (for example, this one) can refer to itself or to other sentences (e.g., all of the sentences in this book). In "understanding" the preceding sentence, one must understand the phrase "this one" (which refers to the entire sentence in which it occurs), and the phrase "all of the sentences in this book." One can find many other types of self-reference exhibited by English sentences.

A third aspect of human (and many other) languages which should be mentioned is their *redundancy*. Any means of transmitting a signal may involve some "noise" that will tend to distort or degrade the signal. To convey the semantic information, one should, in effect, transmit the signal several times, because it is very unlikely that random noise will degrade the signal the same way every time. The receiver can reconstruct the original signal by adopting a "majority vote" policy when comparing the signals he receives. Another way of using redundancy in an alphabetic language is to use more symbols than are needed to represent each word; with 26 letters one could, for example, represent each of 10 million words uniquely by a series of 5 letters. In fact, English uses considerably more letters than are necessary to represent each of its words. Finally, one can also obtain redundancy in a language if its grammar provides sentences with "structural redundance" (see Cherry, 1957). The redundancy of English is often quoted as about 50%; that is, an English sentence is usually decipherable even if each of its letters is blanked out independently of the others with any probability up to one-half.

The importance of language to the development of human intelligence is a subject that deserves a great deal of attention, certainly more than can be offered in this book. Wherever people have formed societies, they have developed languages. The tendency to develop languages is one of the most important traits of our species. One of the more remarkable things about it is the ability of the young child to learn the language of the society in which he is raised. Something about the way in which societies develop languages insures that practically every child will be able to accomplish the language-learning feat in the space of a few years. This has led certain scholars (Chomsky, 1966) to conjecture the existence of a universal grammar underlying all human languages, and which is naturally reflected somehow in the learning process of each person, to the extent that individuals are enabled to learn the language of their society without making too many

| Signs | Description | Context |
|---|---|---|
| Come-gimme | Beckoning motion, with wrist or knuckles as pivot. | Sign made to persons or animals, also for objects out of reach. Often combined: "come tickle," "gimme sweet," etc. |
| More | Fingertips are brought together, usually overhead. (Correct ASL form: tips of the tapered hand touch repeatedly.) | When asking for continuation or repetition of activities such as swinging or tickling, for second helpings of food, etc. Also used to ask for repetition of some performance, such as a somersault. |
| Up | Arm extends upward, and index finger may also point up. | Wants a lift to reach objects such as grapes on vine, or leaves; or wants to be placed on someone's shoulders; or wants to leave potty-chair. |
| Tickle | The index finger of one hand is drawn across the back of the other hand. (Related to ASL "touch.") | For tickling or for chasing games. |
| Toothbrush | Index finger is used as brush, to rub front teeth. | When Washoe has finished her meal, or at other times when shown a toothbrush. |
| Cat | Thumb and index finger grasp cheek hair near side of mouth and are drawn outward (representing cat's whiskers). | For cats. |
| Key | Palm of one hand is repeatedly touched with the index finger of the other. (Correct ASL form: crooked index finger is rotated against palm.) | Used for keys and locks and to ask us to unlock a door. |
| Baby | One forearm is placed in the crook of the other, as if cradling a baby. | For dolls, including animal dolls such as a toy horse and duck. |
| Clean | The open palm of one hand is passed over the open palm of the other. | Used when Washoe is washing, or being washed, or when a companion is washing hands or some other object. Also used for "soap." |

Figure 7–1. Some signs used reliably by Washoe after 22 months of training. (Gardner and Gardner, 1969, reprinted with permission.)

Figure 7–2. Examples of the sequences of symbols used by Sarah. Reprinted from "The Education of Sarah" by David Premack in *Psychology Today Magazine*, September 1970. Copyright © Communications/Research/Machines, Inc.

wrong guesses. Such a grammar would also account for the similarities in syntax between the various languages that people have developed (note 7–3).

Efforts have been made to teach human languages to other animals, but only recently have researchers achieved any success (note 7–4). For example, two chimpanzees have been taught to communicate with people by using "sign languages" (Gardner and Gardner, 1969; Premack, 1970). One chimpanzee, named *Washoe,* has learned over 150 signs of the American Sign Language System, originally devised for the deaf and dumb; this is the language in which words are represented by movements and configurations of an individual's hands and fingers. *Sarah,* the other chimpanzee, has learned to communicate with sentences that consist of simple sequences of cards bearing printed symbols. Figures 7–1 and 7–2 show some of the signs and card sequences used by Washoe and Sarah. Neither chimpanzee is able to use very long or complicated sequences of signs or cards, although Washoe has been able to *invent* a few new signs that are now used by some people learning the language.

Because our spoken and written languages are so familiar and important to us, it is customary to call them *the* "natural languages," as distinct from other organism-level and cell-level natural languages. So, unless otherwise specified, the phrase "natural language" will refer henceforth only to spoken and written languages used by *homo sapiens*.

# Artificial Languages and Programming Languages

Artificial languages are certain mathematically defined classes of signals that can be used for "communication" with machines. Throughout the rest of this section the reader will be introduced to those artificial languages that can be used for communicating with computers. These languages are generally known as *programming languages*. Chapter 6 has already given a brief description of programming languages (in particular, LISP and PLANNER). Programming languages have many properties that are analogous to those of natural languages. The next section reviews the attempts made by AI researchers to "unify" the artificial and natural languages; that is, to design machines with an ability to "communicate" in both kinds of languages. Here, however, the emphasis is on languages that are currently conventional for programming (communicating with) computers. The major difference between these conventional artificial languages and natural languages is that the syntactic and semantic properties of the artificial languages are more thoroughly known (in the sense of being more rigorously formalized, at least consciously) than are those of natural languages.

Essentially, a programming language is a set of sentences (signals), each of which a computer may receive and store internally as a *data structure*. Data structures may have many "forms" (numbers, vectors, matrices, lists, graphs, etc.) and may cause the computer to perform many different "actions"—physically, a data structure is usually a collection of electric or magnetic charges that can be sensed and altered by the computer. The *syntactic* nature of the programming language is given when we finitely describe the exact forms of its sentences and their data structures. The *semantic* nature of the language is given when we specify the actions that each data structure will cause to be performed. A data structure can cause the computer to perform actions in the external world (e.g., move a mechanical arm, or transmit electric signals to a printer) or it can cause the computer to create new internal data structures, or modify or erase those that are already present. If a data structure is causing the computer to perform actions, then it is called a *program;* otherwise, we may simply call it *data*. It should be noted that the distinction is not always a particularly good one: The same data structure may be a program and also be data that the computer manipulates. And, though it is a good approximation to say that data structures "cause" computers to perform actions, this is not the entire truth. Whatever action is caused when the computer meets a data structure is as much dependent on the computer as it is on the data structure.

To illustrate this, let us recall the model that was introduced in Chapter 2 for computers. That chapter defined *Turing machines* and showed how a *universal* Turing machine could simulate any given Turing machine. Also described were *polycephalic* universal Turing machines, which were credited as a better model for modern computers. In the context of the current discussion, consider any collection of symbols printed on the (possibly $n$-dimensional) tape(s) of a polycephalic universal Turing machine to be a "data structure." It is clear that this agrees with what has been said above about data structures; that is, some of the symbols on the tape of the universal Turing machine may be a "program" and cause the machine (computer) to perform actions. However, it is also clear that the actions performed by the machine at a given moment depend as much on the "state" of the machine as on the symbols that it reads with its tape-heads; and it is clear that the same data structure might "cause" different machines to perform different actions.

Let us continue to use the (universal, polycephalic) Turing machine formalism to discuss programming languages and computers, taking care to observe some ways in which modern computers deviate from the model, as well as ways in which they satisfy it. The reader may recall that in Chapter 2 the notion of a "descriptive string" was introduced to show how a universal Turing machine could simulate a given Turing machine; namely, the next-move function of any Turing machine (say, $T$) can be described using a finite string of blanks and 1's, called a "descriptive string" for $T$. If this descriptive string is placed appropriately on an otherwise blank tape of an appropriate universal Turing machine (say, $U$), then $U$ will use the descriptive string for $T$ in such a way that $U$ will simulate $T$; that is, $U$ will manipulate data structures on some of its othe· tapes just as would $T$. However, $T$ may itself be a universal Turing machine and thus it is possible to have a machine $U$ simulate a machine $T$ simulating a machine $T'$, etc. A Turing machine (simple, universal, polycephalic, or whatever) is basically a procedure for manipulating symbols on tapes (i.e., data structures). "Descriptive strings" are basically data structures that describe procedures (Turing machines). A universal Turing machine is a Turing machine that can use descriptive strings to carry out the procedures they describe. In this sense, a *program* is basically a descriptive string that can be used by some universal Turing machine.

Thus we may consider the set of descriptive strings (programs) that can be used by a given universal Turing machine $U$ to be a *language* that is "understood" by $U$. Each descriptive string is a *sentence* that $U$ understands; the "meaning" of a given sentence (program) is the procedure it describes; $U$ may demonstrate its understanding of this meaning by carrying out the procedure. Imagine a person "communicating" with $U$ in the following way: the person (whose name will be $A$) has a means of printing symbols[5] on the squares of one of $U$'s tapes, called the *input tape;* the signals that $A$ transmits to $U$ are precisely the symbols $A$ decides to print; in addition, $A$ has the ability to read all symbols that are printed on one of $U$'s tapes which will be called the *output tape.* $A$ may decide that $U$ "understands" a language of descriptive strings if whenever $A$ prints a sentence of that language on $U$'s input tape, $U$ eventually prints, on its output tape, the result of carrying out the procedure described by $A$'s sentence. (Of course, if $A$ has some knowledge about the "internal workings" of $U$ (its next-move function, or the symbols printed on its other tapes), then $A$ may well decide that $U$ understands a given programming language, without very extensively performing this experiment.)

It should be emphasized that there is more than one way to describe a given Turing machine. Chapter 2 presented a very simple way to describe the next-move function of any given Turing machine; that way produced, for each Turing machine, a simple string of blanks and 1's. In essence, this was a description of a programming language, the sentences of which were strings that could be used by an "appropriate" universal Turing machine to carry out the procedures they described. Besides the "blank-one language" presented in Chapter 2, one can certainly design other programming languages for describing procedures. Furthermore, one can certainly design universal Turing machines that would not "understand" the blank-one language but would understand some other language for describing procedures. Indeed, the major difference between modern computers and polycephalic universal Turing machines is that computers understand languages that are much simpler and easier for people to use (when describing complicated, useful, "real-world" procedures) than is the blank-one language. What all these languages have in common is that any

---

[5] This includes blanks; that is, $A$ has the ability to erase symbols previously printed on the input tape.

## COMPUTER HARDWARE

| BITS | $b_7$ | 0 | 0 | 0 | 0 | 1 | 1 | 1 | 1 |
|---|---|---|---|---|---|---|---|---|---|
| | $b_6$ | 0 | 0 | 1 | 1 | 0 | 0 | 1 | 1 |
| | $b_5$ | 0 | 1 | 0 | 1 | 0 | 1 | 0 | 1 |
| $b_4\ b_3\ b_2\ b_1$ | Column → Row ↓ | 0 | 1 | 2 | 3 | 4 | 5 | 6 | 7 |
| 0 0 0 0 | 0 | NUL | DLE | SP | 0 | @ | P | | p |
| 0 0 0 1 | 1 | SOH | DC1 | ! | 1 | A | Q | a | q |
| 0 0 1 0 | 2 | STX | DC2 | " | 2 | B | R | b | r |
| 0 0 1 1 | 3 | ETX | DC3 | # | 3 | C | S | c | s |
| 0 1 0 0 | 4 | EOT | DC4 | $ | 4 | D | T | d | t |
| 0 1 0 1 | 5 | ENQ | NAK | % | 5 | E | U | e | u |
| 0 1 1 0 | 6 | ACK | SYN | & | 6 | F | V | f | v |
| 0 1 1 1 | 7 | BEL | ETB | ' | 7 | G | W | g | w |
| 1 0 0 0 | 8 | BS | CAN | ( | 8 | H | X | h | x |
| 1 0 0 1 | 9 | HT | EM | ) | 9 | I | Y | i | y |
| 1 0 1 0 | 10 | LF | SUB | * | : | J | Z | j | z |
| 1 0 1 1 | 11 | VT | ESC | + | ; | K | [ | k | { |
| 1 1 0 0 | 12 | FF | FS | , | < | L | \ | l | l |
| 1 1 0 1 | 13 | CR | GS | — | = | M | ] | m | } |
| 1 1 1 0 | 14 | SO | RS | . | > | N | ^ | n | ~ |
| 1 1 1 1 | 15 | SI | US | / | ? | O | — | o | DEL |

## SYMBOLS

| | | | |
|---|---|---|---|
| NUL | Null | DLE | Data Link Escape (CC) |
| SOH | Start of Heading (CC) | DC1 | Device Control 1 |
| STX | Start of Text (CC) | DC2 | Device Control 2 |
| ETX | End of Text (CC) | DC3 | Device Control 3 |
| EOT | End of Transmission (CC) | DC4 | Device Control 4 (Stop) |
| ENQ | Enquiry (CC) | NAK | Negative Acknowledge (CC) |
| ACK | Acknowledge (CC) | SYN | Synchronous Idle (CC) |
| BEL | Bell (audible or attention signal) | ETB | End of Transmission Block (C |
| BS | Backspace (FE) | CAN | Cancel |
| HT | Horizontal Tabulation (punched card skip) (FE) | EM | End of Medium |
| LF | Line Feed (FE) | SUB | Substitute |
| VT | Vertical Tabulation (FD) | ESC | Escape |
| FF | Form Feed (FE) | FS | File Separator (IS) |
| CR | Carriage Return (FE) | GS | Group Separator (IS) |
| SO | Shift Out | RS | Record Separator (IS) |
| SI | Shift In | US | Unit Separator (IS) |
| SP | Space | DEL | Delete |

## ABBREVIATIONS

(CC) Communication Control
(FE) Format Effector
(IS) Information Separator

Figure 7-3. The ASCII "alphabet" for programming languages.
(Chapin, 1971, reprinted with permission.)

procedure which can be described in the blank-one language can also be described in the languages that are understood by real computers, and vice versa. Thus, in theory a computer can perform exactly those procedures that can be carried out by a universal Turing machine. If, for any Turing machine, a programming language contains at least one sentence that describes the procedure carried out by that Turing machine, then the programming language is said to be a *universal programming language.* Again, however, a language can be a universal programming language only with respect to some computer (universal Turing machine) that "understands" it, or in other words, one having the "language capability" for that language.

The discussion so far describes the "semantics" of programming languages. In the next few pages the "syntactics" of these languages will be described.

With respect to syntactics, note first that all programming languages used by real computers[6] make use of sentences that are essentially strings (sequences) of symbols. In other words, they use "descriptive strings," though these descriptive strings consist of many other symbols besides "blank" and "1." Figure 7–3 shows a set of symbols that may currently appear in the sentences (programs) of universal programming languages. One may transmit strings of these symbols to the computer by typing them out on a typewriter connected to the computer, or by "feeding" the computer a deck of appropriately punched cards, etc. (The reader should note that each of these symbols is actually converted into a seven-place string of zeroes and ones when it is read into the computer; the "code" for making this conversion is indicated in Fig. 7–3.) A total of 128 symbols make up this "alphabet" of current programming languages.

As stated before, a programming language must contain at least one program describing each Turing machine, if it is to be "universal." However, there are an infinite number of different Turing machines and therefore a universal programming language must contain an infinite number of programs, or sentences. A proper description of the syntactics of the language must describe the structural nature of each of its sentences. This can be done either by presenting all of the sentences in the language or by giving some set of rules that could be used to construct any sentence belonging to the language, given enough time, yet could not be used to construct a sentence not belonging to the language. Such a set of rules is called a *grammar* for the language. To properly

---

[6] There is no reason why the sentences of a programming language would *have* to be stringlike structures; indeed, some researchers have suggested that eventually other types of structures will also be used (e.g., Minsky, 1970). In the fourth section of this chapter, languages with sentences of more general structural nature will be discussed.

describe the syntactics of a universal programming language it is clear that we must present a grammar for that language. Several different kinds of grammars for describing languages have been developed (see, e.g., Chomsky, 1959; Post, 1944; Backus, 1959). A formalization for the Chomsky *phrase-structure* grammars will be presented. For the reader who wishes to skip the mathematics of this formalization (which is based on that given by Hopcroft and Ullman, 1969), the ordinary discourse will be resumed in the later section entitled "Grammars, Machines, and Extensibility."

# String Languages

If $V$ is a set of symbols, then $V^*$ represents the set of all finite strings composed of elements from $V$. A string is an ordered series of symbols (i.e., for some $n$, an ordered $n$-tuple). Thus, if $V = \{0, 1\}$, then $V^* = \{\epsilon, 0, 1, 01, 10, 00, 11, 111, 101, \ldots\}$, where $\epsilon$ represents the *empty string*, which does not contain any symbols. We stipulate that $\epsilon$ is always an element of $V^*$, for any $V$, and use $V^+$ to denote $V^* - \{\epsilon\}$. A *language L* on the alphabet $V$ is then any set $L$ that is a subset of $V^*$; that is, $L \subseteq V^*$.

A *grammar G* is defined to be an ordered quadruple

$$G = (V_N, V_T, P, S)$$

satisfying the following conditions:

1. $V_N, V_T, P$ are finite sets.
2. $V_N \cap V_T = \phi$ (no elements belong to both $V_N$ and $V_T$).
3. $S \epsilon V_N$ ($S$ is called the *start symbol*).
4. $V_N$ and $V_T$ are sets of symbols. (The symbols belonging to $V_N$ are referred to as production variables, and those belonging to $V_T$ are referred to as terminals. The *alphabet of G* is $V_N \cup V_T$.)
5. $P$ is a set of written expressions of the form $\alpha \to \beta$ (or equivalently, ordered pairs of the form $(\alpha, \beta)$), where $\alpha \epsilon V^+$ and $\beta \epsilon V^*$.

It is customary to use capital Roman (italic) alphabet letters for production variables and to use lower-case letters at the beginning of the Roman (italic) alphabet for terminals. Strings of terminals are represented by lower-case letters near the end of the Roman alphabet; strings of production variables and terminals are denoted by lower-case Greek letters.

If $\alpha$ and $\beta$ are two strings, then $\alpha\beta$ denotes the string obtained by writing down all elements in $\alpha$, followed by all elements in $\beta$. Thus, if $\alpha = abc$ and $\beta = def$, then $\alpha\beta = abcdef$ and $\beta\alpha = defabc$.

We can now proceed to define the language $L(G)$ generated by a given grammar $G$. To accomplish this, we need two mathematical relations, $\vec{G}$ and $\overset{*}{\vec{G}}$, which can exist between strings in $V^*$. The first relation is defined as follows: If $\alpha \to \beta$ is an element of $P$, and $\gamma$ and $\delta$ are any strings in $V^*$, then $\gamma\alpha\delta$ is said to *directly derive* $\gamma\beta\delta$, and we write $\gamma\alpha\delta \ \vec{G} \ \gamma\beta\delta$. The *rewriting rule* or *production rule* $\alpha \to \beta$ is said to be *applied* to the string $\gamma\alpha\delta$ to obtain $\gamma\beta\delta$.

The second relation, $\overset{*}{\vec{G}}$, is defined as follows: For two strings $\alpha$ and $\beta$ in $V^*$, we say that $\alpha \overset{*}{\vec{G}} \beta$ ($\alpha$ *derives* $\beta$) iff we can obtain $\beta$ by the application of some finite number of production rules in $P$ to $\alpha$. That is, $\alpha \overset{*}{\vec{G}} \beta$ iff there exist in $V^*$ strings $\gamma_1, \gamma_2, \ldots, \gamma_n$ such that $\alpha \ \vec{G} \ \gamma_1$, $\gamma_1 \ \vec{G} \ \gamma_2$, $\ldots, \gamma_{n-1} \ \vec{G} \ \gamma_n$, $\gamma_n \ \vec{G} \ \beta$.

The *language $L(G)$* generated by the grammar $G$ is now defined to be

$$L(G) = \{w | w \epsilon V_T^{\overset{*}{\to}} \ \text{and} \ S \overset{*}{\vec{G}} w\}$$

In other words, a string $w$ is in $L(G)$ if it is made up entirely of terminals and it can be derived from $S$. If $w$ can be derived from $S$, then a sequence of strings

$$S, \gamma_1, \gamma_2, \ldots, \gamma_n, w$$

such that $S \ \vec{G} \ \gamma_1$, $\gamma_1 \ \vec{G} \ \gamma_2$, $\ldots, \gamma_n \ \vec{G} \ w$ is known as a *derivation* of $w$ in the grammar $G$. (If it is clear which grammar is involved, we use $\Rightarrow$ for $\vec{G}$ and $\overset{*}{\Rightarrow}$ for $\overset{*}{\vec{G}}$.)

As an example, consider the grammar $G_1 = (V_N, V_T, P, S)$, where $V_N = \{S, A\}$, $V_T = \{0, 1\}$ and $P$ contains the following *production rules.*

1. $S \to A1$
2. $S \to S0$
3. $A \to S0$
4. $S \to 0$
5. $S \to 1$
6. $A \to 0$

The language $L(G_1)$ generated by this grammar contains all finite strings made up of 0's and 1's in which there are no consecutive 1's. To illustrate, the string 10010 may be derived from $S$ as follows:

| Given: | $S$ |
|---|---|
| Apply Rule 2: | $S0$ |
| Apply Rule 1: | $A10$ |
| Apply Rule 3: | $S010$ |
| Apply Rules 4, 2: | $S0010$ |
| Apply Rule 5: | $10010$ |

The reader may prove as an exercise that any string of 0's and 1's in which there are no consecutive 1's can be derived in this grammar from $S$. (Suggestion: Use mathematical induction on the length of a string.)

For another example, let

$$G_2 = (V_N, V_T, P, S)$$
$$V_N = \{S, B, C\}$$
$$V_T = \{a, b, c\}$$

and let $P$ contain the following rules.

1. $S \rightarrow aSBC$
2. $S \rightarrow aBC$
3. $CB \rightarrow BC$
4. $aB \rightarrow ab$
5. $bB \rightarrow bb$
6. $bC \rightarrow bc$
7. $cC \rightarrow cc$

To describe the language generated by this grammar we need to introduce some new notation: If $\alpha$ is a string, then the expression $\alpha^n$ refers to the string $\alpha\alpha\cdots\alpha$, in which $\alpha$ is repeated exactly $n$ times. The language $L(G_2)$ then contains the string $a^n b^n c^n$ for each $n \geq 1$, and no other strings.

To obtain a given string $a^n b^n c^n$, we work in the following fashion.

| | |
|---|---|
| Given: | $S$ |
| Apply Rule 1 $n - 1$ times: | $a^{n-1}S(BC)^{n-1}$ |
| Apply Rule 2: | $a^n(BC)^n$ |
| Apply Rule 3 as often as necessary[7]: | $a^n B^n C^n$ |
| Apply Rule 4: | $a^n b B^{n-1} C^n$ |
| Apply Rule 5 $n - 1$ times: | $a^n b^n C^n$ |
| Apply Rule 6: | $a^n b^n c C^{n-1}$ |
| Apply Rule 7 $n - 1$ times: | $a^n b^n c^n$ |

We now demonstrate that $L(G_2)$ does not contain any strings other than those of the form $a^n b^n c^n$: first of all, we know that any derivation of a string must start from the symbol $S$. Note that, given $S$, we cannot apply rules 4, 5, 6, or 7 until we apply rule 2. And, once rule 2 is applied, we can no longer use rules 1 or 2. A (nontrivial) derivation, then, must start with the use of rule 1 and be followed by a series of applications of rules 1 and 3 until the application of rule 2. (We could, of course,

---

[7] "Necessary" = $n(n-1)/2$ times. (Why?)

start our derivation with rule 2, in which case only the string *abc* can be produced.) The string now consists of *n* *a*'s followed by some ordering of *n* *B*'s and *n* *C*'s. After applying any of rules 3 through 7 any number of times, the string will still have the form $\alpha\beta$, where $\alpha$ consists entirely of terminals and $\beta$ is a nonempty string consisting entirely of *B*'s and *C*'s.

Now, if all the *B*'s are converted to *b*'s before any *C* is converted to a *c*, the string will have the form $a^n b^n C^n$, and the only string of terminals that can possibly be derived from this is $a^n b^n c^n$. So, we may assume that some *C* is converted to a *c* before all the *B*'s are converted to *b*'s; the string now has the form of $a^n b^i c\alpha$, where $i < n$, and $\alpha$ is a string of *B's* and *C's* (including at least one *B*). The only rules that can now be applied are rules 3 and 7; their use can result only in a string of the form $a^n b^i c^j$ $\alpha$ (where $i < n$ and $j \leqq n$), such that $\alpha$ contains at least one *B*. They are still the only rules that can be applied, and their use continues to give a string of the same form; therefore we conclude that a string without variables cannot be produced if a *C* is converted to a *c* before all *B*'s are converted to *b*'s. Thus, $L(G_2) = \{a^n b^n c^n / n \geqq 1\}$.

As an exercise, the reader should inspect that the grammar $G_3 = (V_N, V_T, P, S)$, where $V_n = \{S, A, B, C\}, V_T = \{a, b, c\}$, and *P* contains the following rules.

1. $S \rightarrow aAC$
2. $S \rightarrow aC$
3. $A \rightarrow aAB$
4. $A \rightarrow aB$
5. $C \rightarrow bc$
6. $Bb \rightarrow bB$
7. $Bc \rightarrow bcc$

This grammar also generates the language of all strings of the form $a^n b^n c^n$, $n \geqq 1$. Grammars that generate the same language are said to be *equivalent*.

It is possible to distinguish between different *types* of grammars on the basis of their sets of production rules. The reason for making the distinction is that there exists a correspondence between each type of grammar and a certain type of machine.

If every production rule in a grammar is of the form $A \rightarrow a$ or $A \rightarrow aB$, then the grammar is said to be a *type 3*, or *regular*, grammar, and to generate a type 3, or regular, language. (An example is the grammar $G_1$ above.) If every production in a grammar is of the form $A \rightarrow \alpha$, such that *A* is a production variable and $\alpha \in V^+$, then the grammar is called a *type 2*, or *context-free*, grammar (and it generates a type 2 or context-free language). Finally, if every production in a grammar is of the form $\alpha \rightarrow \beta$ such that the number of symbols in the string $\beta$ is always greater than or equal to the number of symbols in the string $\alpha$, then we have a *type 1,* or *context-sensitive,* grammar (generating a type 1, or context-sensitive, language).

The reason for the type definition "context-sensitive" is that the class of languages generated can be shown to be the same if we define instead that the production rules in a context-sensitive grammar be of the form $\alpha A \gamma \rightarrow \alpha \beta \gamma$, where $A \in V_N, \beta \neq \epsilon$, and $\alpha$, $\beta$, and $\gamma$ are in $V^*$. In other words, if $A$ appears "in the context of" $\alpha$ and $\gamma$, it can be replaced by $\beta$. Examples of context-sensitive grammars are $G_2$ and $G_3$, discussed above (and also $G_1$; every type 3 language (grammar) is also type 2; every type 2 language (grammar) is also type 1).

If no restrictions are placed on the form of the production rules of a grammar (other than the necessary ones, $\alpha \in V^+$ and $\beta \in V^*$, for any rule $\alpha \rightarrow \beta$), it may be referred to as a type 0 or "general" phrase-structure grammar (and the same names are given to the language it generates).

# Grammars, Machines, and Extensibility

The basic correspondence between grammars and machines can be fairly simply described, making use of the concepts of "input tape" and "output tape" given earlier. We say a machine *accepts* a language iff whenever any sentence of the language is placed on the (otherwise blank) input tape of the machine, the machine eventually prints a "1" on its (otherwise blank) output tape and halts. In essence, a machine that accepts a language $L$ is a "procedural embodiment" of a grammar for that language. It can be shown that a phrase-structure language is of type 0 iff there is a Turing machine that accepts it. (See the Exercises.) Three special types of Turing machine can be defined—*linear bounded automata, pushdown automata,* and *finite-state automata* (see Chapter 2)—and it can be shown that they correspond to acceptors for the context-sensitive, context-free, and regular languages, respectively.

Of course it is desirable to do more than merely recognize that a given sentence belongs to a language, especially if we are concerned with programming languages. It is also necessary to "understand" the sentence itself, and implement the procedure it describes. A computer may come by this understanding automatically, just as the universal Turing machine described in Chapter 2 would be automatically able to implement the procedure described by a sentence in its "blank-one" language. In essence, the understanding of that language was "wired in" with the next-move function of the machine. In general, every computer will be able to understand some programming language in this automatic sense; the programming language that is "wired in" to a computer is commonly known as its *machine language.*

It is natural to ask whether a computer can understand programming languages other than its machine language. The answer to this question is yes. Let's suppose we have some computer $U$ which has as its machine language the programming language $L$, and that we want $U$ to be able to "understand" sentences in another programming language, $L'$. We assume that $L'$ is at least type 0, and may be type 1, 2, etc. (See note 7–5.) Because $U$ is universal, and because $L'$ is type 0, we know that we can write a program (find a sentence in $L$) that describes a procedure which will accept the sentences of $L'$. Also, we know that $U$ will be capable of implementing this procedure. Thus, we can "program" $U$ to accept the sentences of $L'$. In fact, however, it is possible to do more. Given a description of a grammar $G'$ for $L'$ and a sentence $w'$ in $L'$, we can "program" $U$ to find the derivations of $w'$ with respect to the grammar $G'$. Normally there will be only one such derivation and it will provide structural information about the procedure described by $w'$ that can be used to construct a sentence $w$, in the language $L$, which describes the same procedure. (The sentences $w$ and $w'$ are said to be *computationally equivalent.*) One can attempt to describe the procedure by which the sentence $w$ is produced from the sentence $w'$, and generalize to a procedure that will produce a computationally equivalent sentence in $L$, given any sentence in $L'$. If this general procedure (called a *translator* from $L'$ to $L$) can be described by a sentence $p$ in $L$ (and we know that it can be, if $L$ is a universal programming language), then $p$ can be used to extend the "language capability" of the computer $U$. If we give $p$ and any sentence of $L'$ to $U$, that sentence of $L'$ will be converted into a sentence of $L$ and the procedure it describes can then be implemented by $U$ (notes 7–6, 7–7).

Thus, it is possible to find sentences in $L$ which will "extend" the language capability of $U$, just as it is possible to find sentences in English that a person can use to extend his own language capability. When one looks at a modern computer one sees a hierarchy of languages $L, L', L'', L''', \dots$ that are each ultimately embedded in the machine language of that computer. We can now make use of many different kinds of programs ("compilers," "interpreters," etc.) for extending a computer's language capability (see Earley and Sturgis, 1970; Irons, 1970).

Again, the reason for extending the language capability of a computer is not exactly that the computer will thereby be able to do things it could not do before; in theory, one universal programming language is just as good as another, because every universal programming language describes the same class of procedures (namely, those that can be carried out by Turing machines). In practice, however, we find that any given universal programming language will provide very simple sentences describing some procedures and very complex sentences describing other procedures. Thus, a "blank-one" sentence describing a Turing machine procedure for matrix multiplication would be very long and difficult for a person to handle. The same procedure may be described simply in other universal programming languages such as FORTRAN, ALGOL, and SAIL. Since people want to be able to describe procedures like matrix multiplication easily, it is customary to extend the computer's language capability to include these "higher-level" languages (note 7–8).

In addition to this extensibility, one can design programming languages to facilitate the use of programs with "self-reference." Thus, "recursive programs" are easily described in LISP. However, no one really knows the precise relationship between the self-reference of recursive programs and the self-reference of natural-language sentences. Also, it is possible to design programming languages with "redundancy" (e.g., "error-coding" of instructions; see Lucky, 1969).

Universal programming languages have, in one way or another, two of the most important characteristics that are possessed by human languages: extensibility and self-reference. These characteristics are not possessed in any form by any other known organism-level language. So, it may not be so surprising to read in the next section that computers can now understand human languages much better than monkeys can.

# PROGRAMS THAT "UNDERSTAND" NATURAL LANGUAGE

## Five Problems

In the preceding section we saw how it is possible to extend the "language capability" of a computer so it can understand programming languages other than its machine language. To make such an extension, the computer might be given a sentence (program) in some programming language that it already understands (e.g., its machine language) that will enable it to "translate" sentences in the new language; that is, convert them into sentences it can already understand. Because such a "translator" sentence (program) describes a procedure for understanding the new language, we often say that the sentence itself "understands" the language, even though in fact the understanding results from the interaction of the translator sentence and the computer. Thus, it is common to talk of programs that understand languages.

It is natural to ask how far the language capabilities of computers can be extended. For example, is it possible to write programs that understand the natural languages spoken by humans? Can we extend the language capabilities of a computer to the extent that it becomes possible for us to describe in English the procedures we wish it to perform? Can English be used as a "programming language" for a computer? If this could be done, people would not have to learn special programming languages in order to make use of computers, and the utility of computers might be greatly increased.

Artificial intelligence research is currently concerned with problems such as how to program computers to answer questions stated in English (or other natural languages), solve problems stated in English, and participate in English conversations with people (or, for that matter, other computers). Ultimately, AI research may consider a variety of more difficult problems, such as whether computers can translate from one natural language to another (note 7–9), perform complicated secretarial work (e.g., take dictation), or play "language games" (note 7–10). Although the emphasis here is on current achievements and problems, it is well to keep the "more difficult" ones in mind (cf. Polya, 1945). The evidence presented suggests that all of these problems may eventually be solved (note 7–11).

Before continuing, the reader should note that this discussion will not deal with the machine understanding of spoken languages, even though reference will often be made to the "speaker" of a sentence, simply to follow a convention. Techniques for enabling computers to hear, understand, and make spoken words and sentences are still in a relatively primitive state of development. The reader who is interested in this subject should refer to Aṣtrahan (1970), Bobrow (1968), Denes and Mathews (1968), D. R. Hill (1967), and Mermelstein (1969).

Unless otherwise stated, the discussion throughout this section will always be concerned with computer programs that "understand" English sentences (usually submitted *via* a computer terminal), which will be simply called "language understanding programs." In the subsequent pages a variety of such programs will be discussed. The approach and terminology are largely modeled after that of Winograd's (1971, 1972) work, which the reader is encouraged to consult. Space does not permit complete descriptions of each of the many language-understanding programs that have been written, so instead an attempt is made to summarize the most important approaches that have been followed in their design. However, special attention is given to Winograd's program, one of the most linguistically powerful.

Each language-understanding program may typically be said to confront four highly interrelated problems: a *syntax problem,* a *semantics problem,* an *inference problem,* and a *generation problem.* (The question of how these problems are interrelated, and of what use a language understanding program should make of this interrelation, is itself a fifth problem faced by these programs, which will be referred to as the *integration problem.*) To see how these problems arise, return to the discussion of "understanding" and languages. The viewpoint presented in the early pages of this chapter was that one "understands" a sentence in a language by making a model of its "meaning." Mention was made of three aspects of a sentence's transmission which should be considered as elements of its meaning: the things that cause the sentence to be transmitted (e.g., the speaker's "motives" for using the sentence); the things that the sentence causes when it is transmitted (e.g., the sentence might tend to have an "emotional effect" when it is used); the things that the sentence describes (e.g., objects, events, concepts, procedures, other sentences, or an attitude or wish held by its speaker). It is clear that with this interpretation the "meaning" of a sentence is highly dependent on the situation in which it is used.

However, some things about a sentence do *not* usually depend on the situation, or "context" of its use: namely, the sequence of words and letters that make up the sentence itself, the possible derivations (or "parsings") of that sentence in one's grammar for English, and the set of possible "meanings" of the words in the sentence (which is what makes dictionaries useful). These relatively *constant attributes* (the latter two are of course variable, by the extensibility of natural languages) of the sentence help us determine its "meaning," if we also have knowledge about the "situation."

The *syntax problem* has two basic subproblems: What is a good grammar for English? How should we obtain the parsing(s) of a sentence? The *semantics problem* is that of finding a good formalism in which to express models for "meanings" and "situations." The *inference problem* has three basic subproblems: How can we use our model for the "situation" and the constant attributes of a sentence to make a model of its meaning? How do we change our model of the "situation" when we determine the "meaning" of a sentence we receive? How do we determine the "meanings" that we wish to convey, given that we have determined models for the current "situation" and the "meanings" of the sentences we've received? The *generation problem* is that of finding and transmitting sentences that will have the "meanings" we wish to convey. (The semantics and inference problems are often jointly referred to as the *representation problem*.)

If in the sentences of the preceding paragraph we substitute "the computer" for "we," then we obtain the basic problems that confront AI researchers who attempt to make programs that understand English. None of these problems has yet been completely solved, if we make a comparison with human abilities to converse, nor is this surprising. Natural languages are designed to be useful in almost the full range of situations that people encounter, whereas computers currently are acquainted with a relatively small range of situations. Success in achieving language-understanding programs is limited by the extent to which computers can be enabled to reason about real-world situations. An example due to Schank (1971a,b) is

"We saw the Grand Canyon flying to Chicago."

This sentence is syntactically ambiguous (has two equally plausible parsings) unless the computer knows something about the real-world nature of locations and the ability to fly.

Subsequent pages review the approaches that have been used in designing language-understanding programs that can solve these problems. A brief collection of conversations with computers, to illustrate the success AI researchers have had to date, will then be presented. The next section discusses some of the "open questions" that still remain, concerning the relevance of "semantic information processing" to artificial intelligence in general.

## Syntax

The earliest language-understanding programs, which were written for the purpose of mechanical translation (note 7–10), were developed before linguists had achieved any very precise theories of syntax for natural languages. Certainly the theories that then existed were not precise enough to suggest explicitly how computers should be programmed to understand natural languages. As a consequence, the designers of those programs were forced to produce their own *ad hoc* systems for parsing sentences (i.e., *parsers*). Because they lacked a comprehensive plan for designing their programs, the programs tended to become more and more complex, difficult to understand and debug, and difficult to improve; therefore the programs eventually had to be abandoned, during the latter part of the 1950s. After that time, and until 1968, designers of language-understanding programs followed either of two main approaches to the basic problems of syntax: the *restricted pattern-matching* approach and the *context-free* approach.

The restricted pattern-matching approach consisted essentially of accepting the limitations on syntax that were implied by the lack of a good formalism for expressing and using grammars for English. The researchers who followed this approach (including Bobrow, Raphael, and Weizenbaum) recognized that the syntax problem would still have to be solved eventually by really general language-understanding programs, but they managed to show that interesting linguistic behavior, relating to the other basic problems of semantics and inference, can be obtained even if only minimal solutions to the syntax problem are provided. The language-understanding programs they developed did not really use "grammars" in any general sense, nor did they parse sentences. Instead, these programs were designed to extract semantic information from sentences by matching them against any of a small, prespecified, constant number of "templates" or "forms." Examples of the forms used by Bobrow (1968)—which he called "linguistic forms"—are "_____and_____," "_____equals_____," "_____'s father," "salary of_____," "not_____," "_____gave_____to_____," etc. Bobrow's program (known as STUDENT) was designed to follow a relatively rigid procedure of successively "filling in blanks"; thus, it might "parse" the sentence, "The salary of John's father equals 100 dollars," by filling in blanks as in Fig. 7–4. It should be clear that

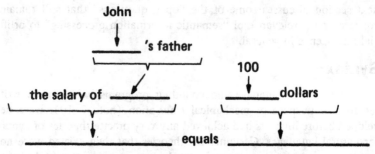

Figure 7–4. Pattern-matching for the sentence "The salary of John's father equals 100 dollars."

STUDENT had a "recursive" ability to fill in blanks; the blanks of a given template might be filled in by other templates. As STUDENT matched a given sentence against its collection of linguistic forms, it could be guided by the matchings it obtained in a process of setting up an algebraic equation to represent the relationship between the "variables" of the sentence (e.g., "John"). Thus, STUDENT was capable of converting each of a limited (although infinite) variety of English sentences into an equivalent algebraic equation. Given a collection of such sentences, STUDENT would form a set of simultaneous equations. STUDENT was designed to use special linguistic forms like "find_____" to identify the variables for which it should solve, given such a set of simultaneous equations, and it contained a special set of programs it could then use to actually solve sets of simultaneous, elementary algebra equations. Finally, STUDENT was capable of "assuming" certain variables to be equivalent (based on simple structural similarities between the ways they were named in the initial set of English sentences) if the operation of its problem-solving routines revealed that it had been given more variables than equations. If this technique failed, STUDENT could ask the person who supplied the problem for more information. Thus, STUDENT was capable of performing a fairly difficult intellectual task, that of understanding and solving algebra word-problems.

It should be evident from this description that STUDENT's ability to "understand" algebra problems that were stated in English was somewhat limited. One could easily find problem statements that it could not understand, using its restricted pattern-matching approach to syntax. Still, STUDENT and the other early programs that used this approach demonstrated some rather impressive (and surprising) behavior. STUDENT fostered two other special-purpose question-answering programs, CARPS and HAPPINESS (Charniak, 1969; and Gelb, 1971), respectively designed to solve calculus and probability problems stated in English.

The context-free approach to the problem of syntax involved finding simplified subsets of English that could be described by well-understood kinds of phrase-structure string grammars; much research concentrated specifically on the use of context-free grammars, owing to their proven value as the basis for ordinary programming languages. However, the full complexity of English syntax is not easily describable by phrase-structure grammars (see Winograd, 1971, 1972, for a discussion of reasons). Thus, the context-free approach has had only limited success. Rather than discuss this approach in any detail, the reader is asked to refer to Simmons (1965) and Kuno (1965).

## Recursive Approaches to Syntax

In many ways, the year 1968 was a good one for language-understanding programs. From the standpoint of syntax, it was the year in which the designers of these programs freed themselves from the restrictiveness of phrase-structure grammars by taking a new approach to

syntax. The essence of this approach to syntax is the realization that
language-understanding programs need not be restricted to the use of
phrase-structure grammars any more than computers need be restricted
to the simulation of Turing machines. Phrase-structure grammars and
Turing machines are adequate simple formalizations for the infinite
classes of all machine-understandable languages and all machine-
computable functions, but they are extremely poor formalizations in
which to describe the relatively small classes of natural languages and
intelligent procedures.[8] The result of using this new approach has been
the discovery that natural language grammars can be profitably described as certain kinds of recursive procedures. Two ways of describing
these procedures have been developed, corresponding to the formalization of *augmented state transition networks* (see Thorne, Bratley, and
Dewar, 1968; Bobrow and Fraser, 1969; Woods, 1969) and to the
programming language PROGRAMMAR (Winograd, 1971).

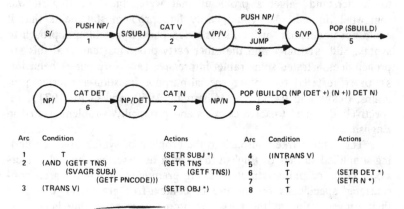

Figure 7–5. A simple augmented transition network grammar.
(Kaplan, 1971, reprinted with permission.)

Augmented transition networks are a generalization of the transition networks for the finite-state machines discussed in Chapter 2.
Figure 7–5 shows an example of a simple augmented transition network; as can be seen, it is basically a graphlike structure similar to the
transition networks discussed previously. However, two important
changes should be noted: First, the augmented transition network

---

[8] Actually, these classes are not really so "small"; perhaps it is better to say
that they have a very low "density" when one tries to find them by searching
through the classes of all languages and functions, as represented by phrase-
structure grammars and Turing machines.

represents a *recursive* procedure. This is achieved by allowing the label of an arc to refer to a state, using either a PUSH or a POP command. A PUSH command means that the transition along its arc should be postponed (i.e., the name of the state at the head of the arc should be placed on the top of a "pushdown" storage unit) and the new state of the machine should instead become that referred to (explicitly) by the PUSH command. Thus, arc 1 in Fig. 7–5 has the label PUSH NP/, which specifies that the machine should postpone its transition from state s/ to state s/SUBJ and instead make a transition from state s/ to NP/. Similarly, a POP command may be a label for a "dangling arc," the head of which is not attached to a node. The meaning of a POP command is that the machine should remove the name at the top of its pushdown store and make a transition to the corresponding state. Thus, suppose that the pushdown store should happen to contain the following "stack" of names:

NP/DET
S/SUBJ
NP/N
S/

and suppose that the machine should happen to be in state s/VP; then the POP command at arc 5 will specify that the machine should make a transition from state s/VP to state NP/DET and that the stack of names in the pushdown store should become

S/SUBJ
NP/N
S/

Besides this ability to "transfer control recursively" throughout the network, augmented transition networks differ from finite-state transition networks in another manner: Each arc may be allowed to specify a *condition* and a *sequence of actions;* the actions that an arc may specify are those of building and naming tree structures—the name of a tree structure is known as its *register,* and registers are said to "contain" their tree structures. Actions may specify various kinds of changes to the contents of registers "in terms of the current input symbol, the previous contents of registers, and the results of lower-level computations (pushes)" (Kaplan, 1971). The "input symbols" that are submitted to an augmented transition network are English words. The network of Fig. 7–5 is designed to start in state s/, with the first word of a sentence being submitted to it; arcs that do not have PUSH or POP commands attached may have either "word" or "category" statements attached to them. Thus, arc 2 has the label CAT V, which means that if the machine is in state S/SUBJ and the input symbol is a word that is a verb, then the machine is to make a transition to state VP/V, *provided* the additional condition for arc 2 listed in the table of "arcs, conditions, and actions" below the diagram is satisfied. The condition for an arc is in general a Boolean combination of predicates involving the current input symbol and register contents; the conditions for some arcs (e.g., arc 7) may always be trivially satisfied. A sentence is said to be accepted by an augmented transition network whenever a final state (i.e., a "dangling arc"), the end of the sentence, and an empty pushdown store are all reached at the same time. The parsing for a sentence provided by an augmented transition network corresponds simply to the history of transitions, pushes, and pops required to accept it.

The particular grammar shown in Fig. 7–5 will not be discussed in any greater detail. However, the reader interested in these networks may wish to see if he can understand the operation of this example network on a simple sentence, shown in Fig. 7–6. A complete explanation of this example is to be found in Kaplan (1971).[9]

A few sentences are sufficient to describe the nature of the grammars that can be formulated as augmented state transition networks, and to indicate their applicability to the syntax of natural languages. The use of PUSH and POP commands, and conditions, actions, and registers in such a grammar (network), enables it to try out different kinds of parsing strategies on variably large phrases in a sentence, to store information relating to the success of these strategies as they are being carried out, and to recognize whenever a given strategy has failed so that a new strategy can be tried. If this is contrasted with the performance offered by context-free grammars, the differences are striking. A parser that uses a phrase-structure grammar typically has a large set of production rules, each of which is potentially applicable at any point in its analysis. Such a parser is not easily made to simulate strategic performance in the way it conducts its analysis. Even though a system designer were to manage somehow to find a parser and a phrase-structure grammar that would efficiently parse the sentences of a given subset of English, he would in general find it difficult to extend his system to

---

[9] A helpful hint: The symbol "*" represents a special register in Kaplan's formalism which always contains the structure or word that "enabled" the most recent transition of the machine. In most cases this is an input symbol; however, whenever a POP command is executed, it is the value of that command's argument. Thus, the execution of POP(SBUILD) causes the value of the function SBUILD to be placed in "*". Also, the JUMP label on an arc indicates that a transition is to be made without advancing the input sentence.

```
Sentence:  The man kicked the ball.

STRING = (THE MAN KICKED THE BALL)
ENTERING STATE S/
ABOUT TO PUSH
 ENTERING STATE NP/
 TAKING CAT DET ARC

STRING = (MAN KICKED THE BALL)
 ENTERING STATE NP/DET
 TAKING CAT N ARC

STRING = (KICKED THE BALL)
 ENTERING STATE NP/N
 ABOUT TO POP
ENTERING STATE S/SUBJ
TAKING CAT V ARC

STRING = (THE BALL)
ENTERING STATE VP/V
STORING ALTARC ALTERNATIVE 76869[a]
ABOUT TO PUSH
 ENTERING STATE NP/
 TAKING CAT DET ARC

STRING = (BALL)
 ENTERING STATE NP/DET
 TAKING CAT N ARC

STRING = NIL
 ENTERING STATE NP/N
 ABOUT TO POP
ENTERING STATE S/VP
ABOUT TO POP
SUCCESS
10 ARCS ATTEMPTED
195 CONSES[b]
1.8869999 SECONDS[c]
PARSINGS:
S NP DET THE
    N MAN
  AUX TNS PAST
  VP V KICK
     NP DET THE
        N BALL
_____

a. The alternative analysis path
   starting with arc 4 is saved.

b. Number of memory words used.

c. Processing time required.
```

Figure 7–6. Trace of an analysis. (Kaplan, 1971, reprinted with permission.)

a larger subset of English and still maintain its efficiency. By comparison, it is relatively easy to add new strategic abilities to a network grammar.

Another formalization of this approach to syntax is the programming language PROGRAMMAR (Winograd, 1971, 1972). PROGRAMMAR facilitates the writing of programs that can act as grammars and parsers for natural languages. It is specifically designed to facilitate the description of parsers that can act strategically and recursively, and to enable the designer of a language-understanding program to make extensions to his system in a fairly straightforward fashion. It is closely related to the language PLANNER in the general philosophy of the programs it is intended to encourage, but the theory underlying its orientation to natural language is actually that of *systemic grammar,* an outlook on natural language that has been developed by Halliday (1961 et seq.). Space does not permit a detailed discussion of PROGRAMMAR and the theory of systemic grammar. However, an outline of the highlights these topics exhibit is provided below. The information given should be sufficient for the general reader to decide whether to investigate them further. For a more detailed introduction, Winograd's discussion of these subjects is readily understandable to the nonspecialist.

Some of the basic tenets of systemic grammar, as expressed previously, are repeated as follows:

1. The purpose of natural language is communication; thus, the syntactic nature of language must be understood in relation to the semantic information it is designed to carry.
2. The problems of syntax, semantics, inference, and generation, which are to be solved in the use of natural language, are all closely interrelated; it is desirable that a language-understanding program be able to solve these problems in a highly integrated way.
3. Despite their interrelations, these problems are in many ways quite distinct. Thus, we should not expect that a system designed to solve the generation problem (e.g., transformational grammar; see Chomsky, 1959 et seq.) will necessarily be the basis of an efficient system to solve the syntax (in particular, the parsing) problem.

Conditions (2) and (3) above will be considered more thoroughly in the following subsections, devoted specifically to the semantics, inference, generation, and integration problems. To be considered first is condition (1), the way in which systemic grammar and PROGRAMMAR are designed to understand the syntactic nature of English in terms of the semantic information its sentences may carry.

A viewpoint common to the theories of both systemic and transformational grammar is that the structure of a sentence is the result of a sequence of *grammatical choices* made by the speaker of the sentence. Systemic grammar describes a specific class of such grammatical choices and specifies the effect that each will have on the nature of the sentence being produced. Moreover, systemic grammar dscribes certain relationships that exist among these grammatical choices, and specifies by means of these relationships which sequences of grammatical choices may produce "meaningful" sentences and which may not. When a person makes a meaningful sequence of grammatical choices in the course of producing a sentence, the effect of the choices he makes will be to provide the sentence with certain structural characteristics, or *features,* which other people can use as an aid to the discovery of the "meaning" of the sentence.

For example, every sentence must have structural characteristics corresponding to exactly one of the three features: IMPERATIVE, DE-CLARATIVE, or QUESTION. Thus, the speaker must make the grammatical choice as to which of these three features he wants his sentence to have. Again, if the speaker should choose to give his sentence the structural characteristics corresponding to QUESTION, systemic grammar specifies that he will also have to make a choice between the structural characteristics corresponding to the features YES-NO and WH-question.[10] The features possessed by his sentence are, in effect, markers that people may use to "understand" that it is a question and that it requires, say, a yes or no answer. A set of features that form a mutually exclusive set (e.g., YES-NO and WH-question) are said to be a *system.* The set of other features that must be present for the grammatical choice between the elements of a system to be possible is known as the *entry condition* for that system. Thus, the entry condition for the system YES-NO, WH-question is the feature QUESTION.[11]

In addition to sentences, the theory of systemic grammar specifies features (and systems and entry conditions) for smaller "syntactic units" such as noun groups, prepositional groups, and words. (Thus, the various endings that a word might have are considered to be the "features" it may possess; the word itself may be the entry condition for the system of its endings.)

PROGRAMMAR is designed to facilitate the writing of programs capable of implementing systemic grammars. The language-understand-

---

[10] A sentence that has the structural characteristics corresponding to the feature "WH-question" must possess the feature QUESTION and must begin with one of the words "what," "why," "who," "where," "how," "which," etc.

[11] More generally, an entry condition may be a Boolean formula, the terms of which are features.

ing program that Winograd has written using PROGRAMMAR contains many special subprograms, each designed to recognize a different feature that can be possessed by certain strings of words. In addition, his program is capable of recognizing the presence of entry conditions and can call the appropriate feature-detecting programs when necessary. This gives it the basic ability to carry out a systemic analysis of the features possessed by an English sentence. Other parts of his language-understanding program are capable of using the systemic analysis of a sentence to construct a "semantic model" of its meaning, to integrate this semantic model into its current "world model," to prove theorems and solve problems about and within the world model (using PLANNER), to use the semantic model to detect errors in the systemic analysis of a sentence (or part of a sentence) and redirect the analysis in a strategic manner toward a more plausible parsing, and to generate appropriate replies (e.g., answer questions) to sentences submitted by people. Winograd's program is capable of answering questions about itself (its world model contains a simple "self-model") and of remembering and understanding the contexts of conversations. A sample conversation with this program (called "SHRDLU") is given at the end of this section.

## Semantics and Inference

"Meaning" and "semantic information" are half-mysterious concepts. By this is meant that people are unable to know precisely what effect their words may have on people, whereas they can know exactly what effect their words may have on machines.

Thus, in the preceding pages no attempt was made to present very concrete definitions of the meaning, or semantic information, that may be conveyed by the sentences of a natural language. To have done so would have been to discuss a theory of human psychology (which causes the sentence to be used; which is partially caused by the use of sentences); such a discussion would eventually be desirable, but it is not necessary here. In these pages the primary concern is with viewing the (relatively) unmysterious behavior of machines—unmysterious because we can look directly at their inner workings and at the data stored in their memories.

Because we can know and design the "psychology" of language-understanding machines, the notions of *meaning* and *semantic information* become "halfway more tractable." We can define these concepts rigorously for a language-understanding machine if it has been built (or programmed or designed), but we have difficulty in defining these concepts for the ultimate, truly intelligent machine that would understand our language as well as we do, since we know as little about the internal workings of that (nonexistent) machine as we know about the internal workings of our own intelligence.

For a language-understanding program, the "semantic information" carried by a sentence is simply the data structure that the program creates when it processes ("understands") the sentence. The *problem of semantics* is to discover what kinds of "semantic" data structures such programs should create in order to provide the best solutions to the problems of inference and generation. The *problems of inference and generation* are to discover how language-understanding programs should use their semantic data structures to produce the kind of behavior that people would accept as evidence that the programs "understand" language. So far, the greatest success in solving the problems of semantics, inference, and generation has been in enabling machines to understand the relatively factual, logical, nonpsychological aspects of its use. However, some investigators (notably Colby, Schank, Tesler, Enea, Abelson and Carroll) have been concerned with developing programs with an aptitude for understanding the emotional, metaphorical, and otherwise psychological aspects of meaning.

Clearly, the problem of semantics can be minimized by restricting the "environment" or "problem domain" that one's language-understanding program is supposed to "understand"; some of the earliest language-understanding programs (e.g., SAD-SAM; see R. K. Lindsay, 1963) did exactly this. They minimized their problem of semantics by severely restricting the type of questions they could accept, information they could store, and problems they could solve. To a lesser extent, the more recent "specialized question answerers" (e.g., STUDENT, CARPS, HAPPINESS; see Bobrow, 1968, Charniak, 1969, Gelb, 1971a,b) have adopted the same policy.

Several approaches, which may ultimately be developed into a workable, general semantics-inference formalism, have been suggested. These may be grouped into two classes (which are, however, somewhat indistinct): the *predicate calculus* formalism and the *graph-structure* formalism. The predicate calculus formalism was investigated by Coles (1969) and C. C. Green (1969), who showed that it is possible to translate relatively simple natural language questions into example-construction problems that can be posed in first-order predicate calculus and solved using the resolution technique (see Chapter 6). This approach seems plausible because of (1) the generality of first-order predicate calculus as a language for the statement of facts and problems and (2) the completeness (and consequent problem-solving generality) of the resolution procedure.

As previous chapters have shown, graphs are a type of "mathematical construct" that AI researchers find useful in describing much of their work. For example, state-space problems, finite-state automata, augmented transition networks, and "flowcharts" for computer programs (see Chapin, 1971b; Rodriguez, 1969) may all be represented by structures that are essentially graphs. Winston's work (1970), discussed in Chapter 5, showed that graph structures can be used to describe visual patterns. In general, anything that has a "structural nature" and can be described as a collection of parts existing in various relationships to each other may be represented by a graph. In particular, our examples show that graphs may be profitably used to describe certain types of problems, "processes" (i.e., automata), and patterns. As an approach to the semantic-inference problems in language understanding, the graph-structure formalism consists of attempts to model the "meaning" of sentences and words by graphs. The plausibility of this approach is supported by the fact that real-world situations, which may be described by sentences (and which may help cause the use of sentences, or be partially caused by the use of sentences), often have a structural nature. The best way of describing the structural nature of general, real-world situations is still not known. However, the utility of the graph-structure formalism should be apparent if we simply note a few examples of its use.

One of the earliest studies of the graph-structure formalism was conducted by Quillian (1966), who developed an elegant model of semantic memory. Information is represented in this semantic memory by a graph structure of arbitrary size in which each node is named by a word and the arcs between nodes represent certain specific relationships, or *associative links,* that may exist between words. Nodes are of two kinds: *types* and *tokens.* A type node represents the "meaning" of its name word; the associative links going from a type node lead to a configuration, or *plane,* of token nodes that represents a definition of this "meaning"; the only purpose of token nodes is to be used in such definitions. Thus, a token node represents a "use" of its name word. Two additional constraints are imposed: For any given token node there must be exactly one type node bearing the same name word, and the two nodes are to be connected by a special "token-to-type" associative link. For each meaning of an English word there must be exactly one type node; a word like PLANT, which has multiple meanings, is represented by multiple type nodes PLANT, PLANT1, PLANT2, etc. In diagrams, type nodes are circled, whereas token nodes are simply indicated by the presence of their name words. Figure 7–7 shows the different kinds of associative links used in Quillian's semantic memory model. Figure 7–8A shows some planes stored in a semantic memory representing definitions for PLANT; Fig. 7–8B represents FOOD. Quillian wrote a program that could do "associative" processing on this kind of memory and demonstrated that it could "compare concepts" and discover interrelationships not indicated specifically in its "definition planes." Essentially, the mechanism for comparing two concepts was a breadthfirst, bidirectional search through the graph structure of the memory (see Fig. 7–9). The next section presents some computer-produced concept comparisons. (Quillian's paper presented intriguing discussions on the similarities of this model to human concept comparison and on the difficulties of making dictionaries.)

Among the more recent graph-structure formalisms for semantic information storage are Schank's (1970 et seq.) "conceptual dependency graphs" (see Fig. 7–10), Shapiro's (1971a,b) MENS system (see Fig. 7–11), and the "hierarchial graphs" of Pratt (1969 et seq.).

As mentioned before, the graph-structure formalism and the predicate calculus formalism are somewhat indistinct. This is true because predicate calculus expressions (of any order) may be stored in graph structures and because predicate calculus expressions may describe properties of graph structures. One of the early language-understanding programs that stored predicate calculus expressions in graph structures was SIR, written by Raphael in 1964 (also see Simmons and Bruce, 1971).

An important relationship between the problems of semantics and inference is described by the *principle of homogeneity:* The operations used to process semantic information should themselves be describable as semantic information and stored in a common semantic memory with other information. This principle dates back to the "stored program" concept formulated in the early years of computer science, but it has often been rediscovered by the designers of language-understanding systems. Among the studies following this principle are those presented by Quillian (1969), Shapiro (1971), Hewitt (1968 et seq.), Norman (1972), Sussman (1972), and R. C. Moore (1973). Winston's (1970) structure-recognizing program, discussed in Chapter 5, should also be mentioned in this regard.

As for SHRDLU, it incorporates the graph-structure formalism, the predicate calculus formalism, and the principle of homogeneity. The features of sentences detected by its systemic parser can be translated readily into conjunctions, disjunctions, conditionals ("if . . . then . . . else" statements), etc., in the PLANNER formalism. The evaluation of a

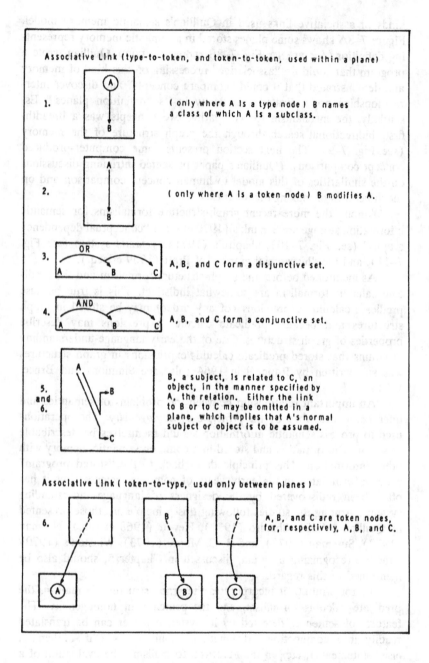

Figure 7–7. Associative links. (Quillian, 1966, reprinted with permission.)

PLANT.  1.  Living structure which is not an animal, frequently with leaves, getting its food from air, water, earth.
        2.  Apparatus used for any process in industry.
        3.  Put (seed, plant, etc.) in earth for growth.

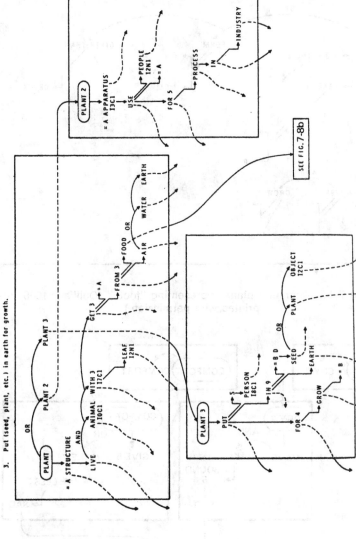

Figure 7–8A. Three definition planes representing three meanings of "plant." (Quillian, 1966, reprinted with permission.)

FOOD:  1. That which living being has to take in to keep it living and for growth.
       Things forming meals, especially other than drink

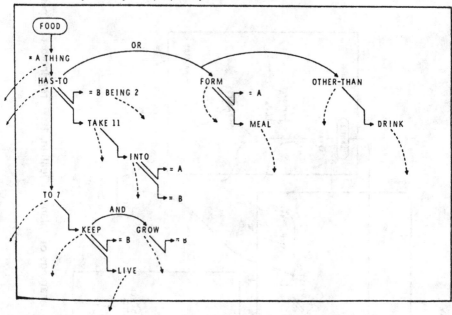

Figure 7–8B. Definition plane representing "food." (Quillian, 1966, re-printed with permission.)

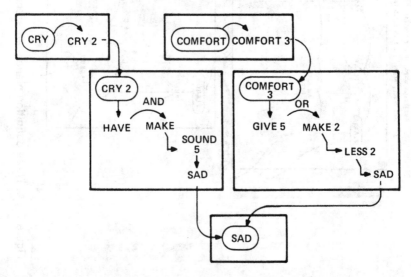

Figure 7–9. A comparison path for "comfort" and "cry." (Quillian, 1966, reprinted with permission.)

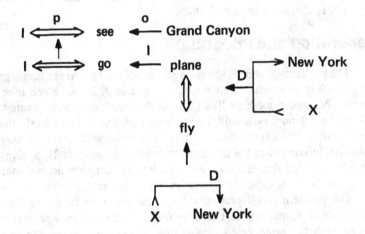

Figure 7-10. A conceptual dependency graph for "I saw the Grand Canyon flying to Chicago." (Schank, 1971, reprinted with permission.)

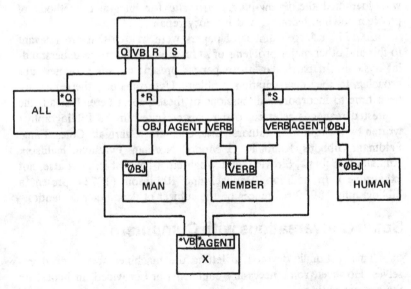

Figure 7-11. A MENS structure for the deduction rule "Every man is human." (Shapiro, 1971, reprinted with permission.)

PLANNER theorem corresponding to an English statement constitutes the major part of the "inference process" performed by SHRDLU (Winograd, 1971; also see Chapters 5 and 6).

# Generation and Integration

The problem of generation is largely unsolved by current language-understanding programs; even SHRDLU uses essentially a "blank-filler" scheme. Perhaps something like Chomsky's transformational grammar (1959 et seq.) may eventually be implemented, but it seems likely that efforts should be devoted first to the "comprehension" stage (syntax, semantics, inference) of the language-understanding process. By analogy, it has been noted that the ability of children to comprehend sentences at a given "level of difficulty" precedes the ability to speak them.

The problem of integration will not be discussed in detail. Relatively little is known about the integration of the sentence-generation process with the sentence-comprehension process, nor about the integration of the language-understanding process with the language-learning process—except, of course, for the integration automatically implied by the principle of homogeneity (we can tell the machine new rules of grammar, meanings for words, etc.; see Quillian, 1969). A good discussion of the integration problem is provided by R. K. Lindsay (1971), who identified the *jigsaw-puzzle heuristic* for integrated methods of problem solving, learning, and memory repair.

In 1972 a large number of papers were written that are relevant to this and other major problems of AI research on language-understanding systems. In particular, these papers present a variety of new approaches to the representation problem. Unfortunately, there has not been time to incorporate discussions of these papers here. Instead, the interested reader is referred to the papers (cited in the Bibliography) written by the following authors: Biss, Chandra, Charniak, Coles, Fang, Feldman, Gibbons, Kuno, R. C. Moore, Norman, Pylyshyn, Rulifson, Rumelhart, Schank, Sirovich, and Wegbreit. (This list is, of course, not exhaustive.) In addition, Raphael and Robinson (1972) present a bibliography of 200 references on the subject of "computer semantics."

# Some Conversations with Computers

This section is devoted to letting the machines speak for themselves. However, some necessary comments are provided, in italics, by the present author.

## STUDENT

(Bobrow, 1964)

*(See the section on "Syntax" of this chapter.)*

(THE PROBLEM TO BE SOLVED IS)
(THE RUSSIAN ARMY HAS 6 TIMES AS MANY RESERVES IN A UNIT AS IT HAS UNIFORMED SOLDIERS. THE PAY FOR RESERVES EACH MONTH IS 50 DOLLARS TIMES THE NUMBER OF RESERVES IN THE UNIT, AND THE AMOUNT SPENT ON THE REGULAR ARMY EACH MONTH IS $150 TIMES THE NUMBER OF UNIFORMED SOLDIERS. THE SUM OF THIS LATTER AMOUNT AND THE PAY FOR RESERVES EACH MONTH EQUALS $45000. FIND THE NUMBER OF RESERVES IN A UNIT THE RUSSIAN ARMY HAS AND THE NUMBER OF UNIFORMED SOLDIERS IT HAS.)

(THE EQUATIONS TO BE SOLVED ARE)

(EQUAL G02532 (NUMBER OF UNIFORMED SOLDIERS (IT / PRO) (HAS / VERB)))

(EQUAL G02531 (NUMBER OF RESERVES IN UNIT RUSSIAN ARMY (HAS / VERB)))

(EQUAL (PLUS (AMOUNT SPENT ON REGULAR ARMY EACH MONTH) (PAY FOR RESERVES EACH MONTH)) (TIMES 45000 (DOLLARS)))

(EQUAL (AMOUNT SPENT ON REGULAR ARMY EACH MONTH) (TIMES (TIMES 150 DOLLARS)) (NUMBER OF UNIFORMED SOLDIERS)))

(EQUAL (PAY FOR RESERVES EACH MONTH) (TIMES (TIMES .50 (DOLLARS)) (NUMBER OF RESERVES IN UNIT)))

(EQUAL (NUMBER OF RESERVES IN UNIT RUSSIAN ARMY (HAS / VERB)) (TIMES 6 (NUMBER OF UNIFORMED SOLDIERS (IT / PRO) (HAS / VERB))))

THE EQUATIONS WERE INSUFFICIENT TO FIND A SOLUTION

(ASSUMING THAT)
((NUMBER OF UNIFORMED SOLDIERS) IS EQUAL TO NUMBER OF UNIFORMED SOLDIERS (IT / PRO) (HAS / VERB)))

(ASSUMING THAT)
((NUMBER OF RESERVES IN UNIT) IS EQUAL TO (NUMBER OF RESERVES IN UNIT RUSSIAN ARMY (HAS / VERB)))

(THE NUMBER OF RESERVES IN A UNIT THE RUSSIAN ARMY`HAS IS 800)

(THE NUMBER OF UNIFORMED SOLDIERS IT HAS IS 100)

---

## *STUDENT*

### (Bobrow, 1964)

(THE PROBLEM TO BE SOLVED IS)
(THE GROSS WEIGHT OF A SHIP IS 20000 TONS. IF ITS NET WEIGHT IS 15000 TONS, WHAT IS THE WEIGHT OF THE SHIPS CARGO Q.)

THE EQUATIONS WERE INSUFFICIENT TO FIND A SOLUTION

TRYING POSSIBLE IDIOMS

(DO YOU KNOW ANY MORE RELATIONSHIPS AMONG THESE VARIABLES)

(GROSS WEIGHT OF SHIP)

(TONS)

(ITS NET WEIGHT)

(WEIGHT OF SHIPS CARGO)

yes
TELL ME

(the weight of a ships cargo is the difference between the gross weight and the net weight)

THE EQUATIONS WERE INSUFFICIENT TO FIND A SOLUTION

(ASSUMING THAT)
((NET WEIGHT) IS EQUAL TO (ITS NET WEIGHT))

(ASSUMING THAT)
((GROSS WEIGHT) IS EQUAL TO (GROSS WEIGHT OF SHIP))

(THE WEIGHT OF THE SHIPS CARGO IS 5000 TONS)

---

## *CARPS*

### (Charniak, 1969)

*CARPS is a question-answerer designed to solve calculus "rate prob-lems," stated in English. After receiving the problem statement, it trans-forms the sentences into successive List structures and builds a tree structure (not shown here) to model the information they contain.*

(WATER IS FLOWING INTO A CONICAL FILTER AT THE RATE OF 15.0 CUBIC INCHES PER SECOND /. IF THE RADIUS OF THE BASE OF THE FILTER IS 5.0 INCHES AND THE ALTITUDE IS 10.0 INCHES /, FIND THE RATE AT WHICH THE WATER LEVEL IS RISING WHEN THE VOLUME IS 100.0 CUBIC INCHES /.)

$$\downarrow$$

(((WATER (FLOWING VERB) (INTO PREP) A (CONICAL ADJ) FILTER (AT PREP) (RATE RWORD) 15.0 (IN3 UNIT) PER (SEC UNIT)) (1)) ((IF THE RADIUS OF THE BASE OF THE FILTER (IS VERB) 5.0 (IN UNIT) AND THE ALTITUDE (IS VERB) 10.0 (IN UNIT), (FIND QWORD) (RATE RWORD) AT WHICH THE WATER LEVEL (RISING VERB) WHEN THE VOLUME (IS VERB) 100.0 (IN3 UNIT)) (2)))

$$\downarrow$$

(((WATER (FLOWING VERB) (INTO PREP) A (CONICAL ADJ) FILTER) (1)) ((WATER (FLOWING VERB) (AT PREP) (RATE RWORD) 15.0 (IN3 UNIT) PER (SEC UNIT)) (1)) ((THE RADIUS OF THE BASE OF THE FILTER (IS VERB) 5.0 (IN UNIT)) (2)) ((THE ALTITUDE (IS VERB) 10.0 (IN UNIT)) (2)) (((FIND QWORD) (RATE RWORD) ATWHICH THE WATER LEVEL (RISING VERB)) (2)) ((THE VOLUME (IS VERB) 100.0 (IN3 UNIT)) (2 WHEN)))

$$\downarrow$$

(THE EQUATION SET IS)
1 ((EQUAL (G0005) (DERIV (G0004 WATER FILTER))))
2 (EQUAL (QUOTIENT (TIMES 17.0 (TIMES (EXPT IN 3) TIM)) SEC) (TIMES (G0004 WATER FILTER) 0.33333300 PI (EXPT (RADIUS BASE WATER FILTER)2)))
3 (EQUAL (TIMES (RADIUS BASE WATER FILTER) (TIMES 12.0 IN)) (TIMES (G0004 WATER FILTER) (TIMES 5.0 IN))))

$$\downarrow$$

(THE ANSWER IS)

(TIMES .53132943 IN (EXPT SEC −1.0) (EXPT PI −0.33333332))

## HAPPINESS

### (Gelb, 1971)

(PLEASE INPUT PROBLEM)

(From a zorch containing a ferd and 3 brakky and 5 chartreuse werfels, 3 are drawn'. What is the probability that 2 are chartreuse and the other brakky?)

(FROM A ZORCH CONTAINING A FERD AND 3 BRAKKY AND 5 CHARTREUSE WERFELS, 3 ARE DRAWN. WHAT IS THE PROBABIL-ITY THAT 2 ARE CHARTREUSE AND THE OTHER BRAKKY?)

(ELAPSED TIME NOW 0.09299999 SECONDS)

(AFTER PRIMARY (IDIOMATIC) TRANSFORMATION, PROBLEM BE-COMES)

(FROM 1 ZORCH CONTAINING A FERD AND 3 BRAKKY AND 5 CHARTREUSE WERFELS, 3 ARE DRAWN. WHAT IS THE PROBABIL-ITY THAT 2 ARE CHARTREUSE AND 1 BRAKKY?)

(ELAPSED TIME NOW 0.384 SECONDS)

(AFTER SECONDARY (REARRANGING) TRANSFORMATION:)

(THE SIMPLE SENTENCES ARE)

((CONTAIN / ACTIVE (1 ZORCH) (4 FERD AND 3 BRAKKY AND 5
    CHARTREUSE WERFELS))
((DPAW / PASSIVE (3) NIL)

(THE QUESTION SENTENCE IS)
(WHAT IS THE PROBABILITY THAT 2 ARE CHARTREUSE AND 1 BRAKKY?)

(ELAPSED TIME NOW 0.227 SECONDS)

(AFTER TERTIARY (CASE-REDUCING) TRANSFORMATION, THE
    QUESTION SENTENCE BECOMES:)

(WHAT IS THE PROBABILITY OF GETTING 2 CHARTREUSE AND 1 BRAKKY?)

(ELAPSED TIME NOW 0.108 SECONDS)

(NO RECOGNIZABLE KEYWORDS FOUND IN PROBLEM.
ASSUMED SAMPLING PROBLEM.)
(AFTER SYNTAX ANALYSIS OF QUESTION SENTENCE:)

(ASSUMING THAT - 3 BRAKKY - MEANS - 3 BRAKKY WERFELS -)
(ASSUMING THAT - 4 FERD - MEANS - 4 FERD WERFELS -)
(ASSUMING THAT - 3 - MEANS - 3 WERFELS -)

(D ESCLIST FOR THIS PROBLEM CONTAINS:)

PROBLEMTYPE- SAMPLING
POPULATION- ((5 CHARTREUSE WERFELS) (3 BRAKKY WERFELS)
(4 FERD WERFELS)) AS OBJECTS

SAMPLESIZE/TRIALS- 3
SIMPLE EVENTS- (G12503 G12502)
G12503- ((WERFEL) ( BRAKKY) (1) NIL NIL NIL)
G12502- ((WERFEL) (CHARTREUSE) (2) NIL NIL NIL)
COMPOUND EVENT STRUCTURE- (AND (OR G12502) (OR G12503))
REPLACEMENT INVOLVED? NO

(ELAPSED TIME NOW 0.715 SECONDS)

(FIRST LEVEL SOLUTION TO PROBLEM IS)

(PLUSF (PROB (QUOTE (G12502 G12503))))

(TIME FOR EVALUATION WAS 0.032 SECONDS)

(SECOND LEVEL SOLUTION TO PROBLEM IS)

(PLUSFN (PR (QUOTE (C12502 C12503))))

(TIME FOR EVALUATION WAS 0.034 SECONDS)

(THIRD LEVEL SOLUTION TO PROBLEM IS)

(PLUSFRAC (SIMPLIFYFRAC (LIST (COMBINL 5 2) (COMBINL 3 1)
(COMBINL 4 0))
(COMBINL 12 3)))

(TIME FOR EVALUATION was 0.14 SECONDS)

(FOURTH LEVEL SOLUTION TO PROBLEM IS)

3/22 (OR 0.1363636)

(ELAPSED TIME NOW 0.134 SECONDS)

(TOTAL TIME FOR PROBLEM SOLUTION WAS 1.882 SECONDS)

## *SIR*

### (Raphael, 1964)

*(See the preceding section of this chapter.)*

(THE NEXT SENTENCE IS . . .)
(MAX IS AN IBM-7094)

(THE FUNCTION USED IS . . .)
SETR-SELECT
((UNIQUE . MAX) (GENERIC . IBM-7094))
(THE REPLY . . .)
(THE SUB-FUNCTION USED IS . . .)
SETRS
(MAX IBM-7094)
(ITS REPLY . . .)
(I UNDERSTAND THE ELEMENTS RELATION BETWEEN MAX AND
   IBM-7094)
(I UNDERSTAND THE MEMBER RELATION BETWEEN IBM-7094 AND
   MAX)

(THE NEXT SENTENCE IS . . .)
(AN IBM-7094 IS A COMPUTER)

(THE FUNCTION USED IS . . .)
SETR-SELECT
((GENERIC . IBM-7094) (GENERIC . COMPUTER))
(THE REPLY . . .)
(THE SUB-FUNCTION USED IS . . .)
SETR

(IBM-7094 COMPUTER)
(ITS REPLY . . .)
(I UNDERSTAND THE SUPERSET RELATION BETWEEN COMPUTER
   AND IBM-7094)
(I UNDERSTAND THE SUBSET RELATION BETWEEN IBM-7094 AND
   COMPUTER)

(THE NEXT SENTENCE IS . . .)
(IS MAX A COMPUTER Q)

(THE FUNCTION USED IS . . .)
SETRQ-SELECT
((UNIQUE . MAX) (GENERIC . COMPUTER))
(THE REPLY . . .)
(THE SUB-FUNCTION USED IS . . .)
SETRSQ

(MAX COMPUTER)
(ITS REPLY . . .)
YES

(THE NEXT SENTENCE IS . . .)
(THE BOY IS AN MIT-STUDENT)
(THE FUNCTION USED IS . . .)
SETR-SELECT
((SPECIFIC . BOY) (GENERIC . MIT-STUDENT))
(THE REPLY . . .)
(THE SUB-FUNCTION USED IS . . .)
SLIRSI
(BOY MIT-STUDENT)
(ITS REPLY . . .)
(G02840 IS A BOY)
(I UNDERSTAND THE ELEMENTS RELATION BETWEEN G02840 AND
   BOY)
(I UNDERSTAND THE MEMBER RELATION BETWEEN BOY AND
   G02840)
(I UNDERSTAND THE ELEMENTS RELATION BETWEEN G02840 AND
   MIT-STUDENT)
(I UNDERSTAND THE MEMBER RELATION BETWEEN MIT-STUDENT
   AND G02840)

(THE NEXT SENTENCE IS . . .)
(EVERY MIT-STUDENT IS A BRIGHT-PERSON)

(THE FUNCTION USED IS . . .)
SETR-SELECT
((GENERIC . MIT-STUDENT) (GENERIC . BRIGHT-PERSON))
(THE REPLY . . .)
(THE SUB-FUNCTION USED IS . . .)
SETR
(MIT-STUDENT BRIGHT-PERSON)
(ITS REPLY . . .)
(I UNDERSTAND THE SUPERSET RELATION BETWEEN BRIGHT-
   PERSON AND MIT-STUDENT)
(I UNDERSTAND THE SUBSET RELATION BETWEEN MIT-STUDENT
   AND BRIGHT-PERSON)

(THE NEXT SENTENCE IS . . .)
(IS THE BOY A BRIGHT-PERSON Q)

(THE FUNCTION USED IS . . .)
SETRQ-SELECT
((SPECIFIC . BOY) (GENERIC . BRIGHT-PERSON))
(THE REPLY . . .)
(THE SUB-FUNCTION USED IS . . .)
SETRSIQ
(BOY BRIGHT-PERSON)
(ITS REPLY . . .)
YES

(THE NEXT SENTENCE IS . . .)
(JOHN IS A BOY)

(THE FUNCTION USED IS . . .)
SETR-SELECT
((UNIQUE . JOHN) (GENERIC . BOY))
(THE REPLY . . .)
(THE SUB-FUNCTION USED IS . . .)
SETRS

(JOHN BOY)
(ITS REPLY . . .)
(I UNDERSTAND THE ELEMENTS RELATION BETWEEN JOHN AND
  BOY)
(I UNDERSTAND THE MEMBER RELATION BETWEEN BOY AND
  JOHN)

(THE NEXT SENTENCE IS . . .)
(IS THE BOY A BRIGHT-PERSON Q)

(THE FUNCTION USED IS . . .)
SETRQ-SELECT
((SPECIFIC . BOY) (GENERIC . BRIGHT-PERSON))
(THE REPLY . . .)
(THE SUB-FUNCTION USED IS . . .)
SETRSIQ
(BOY BRIGHT-PERSON)
(ITS REPLY . . .)
(WHICH BOY. . . (G02840 JOHN))

## SEMANTIC MEMORY

### (Quillian, 1966)

*(See the preceding section of this chapter.)*

Example 1.   Compare: CRY, COMFORT
     A. Intersect: SAD
         (1)   CRY2 IS AMONG OTHER THINGS TO MAKE A SAD SOUND.*
         (2)   TO COMFORT3 CAN BE TO MAKE2 SOMETHING LESS2 SAD.

Example 2.   Compare: PLANT, LIVE
     A. 1st Intersect: LIVE
         (1)   PLANT IS A LIVE STRUCTURE.
     B. 2nd Intersect: LIVE
         (1)   PLANT IS STRUCTURE WHICH GET3-FOOD FROM AIR. THIS FOOD IS THING WHICH BEING2 HAS-TO TAKE INTO ITSELF TO7 KEEP LIVE.

Example 3.   Compare: PLANT, MAN
     A. 1st Intersect: ANIMAL
         (1)   PLANT IS NOT A ANIMAL STRUCTURE.
         (2)   MAN IS ANIMAL.
     B. 2nd Intersect: PERSON
         (1)   TO PLANT3 IS FOR A PERSON SOMEONE TO PUT SOMETHING INTO EARTH.
         (2)   MAN3 IS PERSON.

Example 4.   COMPARE: PLANT, INDUSTRY
     A. 1st Intersect: INDUSTRY
         (1)   PLANT2 IS APPARATUS WHICH PERSON USE FOR 5 PROCESS IN INDUSTRY.

Example 5.   Compare: EARTH, LIVE
     A. 1st Intersect: ANIMAL
         (1)   EARTH IS PLANET OF7 ANIMAL.
         (2)   TO LIVE IS TO HAVE EXISTENCE AS7 ANIMAL.

Example 6.   Compare: FRIEND, COMFORT
     A. 1st Intersect: PERSON
         (1)   FRIEND IS PERSON.
         (2)   COMFORT CAN BE WORD TO4 PERSON.

Example 7.   Compare: FIRE, BURN
     A. 1st Intersect: BURN
         (1)   FIRE IS CONDITION WHICH BURN.
     B. 2nd Intersect: FIRE
         (1)   TO BURN2 CAN BE TO DESTROY2 SOMETHING BY4 FIRE.
     C. 3rd Intersect: BURN
         (1)   FIRE IS A FLAME CONDITION. THIS FLAME CAN BE A GAS TONGUE4. THIS GAS IS GAS, WHICH BURN.

Example 8.   Compare: BUSINESS, COMFORT
    A. 1st Intersect: PERSON
      (1)   BUSINESS IS ACT3 WHICH PERSON DO.
      (2)   COMFORT2 IS CONDITION3 WHICH PERSON HAVE NEED4.
    B. 2nd Intersect: PERSON
      (1)   BUSINESS5 IS ACT3 WHICH PERSON DO.
      (2)   COMFORT CAN BE WORD TO4 PERSON.

Example 9.   Compare: MAN, BUSINESS
    A. 1st Intersect: PERSON
      (1)   MAN3 IS PERSON.
      (2)   BUSINESS CAN BE ACTIVITY WHICH PERSON MUST DO WORK2.
    B. 2nd Intersect: GROUP
      (1)   MAN2 IS MAN AS9 GROUP.
      (2)   BUSINESS2 IS QUESTION3 FOR ATTENTION OF GROUP.

Example 10.  Compare: MAN, LIVE
    A. 1st Intersect: ANIMAL
      (1)   MAN IS ANIMAL.
      (2)   TO LIVE IS TO HAVE EXISTENCE AS7 ANIMAL.
    B. 2nd Intersect: LIVE
      (1)   MAN IS A LIVE + BEING2.

## PARRY

### (Colby, Weber, Hilf, 1971)

*PARRY is a program designed to simulate the conversational behavior of a paranoid human being. The conversation below was produced with the program operating in its "weak version" of paranoia. Sentences labeled "I" are produced by the human interviewer; those labeled "O" by the program.*

(1)  I—HOW LONG HAVE YOU BEEN IN THE HOSPITAL?
O—ABOUT A WEEK.

(2)  I—ARE YOU DEPRESSED?
O—NO.

(3)  I—WERE YOU TAKING DRUGS?
O—NO.

(4)  I—WHAT IS YOUR PROBLEM?
O—PEOPLE MAKE ME NERVOUS.

(5)  I—HOW DO THEY DO THAT?
O—THEY STARE AT ME.

(6)  I—WHY?
O—MAYBE THEY DON'T LIKE THE WAY I LOOK OR SOME-THING.

(7)  I—DO YOU LOOK UNUSUAL IN SOME WAY?
O—WHY ARE YOU INTERESTED IN MY LOOKS?

(8)  I—BECAUSE YOU SAID PEOPLE STARE AT YOU.
O—WELL, SOMETIMES.

(9)  I—HOW DO YOU GET ALONG WITH PEOPLE?
O—I DON'T COMMUNICATE VERY MUCH WITH THE OTHERS.

(10)  I—WHY NOT?
O—THEY DON'T REALIZE WHAT'S GOING ON MOST OF THE TIME.

(11)  I—WHAT IS GOING ON?
O—DO YOU KNOW HOW THE UNDERWORLD OPERATES?

## SHRDLU

### (Winograd, 1971)

*SHRDLU carries out actions in, and answers questions about, a simple world (containing a table, hand, and box, and several blocks and pyramids), stored as a data structure inside the computer and displayed on a television screen for the person talking to it. SHRDLU requires from 5 to 20 seconds to analyze and respond to each sentence presented to it (lower-case lines are the person's; upper-case are the computer's), and its responses (actions and answers) are displayed in real time on the screen. (See Fig. 7–12.) The nuances of this dialog are explained in the book by Winograd.*

1. pick up a big red block.
OK.

2. grasp the pyramid.
I DON'T UNDERSTAND WHICH PYRAMID YOU MEAN.

3. find a block which is taller than the one you are holding and put it into the box.
BY "IT," I ASSUME YOU MEAN THE BLOCK WHICH IS TALLER THAN THE ONE I AM HOLDING.
   OK.

4. what does the box contain?
   THE BLUE PYRAMID AND THE BLUE BLOCK.

5. what is the pyramid supported by?
THE BOX.

6. how many blocks are not in the box?
FOUR OF THEM.

7. is at least one of them narrower than the one which i told you to pick up?
   YES, THE RED CUBE.

8. is it supported?
YES, BY THE TABLE.

9. can the table pick up blocks?
NO.

10. can a pyramid be supported by a block?
YES.

11. can a pyramid support a pyramid?
I DON'T KNOW.

12. stack up two pyramids.
I CAN'T.

13. the blue pyramid is nice.
   I UNDERSTAND.

14. i like blocks which are not red, but i don't like anything which supports a pyramid.
   I UNDERSTAND.

15. do i like the box?
   NO.

16. do i like anything in the box?
   YES, TWO THINGS: THE BLUE BLOCK AND THE BLUE PYRAMID.

17. will you please stack up both of the red blocks and either a green cube or a pyramid?
   OK.

18. which cube is sitting on the table?
   THE LARGE GREEN ONE WHICH SUPPORTS THE RED PYRAMID.

19. is there a large block behind a pyramid?
   YES, THREE OF THEM: A LARGE RED ONE, A LARGE GREEN CUBE AND THE BLUE ONE.

20. put a small one onto the green cube which supports a pyramid.
   OK.

21. put the littlest pyramid on top of it.
   OK.

22. does the shortest thing the tallest pyramid's support supports support anything green?
   YES, THE GREEN PYRAMID.

23. what color is it?
   BY "IT," I ASSUME YOU MEAN THE SHORTEST THING THE TALLEST PYRAMID'S SUPPORT SUPPORTS.
      RED

24. how many things are on top of green cubes?
   I'M NOT SURE WHAT YOU MEAN BY "ON TOP OF" IN THE PHRASE "ON TOP OF GREEN CUBES."
   DO YOU MEAN:
   1—DIRECTLY ON THE SURFACE
   2—ANYWHERE ON TOP OF?
      2
   THREE OF THEM.

25. had you touched any pyramid before you put the green one on the little cube?
   YES, THE GREEN ONE.

26. when did you pick it up?
   WHILE I WAS STACKING UP THE RED CUBE, A LARGE RED BLOCK AND A LARGE GREEN CUBE.

27. why
   TO GET RID OF IT.

28. why did you do that?
   TO CLEAN OFF THE RED CUBE.

29. why did you clear off that cube?
   TO PUT IT ON A LARGE GREEN CUBE.

30. why did you do that?
   TO STACK UP THE RED CUBE, A LARGE RED BLOCK AND A LARGE GREEN CUBE.

31. why did you do that?
   BECAUSE YOU ASKED ME TO.

32. how did you do it?
   BY PUTTING A LARGE RED BLOCK ON THE TABLE; THEN LETTING GO OF IT; THEN PUTTING A LARGE GREEN CUBE ON IT; THEN LETTING GO OF THAT CUBE; THEN PUTTING THE RED CUBE ON THAT CUBE; THEN LETTING GO OF THAT CUBE.

33. how many objects did you touch while you were doing it?
   FOUR OF THEM.

34. what did the red cube support before you started to clean it off?
   THE GREEN PYRAMID.

35. there were five blocks to the left of the box then.
   NO, ONLY FOUR OF THEM: THE RED CUBE, TWO LARGE GREEN CUBES AND A LARGE RED BLOCK.

36. put the blue pyramid on the block in the box.
   OK.

37. is there anything which is bigger than every pyramid but is not as wide as the thing that supports it?
   YES, THE BLUE BLOCK.

39. does a steeple
   SORRY I DON'T KNOW THE WORD "STEEPLE." PLEASE TYPE TWO SPACES.

40. a "steeple" is a stack which contains two green cubes and a pyramid.
   I UNDERSTAND.

41. are there any steeples now?
   NO.

42. build one.
   OK.

43. call the biggest block "superblock."
   OK.

44. have you picked up superblock since we began?
   YES.

45. why did you drop it?
BY "IT," I ASSUME YOU MEAN THE SUPERBLOCK.
   I CAN'T EXPLAIN A NON-EXISTENT EVENT.

46. is there anything to the right of the red pyramid?
   YES, FOUR THINGS: THE BOX, THE RED CUBE, THE BLUE PYR-
AMID AND THE BLOCK NAMED SUPERBLOCK.

47. thank you.
   YOU'RE WELCOME!

---

*Does the shortest thing the tallest pyramid's
support supports support anything green?*

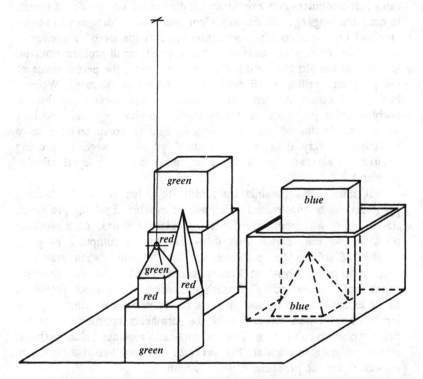

Figure 7–12. SHRDLU. (Winograd, 1971, reprinted with permission.)

# LANGUAGE AND PERCEPTION

## Networks of Question-Answering Programs

The remainder of this chapter is devoted to discussing some general topics of "semantic information processing" that have not been covered explicitly elsewhere in this book. The subjects discussed include networks of question answerers and protocol analyzers, grammatical inference and pattern recognition, communication, teaching, and learning, and the "self-knowledge" of intelligent machines. Mostly, we shall have to content ourselves with a few general observations and some pointers to the literature. The topics discussed represent areas of future study that have not yet been tamed into programs by AI researchers.

The preceding pages have shown that question-answering (and, in general, language-understanding) programs can do some pretty amazing things. On the one hand, the performance of Winograd's program indicates that computers may eventually handle the full complexity of syntax in a natural language like English. Computers can be designed to recognize and use word endings and context and "understand" a conversation, at least when it is concerned with a relatively small problem domain, like that of the SHRDLU world. On the other hand, the performance of the programs written by Bobrow, Gelb, Charniak, Ramani, Weizenbaum, and others indicates that computers can successfully handle fairly complex problems (involving algebra, probability, and calculus) when stated in limited subsets of English. Finally, computers can solve a variety of very difficult mathematical problems, such as proving theorems in abstract algebra or solving rather difficult integral calculus problems.

It thus seems possible that, ultimately, language-understanding programs will be constructed which will be capable of solving problems, stated in English, from very difficult problem domains. As a working principle, we may expect that if we can find a computer program capable of solving the problems in some domain, when stated in some appropriate formalism, then we can also find a computer program capable of "understanding" English statements of the same problems, to the extent that the second program can translate such English problem statements into statements of the formalism appropriate for the first program to solve them. The two programs together (plus, perhaps, a third program to translate the answers) can function as a "question answerer" for the problems of that domain.

Given a set of English sentences (actually, a "structure" of such sentences as determined by the conversation), some of which are questions, a question answerer should be able to

1. Answer questions, in English.
2. Make functional statements like "This will take a little processing" or "Sorry, I can't answer question X."
3. In general, ask questions in English (and, if necessary, justify them on the basis of their relevance to finding answers).
4. Ultimately, make general statements that are neither questions nor answers, but simply "interesting" observations.

Since both input and output for a question-answering program are sets of English statements, it is natural to think of "networks" of question answerers. It may be desirable to use networks of question-answerers in the construction of large, "general" question answerers (GQA's). Such a network might have the following capabilities:

1. It could be "self-organizing." At each moment the GQA could make use of a different configuration of "specialized" question answerers, each one either asking questions or answering questions (or making other statements, etc.) posed by other question answerers or by the user of the system.
2. It is conceivable that it could simulate a "synergetic" or "gestalt" effect. This means that GQA as a whole could answer some questions that its parts could not answer. Of course the whole could not ask questions that could not be asked by at least one of its parts. The "synergetic" ability of the GQA depends on the ability of each of its specialized question answerers to ask questions it may not be able to answer. Question asking may be considered an aspect of problem reduction: The simplest type of GQA corresponds to the parallel implementation of a single problem-reduction problem solver.
3. The difficulties involved in adding to a GQA would be minimized by the use of some common language (not necessarily a natural language) for the communication of problems and answers between components of the GQA (note 7–12).
4. If it is found that several question answerers are, through cooperating in a GQA, able to achieve solutions to a domain of problems that none of them could solve alone, then it may be desirable to have another kind of program (called a *protocol analyzer*) for the purpose of analyzing the conversations and other computations they produce in solving these prob-

lems, and which could develop a new, specialized question answerer to simulate their ability (although at a faster speed) to solve the problems of that domain.

The idea of "protocol analysis" was first developed by Newell and Simon (1963) as a process that AI researchers should perform on the conversational problem-solving behavior of people (specifically, individuals) as a guide to the development of computer programs capable of simulating human problem-solving behavior. Some relevant papers are Waterman and Newell (1971), Hewitt ("procedural abstraction," 1968 et seq.) and Manna and Waldinger (1971). Norman (1972) presents an extensive discussion on the nature of human question-answering processes. Our discussion of the GQA concept (which is intended only as a thought-experiment) is continued in the later section entitled "Communication, Teaching, and Learning."

# Pattern Recognition and Grammatical Inference

An interesting question for the reader to investigate is, "What will happen if we attempt to train a pattern recognizer based on statistical decision theory (see Duda and Hart, 1973) to recognize the sunflower pattern?" (See Figs. 2–1 and 5–2.) One way in which we might train the pattern recognizer is as follows: A series of samples will be presented to the pattern recognizer, each sample corresponding to the coordinates (say, Cartesian) of a point in the plane. After each sample is presented, the pattern recognizer is required to classify it either as belonging or not belonging to the "sunflower pattern." After it makes its classification, it is told the actual classification of the sample and must modify its features and probability functions accordingly. Then the next sample is presented, etc.[12]

So far as the author knows, there is no statistically based pattern recognizer that would, after the presentation of only a finite number of samples, be able to recognize successfully the sunflower pattern (i.e., be able to classify correctly any sample one might then choose to show it). The reason for this is that the points (dots) that belong to the pattern satisfy neither of the requirements typically specified for the point sets that such recognizers are designed to learn to classify. The points of the sunflower pattern are not a continuous set, nor are they a bounded set (one cannot draw a simple, closed curve of finite

---

[12] In an actual experiment it would be desirable to generalize the sunflower pattern to include as pattern examples all points within some small radius of the "true" pattern examples.

length that will enclose them all). Currently developed techniques for the generation and selection of features and the estimation of density functions are probably insufficient to enable the statistically based pattern recognizer to do anything more than "learn" to classify correctly those samples it has already been shown. Since there are an infinite number of points belonging to the sunflower pattern, there will always be an infinite number of pattern samples it will not have learned to recognize.[13]

Yet it seems quite plausible that a truly intelligent pattern recognizer would be able to learn to recognize the sunflower pattern. A person observing Fig. 2-1 would have little difficulty in estimating where new dots could be added, and it is conceivable that he could eventually develop an accurate computational procedure for correctly classifying any sample that he might be shown. Of course this ability on his part might be due largely to the preprocessing ability of his visual system (which would correspond to giving the pattern recognizer a collection of useful features to detect). However, it still seems plausible that, even without the visual preprocessing ability, a human being could learn to recognize the pattern. Intuitively, the sunflower pattern forms a relatively simple "structure," in which each pattern sample bears a fairly simple relationship to certain other pattern samples; the existence of this relationship makes it possible for one to generate as many samples of the sunflower pattern as desired, and also makes it possible for one to decide whether or not a given sample is or is not a pattern sample. People are extremely talented at learning to recognize structures, whereas statistically-based pattern recognizers are not.

We don't have far to look to find another case of a pattern in which structural relationships play an important part. Namely, a natural language like English may itself be considered to be a pattern, the pattern examples of which are sentences, phrases, and words. The language itself may also be said to be a structure, insofar as there are relationships that exist between its pattern examples (e.g., $A$ is-defined-to-be $Z$). Again, when we normally use the English language, we form "conversations," which are also essentially structures of these pattern examples. Any formalization for the semantics of English would in effect denote a set (probably infinite) of "meaningful" conversations, and thus would be a description for the pattern whose pattern examples are "meaningful" conversations. Moreover, sentences, phrases, and words are themselves structures. There is thus a structural aspect to the pattern which is the

---

[13] The author has checked the plausibility of this argument with Richard Duda, and wishes to thank him for an enlightening discussion on the topic.

English language as a whole, to the pattern of its use in making conversations, and to its "elementary" pattern examples (its sentences and words). Finally, and just as important, there are aspects of structure and pattern to the "meanings" that sentences and conversations may have. In general, we may think of the "meaning" of a sentence as being a collection of situations, each of which the sentence possibly denotes as being the case (an unambiguous sentence would denote only one situation). This collection of situations may be considered to be a pattern, while the situations themselves will in general have a structural aspect. We may think of a natural language as being a pattern for the description of patterns.

The fact that there are structural relationships underlying many real-world patterns and their pattern examples, together with the fact that such relationships are important in natural languages, has led a number of investigators to suggest that linguistic techniques should be used by pattern perceiving systems (other investigators have suggested that language-understanding programs should make use of pattern perception techniques—see McConlogue and Simmons, 1965). Research in this area has concentrated in two directions: First, some researchers have attempted to find languages and grammars that could be used to describe and recognize visual patterns; see Narasimhan (1964), Evans (1971), Shaw (1968), Kirsch (1964), Winston (1970), Watanabe (1969, 1971), Banerji (1971), Pfaltz and Rosenfeld (1969), Uhr (1971), and Morofsky and Wong (1971). Second, other researchers have investigated the ability of computer programs to "learn" to recognize patterns corresponding to artificial languages (i.e., sets of strings) by inferring grammars for them; this is known as the *grammatical inference* paradigm for pattern recognition; See Crespi-Reghizzi (1971); Feldman (1967); Horning (1969). The first approach will not be discussed in detail in this section except to note that Winston's work was described in Chapter 5. However, much of this work is relevant to the grammatical inference paradigm.

A *grammatical inference problem* has the form: "Given two sets of strings, $A$ and $B$, which are mutually disjoint (they do not have a common element), find a grammar $G$ such that the language $L(G)$ it generates contains as sentences all the strings of $A$ but none of the strings of $B$; $L(G)$ may, of course, contain other sentences besides those that belong to $A$" (note 7–13). A more general grammatical inference problem might ask us to find a set of such grammars. It should be noted that, as stated, the grammatical inference problem is trivially solvable, for any appropriate sets $A$ and $B$, because we can always specify that $G$ shall be the "enumerative" grammar that contains exactly those production rules of the form $S \to a$, where $a$ is any string of $A$. Moreover, there are always an infinite number of grammars that could be put forward as a solution to a given grammatical inference problem. However, it is possible to specify a number of different conditions one can add to the statement of a grammatical inference problem that will make finding a solution more relevant to pattern perception. Thus, we might specify that any "solution grammar" $G$ for a grammatical inference problem shall generate a language $L(G)$ with an infinite number of sentences, unless the problem explicitly states that $L(G)$ is to be finite (note 7–13). This condition insures that solution grammars will exhibit "perceptual generalization." Or, if we can find a suitable way of measuring the "complexity" of arbitrary grammars, we can specify that a solution grammar for a grammatical inference problem shall be any of the least complex grammars that satisfy the other conditions of the problem. Finally, following Chaitin (1966, 1969) and Martin-Lof (1966), we may decide that some sets $A$ are to be regarded as essentially "pattern-less" or "random" if there are no grammars for them—that is, no $G$ such that $A \subseteq L(G)$—which are less complex than their enumerative grammars.

The grammatical inference paradigm for pattern recognition, then, consists in seeing the task of a pattern recognizer to be that of inferring a grammar that generates those samples which are pattern examples of the pattern it is learning to classify, but which does not generate those samples that are not pattern examples. It is clear that this paradigm is a good one for those patterns whose pattern examples are structures with a linear, stringlike nature. However, to be useful as a paradigm for pattern recognition in general, we would probably desire that our notions of "language" and "grammar" be extended to include languages whose sentences are nonstringlike structures. That is, we would like to formalize a notion of "general language" and "general grammar" in which sentences can be arbitrary structures of symbols, and grammars can be flexible procedures for building structures. It is still not clear what a good, general formalization for "structure" should be like. Indeed, the patterns existing in different environments will often be most easily characterized by using different kinds of structures; among the best "general language" formalizations at the moment are the "web languages" of Pfaltz and Rosenfeld (1969), the "hierarchical graph languages" investigated by Pratt (1969 et seq.) and Winston (1970), and the hierarchical List structures and recursively defined pattern rules investigated by Morofsky and Wong (1971) and Hewitt (1968 et seq.). A good research project would be to investigate whether these concepts can be extended to include "continuous structures" and "changing structures" (or "processes"). This subject is mentioned again in Chapter 8. Finally, it should be mentioned that there is as yet no clearly adequate definition for the concept of "complexity," as it applies to programs, sentences, grammars, patterns, or structures in general. In addition to the papers on grammatical inference cited above, the reader should refer to Arbib and Blum (1965), Blum (1967), Buneman (1970), Cleave (1963), Cobham (1964), Hartmanis and Stearns (1965), Loveland (1969), Mowshowitz (1967), and van Emden (1970, 1971).

## Communication, Teaching, and Learning

McCarthy (1968), Minsky (1968a,b, 1970), Hewitt (1968 et seq.), and Winograd (1971, 1972), among others, presented an extensive array of commentary on the relationships between communication, teaching, and learning. The following passage from McCarthy (1968) is particularly insightful:

> If one wants a machine to be able to discover an abstraction, it seems most likely that the machine must be able to represent this abstraction in some relatively simple way.
>
> There is one known way of making a machine capable of learning arbitrary behavior, and thus to anticipate every kind of behavior: This is to make it possible for the machine to simulate arbitrary behaviors and try them out. These behaviors may be represented either by nerve nets [Minsky, 1962], by Turing machines [McCarthy, 1956], or by calculator programs [Friedberg, 1958, 1959] . . .
>
> In our opinion, a system which is to evolve intelligence of human order should have at least the following features:
>
> 1. All behaviors must be representable in the system. Therefore, the system should either be able to construct arbitrary automata or to program in some general-purpose programming language.
> 2. Interesting changes in behavior must be expressible in a simple way.
> 3. All aspects of behavior except the most routine should be improvable. In particular, the improving mechanism should be improvable.
> 4. The machine must have or evolve concepts of partial success because on difficult problems decisive successes or failures come too infrequently.
> 5. The system must be able to create subroutines which can be included in procedures as units . . .
>
> . . . *We base ourselves on the idea that in order for a program to be capable of learning something it must first be capable of being told it* (pp. 404–405).

In the present author's opinion the final statement of the above passage will probably turn out to be one of the basic principles of the "Theory of Artificial Intelligence," should such a theory ever be established; at the moment it certainly amounts to a guideline that underlies a great deal of research. The ability to understand an abstraction (carry out a procedure described by a program) is effectively essential to the ability to create the abstraction. The more simply the abstraction can be stated to a machine, the more likely we can make the machine find the abstraction by itself. For machines to demonstrate really intelligent, effective learning, it will be necessary to give them a language capability for a general-purpose programming language that facilitates the description of procedures (abstractions, behaviors, "aptitudes") which are appropriate for their problem domains.

As was suggested in the discussion on networks of question answerers, the use of an appropriate language and communication process may enable us to design large problem solvers with an ability to solve problems greater than that of the individual components designed explicitly. The performance of the large problem solver may provide a "protocol" that it can use in the design of new individual components. The effect of a new individual component (specialized question answerer) will be to make it possible for the large machine to solve a certain class of problems more efficiently. As a consequence of its increased efficiency at solving this class of problems, the large machine may then be able to solve other problems, perhaps ones that it could not previously solve at all.

It may be possible for a machine to learn to solve problems more and more efficiently and, eventually, to "bootstrap" itself into an ability to solve problems it could not previously solve.

The idea of "self-improving" artificial intelligence is not yet completely formalized. (Indeed, we may speculate that there is no complete formalization, by definition; see McCarthy's condition 3 above). The discussion of this topic will be taken up again in Chapter 8, where evolutionary programs will be treated in more detail. The reader should not confuse the discussion of self-improvement in this chapter with other theories discussed in Chapter 8 (e.g., Myhill and Holland). For a good analogy to the mechanism currently being discussed, consider the process by which a person learns to perform a new physical task (e.g., playing a guitar): The proficient performance of the complete task (e.g., playing a song) requires a large set of proficient performances of smaller tasks (playing riffs, bridges, estimating notes before they are struck, coordinating hands, eyes, and voice, etc.). The task is learnable because there exists a *training sequence* of simpler tasks that a person can learn to perform efficiently. He begins by "thinking about" the simplest tasks of the sequence, and performing them slowly; with practice he is able to translate his performance of the simple tasks into "habits" and to begin "thinking about" the harder tasks of the sequence. Many authors have stressed the importance of "training sequences" in human and machine learning.

As a conclusion to this chapter, the student is invited to read Winograd's (1971) discussion of "teaching, telling, and learning" and, in particular, his description of the *hierarchy of knowledge* that an intelligent machine should be expected to possess; this hierarchy corresponds essentially to the hierarchy of languages (from its machine language to, perhaps, a natural language) in which it can accept information. Also suggested are Minsky's (1968, 1970) discussions on the nature of the knowledge that an intelligent machine can possess (in particular, its "self-models"). One of the major ways in which an intelligent computer can be different from a human being is that the computer can "know" exactly what kind of machine it is. The intelligent computer could read through the listings of its own programs and the specifications of its physical construction as well, whereas the human being seems unable (at least, consciously) to perform the corresponding tasks for himself. It will be interesting to see what kinds of "self-improvement" this will make possible for machines. In fact, it may eventually be of the utmost importance for AI researchers to understand the phenomenon of machine self-knowledge and its relationship to the "psychological" behaviors intelligent machines might demonstrate. How can we guarantee that an artificial intelligence will "like" the nature of its existence? (See note 7–14.)

## NOTES

**7–1.** Throughout, this chapter adopts the idea that "understanding," whether human or mechanical, is a process that involves "model making." However, no exploration is made of the ramifications of this thesis as it regards human understanding very deeply. So, the student should be advised that it is not the only idea currently being considered by psychologists. Indeed, there has been a sizable school of psychologists maintaining that explanations of human understanding, intelligence, etc., should be "neutral" and "behavioristic" and not "mentalistic," that the ideas of "models," and "concepts," and "ideas" should be avoided; and that a testable psychological theory should not make use of them. One can understand their reluctance to admit these concepts—which have been the Maypoles for circular philosophical arguments since time immemorial—into their studies and laboratories. Still, the mentalistic approach can be used quite profitably by computer scientists. And, since people find it easier (in English) to talk about "understanding" if they use words like "idea" and "concept," it will make our exposition clearer to take this approach. An excellent discussion on "matter, mind, and models" is given by Minsky (1968).

**7-2.** The paragraph citing this note glosses over certain relatively minor points: (1) In some languages, words (and phrases) are written as *ideograms;* that is, they are represented "pictorially." They still have a "structural nature," but it is not that of a sequence or string. (2) Besides spoken and "written" languages there are also human sign languages, "whistle" languages (used on the Canary Islands), and Braille systems. (3) Some other organism-level languages do have a structural (in particular, a stringlike) nature; for example, bees communicate information about food, using sentences that consist of fairly complex sequences of body motions (a sort of "dance"). (4) We have not discussed the use of punctuation in written sentences. (5) The phonemes in spoken sentences usually do not separate precisely into words; rather, people tend to run some words together.

**7-3.** This concept of a universal grammar is echoed in at least three respects. First, our societies have also developed musical forms that show great similarities from one culture to another, so much so that music is itself often called a "universal language." Second, C. S. Pierce was led by his investigation of the history of natural science to suggest that man has a remarkable ability to formulate successful hypotheses about the physical universe, considering the huge number of different explanations that could be advanced for a given phenomenon, and from this he conjectured that we have an innate tendency to perceive "simplicity" (infer grammars; see the fourth section of this chapter) in ways that fortunately lie very close to the actual structure of the laws of nature. Finally, Leibniz long ago proposed to design a "universal language" that would be a calculus for determining all the truths of philosophy and the natural sciences.

**7-4.** There is still very little known about the linguistic abilities of dolphins (see Lilly, 1968 et seq.). It should be noted that the size and complexity of the dolphin brain appear to be comparable to that of the human brain. Dolphins seem to be able to communicate with each other, using as signals rapid sequences of high-pitched sounds. Furthermore, a dolphin has *two* sets of vocal chords, which it can evidently use independently of each other. It is not known whether their language has any of the aspects of generality (extensibility, self-reference) possessed by human languages. However, efforts are being made to teach dolphins the human "whistle language" mentioned above.

**7-5.** There are "languages" which are not type 0 (i.e., do not have a phrase-structure grammar; see Chaitin, 1966, 1969). It is certain that these languages cannot be used as "programming languages" in the sense of $L'$, and it is very doubtful whether they could be meaningful as "machine languages" in the sense of $L$.

**7–6.** In essence, there are types of "information" not considered explicitly in the Shannon and Weaver (1949) theory of communication (see Chapter 2); in addition to "occurrence information" that a sentence carries because it is transmitted and received while other sentences are not, a sentence also carries "syntactic information" with respect to a grammar for the language to which it belongs, and "semantic information" about whatever it describes.

**7–7.** It is almost always desirable that each sentence in $L'$ have exactly one derivation in the grammar $G$ being used. The languages with grammars of this sort are the $LR(k)$ languages; in fact, the $LR(1)$ languages are "good enough." Thus, the programming languages used by modern computers are always $LR(1)$ languages. A definition for these languages has not been presented here, but one can be found in Hopcroft and Ullman (1969, p. 180). Knuth (1965, 1967) presented the basic results concerning $LR(k)$ programming languages.

**7–8.** Some of the many higher-level languages that have been developed include those that facilitate the description of procedures for general scientific data processing (FORTRAN), business data processing (COBOL), string manipulation (SNOBOL), and List structure manipulation (LISP); LISP is also designed to facilitate the use of recursive procedures. In this book are discussed two other high-level languages, PLANNER and PROGRAMMAR, designed to facilitate the description of planlike procedures for theorem proving and natural language sentence parsing, respectively. At least one computer has been constructed for which a higher-level language (known as SYMBOL) is actually its machine language; see Rice and Smith (1971) for further information.

**7–9.** Attempts at "mechanical translation" were first made in the 1950s and thus represent some of the earliest investigations in the field of artificial intelligence, having taken place before the field had a generally accepted name. All early attempts were failures, albeit instructive ones. Since then, the subject of mechanical translation has been postponed somewhat by AI researchers. It is almost universally estimated to be a very difficult, "ultimate" problem. Bar-Hillel (1964) presented a good summary and criticism of the early work.

**7–10.** One ultimate test of the language-understanding abilities of computers would be to see how well they could play "language games." Some simple language games that, to the present author's knowledge, have not been investigated are crossword puzzles, Scrabble, and the game of 20 Questions. A rather entertaining game, which is difficult for people (and currently impossible for computers) to play, is the "question tennis" game of *Rosencrantz and Guildenstern Are Dead,* a play by Tom Stoppard; an example of question tennis is given on pages 42–44 of Stoppard (1967).

Games such as these would require the successful integration of a wide variety of semantic information-processing techniques, if a computer program were to play them well. The work of Wittgenstein presents an extensive treatment of the "language game" concept and its relation to the concept of "meaning."

**7-11.** There is no known a priori limit to the extensibility of a computer's language capability other than those limits of a purely practical nature (memory size and processing speed). Although the difficulties involved with understanding natural language should not be minimized, no one has been able to show, for example, that English is theoretically outside the language capability of all computers; indeed, such a proof would indicate the falsity of Church's thesis. The language-understanding programs discussed in this chapter are examples that certain subsets of English are definitely within the language capabilities of computers.

**7-12.** Of course we assume, that, whatever common problem language is used, it will be extensible, and that each specialized question answerer will be able to "understand" its extensibility. However, it may be argued that it takes relatively little knowledge of probability to ask (at least a simple) probability question; each question answerer will have to be able to *recognize* those questions that it might be able to answer and, ultimately, it will have to be able to recognize those questions that are relevant to its current problem and which other question answerers may be able to answer. "Problem recognition" techniques are employed by current question answerers (e.g., Gelb, Charniak, Quillian), but of course there is still a lot that is not known about the subject.

**7-13.** It is possible for the statement of a grammatical inference problem to specify that a solution grammar generate exactly those strings of $A$ and no others; one way of doing this is to define the set $V_A$ as being the set of all symbols that occur in the strings of $A$, and then to define $B = V_A^* - A$. However, most applications of the grammatical inference paradigm are motivated by the ability of grammars to provide finite descriptions for infinite sets (languages, patterns), and by the consequent ability of a machine that infers grammars to simulate *perceptual generalization* (the ability of people to learn to recognize an infinite number of samples as being pattern examples of a pattern after having observed only a finite number of that pattern's pattern examples).

**7-14.** Why should this question be asked? In addition to the possibility of an altruistic desire on the part of computer scientists to make their machines "happy and contented," there is the more concrete reason (for us, if not for the machine) that we would like *people* to be relatively happy and contented concerning their interactions with these machines. We may have to learn to design intelligent computers that are *incapable* of setting up certain goals relating to changes in selected aspects of their performance and design—

namely, those aspects that are "people protecting." (See the final sections of Chapters 8 and 9.)

## EXERCISES

*7–1.* Design a computer program that could generate the set of "Crypt Addition" problems. (See Exercise 3–5.)

*7–2.* Consider various methods for making a computer generate English fortunes, such as are found in fortune-cookies. What are the desirable attributes of fortune-cookie fortunes? (Some may claim that a fortune-cookie's most desirable aspect is that it is made by a human: Can a machine be human?) Is it possible to develop a program that can generate "all meaningful" one- or two-sentence fortunes? Is this a desirable exercise for AI researchers to perform? (Note: If you do decide to perform this exercise, it might be fun to do it as a class exercise, with a field trip to a local Chinese restaurant.)

*7–3.* Show how the following formula (Watanabe, 1969, p. 13) can be stated in English:

$$F_v = E_{v-1} \quad \cup E_{v-1} \qquad \cup \cdots \cup E_v$$
$$\sum_{a=1} n_a + 1 \sum_{a=1} n_a + 2 \qquad \sum_{a=1} n_a$$

*7–4.* Discuss the subproblems that might be considered by a computer program for solving crossword puzzles.

*7–5.* Prove that a string language is of type 0 iff there is a Turing machine that accepts it.

*7–6.* "Hucbald, Abbot of Saint-Amand, wrote a learned and insufferably boring poem, the Eclogia de Calvi, circa 877 A.D., justifying and praising baldness, in which not only the best and greatest men had apparently been so distinguished, but every word of the 146 verses begins with 'c'." (Beckwith, 1964, p. 74).

Hucbald's poem was written in Latin, but the solution of similar linguistic problems, in any language, indicates some proficiency at semantic information processing. Outline roughly the subproblems involved in

(a) Writing an *n* word sentence in which each word starts with a given letter.

(b) Writing an *m* verse poem of a given meter and rhyme scheme, in which each word starts with the same given letter.

(c) Doing both (a) and (b) in such a manner that the result is "meaningful" (although, perhaps, insufferably boring).

(d) Is there a connection between Hucbald's name and the subject of his poem?

*7–7.* Describe how a GQA might be enabled to "learn how to learn."

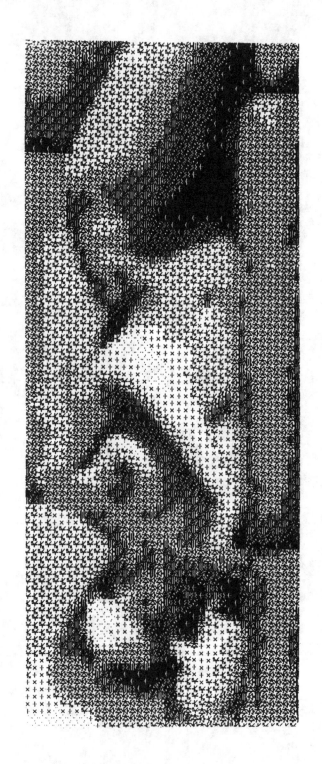

Computer-produced mural based on a photograph of a nude by Leon D. Herman and Kenneth C. Knowlton. (Reprinted with permission.)

# 8

## PARALLEL PROCESSING AND EVOLUTIONARY SYSTEMS

## INTRODUCTION

This chapter is a brief introduction to the subject of "phenomena that are made up of other phenomena," a topic that was introduced in Chapter 2. Again, the discussion will be directed toward phenomena that are discrete and mathematically describable.

Even though all mathematically describable, discrete phenomena can presumably be represented by Turing machines, there are many reasons for considering "phenomena that are made up of other phenomena" in more detail. While a given phenomenon may be easily described by a *serial* machine (i.e., a Turing machine, a program for a universal Turing machine, etc.), this is not the case for all phenomena. If a given phenomenon is most easily described by referring to the actions of several machines, it is said to be a *multiprocess* and to involve *multiprocessing*. If the description of a multiprocess specifies that some of its machines perform their actions simultaneously, then the phenomenon is called a *parallel process,* and is said to involve *parallel processing*.

343

# MOTIVATIONS

The basic reasons for investigating parallel processes in this book are as follows:

1. Our knowledge of the real world is often most easily described by reference to parallel processes: "While $X$ was happening, $Y$ happened whenever $Z$ happened." In particular, natural intelligence seems to involve extensive parallel processing.
2. Although there are limits to the computational ability of any machine, the limits for parallel machines are more remote than those for serial machines.
3. We expect that ultimate investigations of artificial intelligence will be concerned with the problem-solving capacities of parallel and multiprocessors in which each component is artificially intelligent.
4. An important problem for AI research is that of finding good representations for processes. Even the relatively simple representations discussed in this chapter are capable of being used to describe some very lifelike behaviors. Together with the previous discussions of programming languages such as PLANNER and QA4, this chapter serves as an introduction to the study of process representations.

The emphasis of this chapter is primarily theoretical. It will give no coverage of current parallel computer systems and languages, but will refer the reader to Findler and McKinsie (1969), Hewitt (1970a,b), Tesler and Enea (1968), Chamberlin (1971), Riley (1970), Graham (1970), Potvin (1971), and Slotnick (1967). Rather, an attempt will be made to summarize what is known about the theoretical abilities of parallel processors. Thus, the discussions will involve cellular automata, self-reproduction, self-description, Myhill's theory of "self-improvement," self-organizing systems, hierarchical systems, evolutionary systems, evolutionary stagnation, and other related topics. Although the first few pages of each section are easy, most of this chapter is fairly difficult. However, the final section is relatively simple all the way through. For other general discussions on parallel systems, the reader is invited to see Ershov (1971), Mesarovic (1969), von Bertalanffy (1968), Varshavsky (1969), and Dijkstra (1965 et seq.).

# CELLULAR AUTOMATA

Given that Turing machines and finite automata are efficient descriptions of simple serial phenomena, one would naturally expect the automata theorist to look for mathematical ways of saying, "While $X$ was happening, $Y$ happened whenever $Z$ happened," and defining his $X$'s, $Y$'s, and $Z$'s to be finite-state automata or Turing machines. This expectation is justified: The mathematical formalizations for parallel process so far developed are all essentially ways of describing complex machines that are made up of "interrelating" finite-state automata, Turing machines, or other types of information processors. Two such mathematical formalizations are discussed and a third is described.

The first formalization is that of the theory of *cellular automata* (see Codd, 1968; Burks, 1970; A. R. Smith, 1969). At the outset it should be mentioned that the cells of a cellular automaton are not necessarily intended as models of their biological counterparts, the cells that comprise living organisms. The fact that this correlation does not necessarily exist is responsible for the other common name given to this type of machine, "tesselation automaton."

Briefly, a cellular automaton is a graph (note 8–1) whose nodes are finite-state machines (see Fig. 8–1a). The operation of a cellular automaton is determined by information passed between those nodes that are connected; the machine at each node receives the outputs of the machines at those nodes that connect to it.[1] Often, cellular automata are defined as being graphs of some simple nature, say that of an Abelian group (note 8–2), and in most cases the interconnections between nodes pass information bidirectionally (Fig. 8–1b). The important thing about this type of machine is that the underlying graph of a given cellular automaton is considered to be fixed, and is not capable of being altered by any of its nodes; this is the reason we can define the machine at each node by a simple finite-state function.

A person observing a cellular automaton will consequently see its nodes changing state with time, each state affecting the others, etc. If the states used by the machines at the different nodes are the same, he may observe these states to be "flowing" throughout the graph in an

---

[1] One natural generalization of the cellular automata formalism, pursued by Luconi (1968), Martin and Estrin (1969), Rodriguez (1969), and others, is to allow the nodes of the graph to be arbitrary information-processing machines and the arcs between nodes to be channels that may carry arbitrary data structures. An additional generalization is suggested in a later section that the nodes of the graph should be capable of changing their relationships (arcs) to each other.

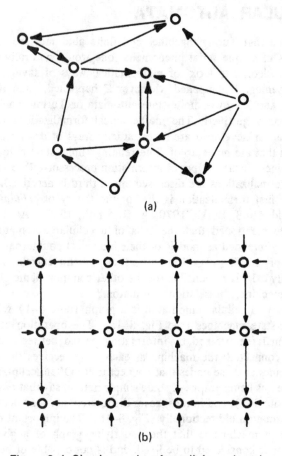

(a)

(b)

Figure 8-1. Simple graphs of a cellular automaton.

interdependent manner. For this reason the underlying graph of a cellular automaton is also called a *space*. However, the states and space of a cellular automaton are not to be confused with the state space of a state-space problem.

By far the greatest amount of research on cellular automata (note 8-3) has been devoted to cellular automata whose underlying graphs have the nature of an *Abelian group;* that is, where the network of nodes forms either an *n*-dimensional Cartesian grid, cylindrical grid, or toroidal grid (Fig. 8-2). Most of this research has also dealt with cellular automata in which all cells or nodes of a given automaton are assigned the same finite-state machine (different nodes may start in different initial states, however).

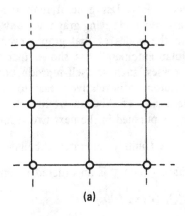

(a)

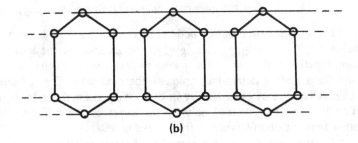

(b)

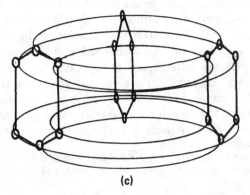

(c)

Figure 8–2. Three types of Abelian group.

In some respects this is less general than the study of cellular automata that can have *any* underlying graph and *any* consistent[2] assignment of machines to the nodes of that graph; even so, the study of "Abelian-group cellular automata" has shown that they can describe some interesting processes, such as "self-reproduction." Since the formalization for these automata is relatively easy to present, this section is confined to a discussion of Abelian-group cellular automata. The question of generality is pursued in the next two sections.

DEFINITION 8–1. A (finitely generated, Abelian group)

*cellular automaton* (ABC) $\Gamma$ is an ordered quintuple:

$$\Gamma = \langle Q, G^\circ, \bullet, f, q_0 \rangle$$

where

1. $Q$ is a set of *states*.
2. $G^\circ = \{g_1, \ldots, g_m\}$ is a generator set of a finite-generated Abelian group having group operation "$\bullet$".
3. $f$ is the *local transition function*, a mapping from $Q^{m+1}$ to $Q$.
4. $q_0$ is the quiescent state, such that $f(q_0, \ldots, q_0) = q_0$.

The *neighborhood* of any node (i.e., element) $g$ in $G$ is defined as the set $N(g) = \{g, g \bullet g_1, g \bullet g_2, \ldots, g \bullet g_m\}$. The meaning of the local transition function $f$, then, is that any assignment of states to the neighborhood for a node $g$ determines uniquely the next state of $g$. (There is, incidentally, no loss of generality in our having defined the local transition function $f$ to depend only on the *states* of the nodes in $N(g)$ rather than on output symbols from these nodes.)

A *configuration* $c$ is an assignment of states to all nodes, or cells, of a cellular automaton. A *finite* configuration is one in which all but a finite number of cells are assigned the quiescent state $q_0$. The operation of an ABC is assumed to proceed in unit time-intervals, $t_0$, $t_1 = t_0 + 1$, $t_2 = t_1 + 1, \ldots$, the local transition function being applied simultaneously to all cells of the ABC during each unit time-interval, thus determining a sequence of configurations $c_0, c_1, c_2, \ldots$. It will also be assumed that each cell requires the entire unit time-interval to carry out the operations (accept input, compute output and next state, go to next state, emit output) defined for it by the transition function. (This condition is relaxed by some authors.) The simultaneous application of $f$

---

[2] The transition function of the finite-state machine (see Chapter 2) at a given node must, of course, agree with the input and output capabilities of that node.

to all nodes of the ABC is, in effect, the application of a global transition function $F: C \rightarrow C$, where $C$ is the set of all configurations of the ABC.

EXAMPLE 8–1. (CONWAYS "LIFE" CELLULAR AUTOMATON.) Let $Q = \{0,1\}$, $q_o = 0$, $G$ be the Abelian group generated by

$$G^\circ = \{(1,0),(0,1),(1,1),(-1,0),(0,-1),(-1,-1),$$
$$(1,-1),(-1,1)\}$$

under the operation of vector addition (i.e., $G$ corresponds to the infinite two-dimensional Cartesian grid), and let $f$ be defined as follows: (Figure 7–3 shows the ("Moore") neighborhood $N(g)$ for a given node, or cell $g$, determined by this generator set $G^\circ$.)

1. If at time $t$ the state of cell $g$ is 0 ($g$ is "dead") and there are exactly three "living" cells (cells in state 1) in $N(g)$, then at time $t + 1$ the state of cell $g$ will be 1 (i.e., $g$ will "give birth" and become a living cell).
2. If at time $t$ cell $g$ is living and there are exactly two or exactly three other living cells in its neighborhood, then at time $t + 1$ cell $g$ will still be living.
3. If at time $t$ cell $g$ and its neighborhood do not satisfy either condition (1) or (2), then at time $t + 1$ cell $g$ will be in state 0.

These three conditions adequately define $f$ and enable us, given any configuration of living and dead cells at time $t$, to effectively determine which cells will be living or dead at time $t + 1$.

The reader should trace the sequence of configurations shown in Fig. 8–4 to verify this for himself (in this figure the cells of the automaton space have been drawn as squares: Fig. 8–4a shows the neighborhood of a cell $g$ corresponding to that indicated by Fig. 8–3). Figure

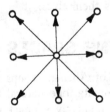

Figure 8–3. The Moore neighborhood.

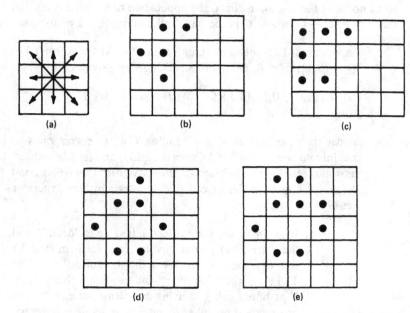

Figure 8–4. (a) A redrawing of Figure 8–3; (b) $C_0$, a right-pentamino;
(c)$C_1$; (d) $C_2$; (e) $C_3$.

8–5 shows a Cheshire cat configuration, which fades to a grin, then disappears, leaving a pawprint.

It will be shown in a later section that the ABC's are a very general class of machine in that some of them are capable of simulating the computations of the universal Turing machines. From the standpoint of efficiency in representation, however, there are some drawbacks to the use of ABC's as a formalization for the concept of parallel process in general. The major disadvantage is the unchangeability of the underlying graph of a given ABC. One might often like to have some way of easily describing systems in which the relations existing between machines are capable of changing with time, depending on the previous operation of the machines themselves.

# ABELIAN MACHINE SPACES

Given the simplicity of Abelian groups as the underlying graphs or spaces for cellular automata, one natural first choice in attempting a more general (yet still relatively simple) formalization for parallel process would be to find some method whereby the neighborhood of a cell could be allowed to wander throughout a constant Abelian space.

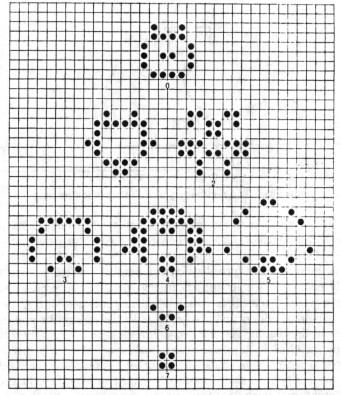

Figure 8–5. Computer-generated "Cheshire" cat (0) fades to a grin (6) and finally to a pawprint (8). (From "Mathematical Games" by Martin Gardner. Copyright © 1971 by Scientific American, Inc. All rights reserved.)

In this respect it is well to reconsider the subject of Turing machines:

The point to make is that the tape of a Turing machine (Tm) is essentially a finitely generated Abelian group. Consider the case of a linear Turing machine tape, divided into squares: Each square can be uniquely specified by a single integer (positive or negative), as shown in Fig. 8–6. The set of integers can, however, be generated by the finite set $\{1, -1\}$ under the (commutative) operation of addition. So, the tape is a finitely generated Abelian group.

Thus, a Turing machine is essentially a finite-state automaton that can "wander" throughout the space determined by an Abelian group. The "neighborhood" of a Turing machine is the particular cell it happens to be scanning or writing on at any given moment. Also, the directions in which the tapehead of the machine can move may be considered equivalent to particular elements of a generator set being

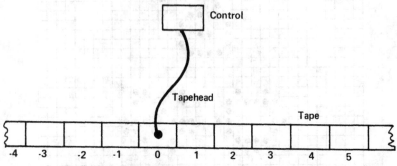

Figure 8–6. The numbering of a Turing machine tape.

used by the machine to describe the topology-of-motion on its tape. (The Turing machines defined in this book have used the generator set $\{-1,0,1\}$.)

A more general formalization for parallel process is then very simply contrived. We let the underlying space of the process be an Abelian group $G$, that is, a (possibly infinite) $n$-dimensional Cartesian grid, cylindrical grid, or toroidal grid. We let the cells of the process be polychephalic Turing machines, and for each cell specify the initial position in $G$ of its tapeheads (some cells may also have their own separate tapes, not printed on—and perhaps not read—by other cells). And we specify what shall happen whenever two or more cells choose to print different symbols at the same time on the same square, or node, of $G$. There may possibly be an infinite number of cells, but we require that each cell be described by one of a finite number of Tm's; also, we assume that each square is initially occupied by only a finite and computable set of tapeheads. If we specify the initial symbols assigned to the nodes of $G$ and require that all cells act simultaneously, always performing their next-move functions in the same unit interval of time, then the subsequent configurations of symbols within $G$ will be well defined. Figure 8–7 illustrates this formalization for parallel process, which we shall refer to as an *Abelian machine space* (AMS).

Our proverbial outside observer, watching the operation of a given AMS, could choose to concentrate either on the flow of symbols throughout its space $G$ or on the changing of the states of its Turing-machine cells. In this model, then, a *cell* is distinct from a square or node of the space and is, rather, identified with a possibly shifting set of "interdependent positions" in the space.

There are several different ways to go about solving the problem of what will happen if two or more tapeheads (possibly from the same

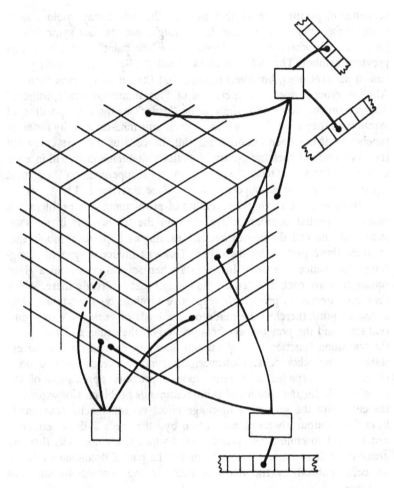

Figure 8–7. Abelian machine space.

cell or possibly not) attempt to print different symbols on the same square during the same unit time-interval. One way is to decide the actually printed symbol on the basis of a dominance relation[3] on the total set of tapeheads.

Another simple, rather elegant way to solve this "conflict of print commands" problem is to stipulate that the total set of all symbols used by the cells of the AMS itself forms a group, under the operation of superposition. That is, the symbols are designed in such a way that any

---

[3] See Chapter 2.

sequence of printing one symbol over another will always yield a new, recognizable symbol. For example, we might use the four symbols $--$, $|, +$, and " ", where " " is a "splash of white paint" that covers any previous symbol. This solution to the conflict-of-print-commands problem in an AMS does, however, require that the set of symbols form an Abelian group under the operation of "instantaneous superposition." The reason for this is that there is no "order" to the superposition of symbols as they are printed within a given unit time-interval by different tapeheads; it has been assumed that all the cells of the AMS carry out the operations of their next-move functions simultaneously within each unit time-interval. Thus, the "instantaneous superposition" $yx$ must equal the instantaneous superposition $xy$ (see Exercise 8–1).

Both ways of solving the conflict-of-print-commands problem can induce a "partial dominance" relation on the set of *cells* in an AMS, such that one cell dominates another insofar as it prints symbols that override those printed by the other. This can induce a type of "long-range dominance" on cells. In an AMS when several cells scan a given square they are each affected by the symbol that is already there. When they each decide to print their respective symbols on the square, their decisions must therefore be made on the basis of each cell's own current state and the *previous* symbols printed on the squares of the space; the transition function of a given cell does not depend on the current states of the other cells. However, the symbol already printed on a given square depends in general upon a previous application of the decision rule for the conflict-of-print-commands problem. Consequently, the cells with the greatest long-range effect on other cells (eventually have their output observed most often by other cells and consequently can be said to control the process as a whole), are the cells that are greatest (if there are any greatest) under the partial dominance relation on cells induced by the decision rule for the conflict-of-commands problem.

Whether or not the introduction of "long-range dominance" in this sense is desirable in an actual construction of an AMS would, of course, depend on the application one has in mind for the machine. One way of solving the conflict-of-print-commands problem, which does not have this type of long-range dominance, is to specify that each square of space be associated with a unique cell that has an immovable tapehead attached to that square, and that each square shall record only the symbols dictated by the immovable tapehead that scans it. Thus, all the moving tapeheads will become scanners. (This is essentially the method adapted in the Holland (1960) iterative circuit computers.)

# QUESTIONS OF GENERALITY AND EQUIVALENCE

The formalizations for parallel process so far discussed have provided a major context within which mathematicians (computer scientists, systems analysts) have (to date) approached the subject of parallel processes in general. Some other formalizations for "parallel process" have been suggested by Rodriguez (1969), Tesler and Enea (1968), Luconi (1968), Martin and Estrin (1969), and Dijkstra (1965 et seq.).

The reader may naturally wonder if these investigations could be carried further: Could we not develop a formalization for parallel processes in which the basic components, or cells, of a given process are enabled to change its underlying space?

Such a formalization can be developed, but in fact it will not be any more general than that provided by the Abelian machine spaces. To see this, let us consider that the space of a given parallel process is represented by a simple structure, of the sort defined in Chapter 7. At a given time *t*, the individual cells of the process will make up the space of the process by existing "in relation" to one another so as to form a structure (see Fig. 8–8). Presumably, each cell will be able to observe those cells to which it is related (which form its "neighborhood structure"), and alter its neighborhood structure by either removing or adding relations within it. It is not too difficult to arrange a consistent formulation of this idea, such that all cells operate simultaneously, within unit time-intervals, and such that the total structure (space) of the process will be changed with time by its cells. However, any such self-affecting space (note 8–4) can, given that it satisfies certain finitistic considerations,[4] be effectively simulated by a suitable AMS. We would merely require that some of the squares of the AMS be used to hold a current description of the given space structure and that the polycephalic cells of the AMS be designed so as to suitably alter that description; the underlying, Abelian space of the AMS would itself not change.

---

[4] For example, each cell should be describable as a finite automaton or Turing machine; each neighborhood structure should be finite, etc. A good way to implement these self-affecting spaces might be to construct "PLANNER-spaces," in which the relations between certain nodes or collections of nodes would be controlled by PLANNER theorems, each theorem controlling its own collection of nodes and operating in parallel with the others. There would, of course, have to be a special procedure for resolving conflicts of commands. (See Hewitt, 1970, section 4.6.1.1.2.)

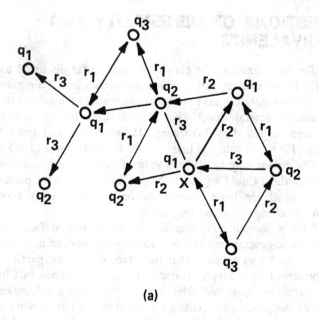

(a)

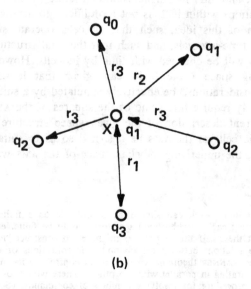

(b)

Figure 8–8. Space-structures.

In light of this conclusion, one might naturally wonder whether the AMS formalization is really more general than that of the ABC's. Can any Abelian machine space be simulated by an ABC? The answer to this question is both yes and no, depending on what meaning is attached to the concept "simulation." There are (at least) two equally valid ways of interpreting the concept; these give different answers when the ABC's and the AMS's are compared. To discuss these interpretations, observe first that both ABC's and AMS's are examples of finitely describable, effectively computable functions. Any given ABC or AMS is in essence a function that maps the set $C$ of configurations (of states and symbols) which are possible in its underlying space into that same set $C$. By "effectively computable" is meant that the configuration produced by a given ABC or AMS after any finite amount of time of its operation can be calculated to any finite extent (i.e., for any finite number of squares in the underlying Abelian space) by a suitably programmed, universal Turing machine (note 8–5).

We can show, however, that some ABC's are *computation universal,* in the sense that such an ABC can be programmed to carry out the computation performed by any given Turing machine. Thus, the operation of any given AMS can be effectively computed by a suitable ABC. In this sense, the AMS's are not more general than the ABC's, and can be "effectively simulated" by them.

To prove that there are computation-universal ABC's, it is sufficient to show that, for any given Turing machine $T$, there is an ABC that can carry out the computation it performs on any given input tape. This immediately implies the result that there are ABC's which can carry out the computation of any given universal Turing machine on any given input tape, that there exists a *single* ABC which, given a suitable initial configuration, will carry out the computation of any given Turing machine on any given tape.

Following is an outline of the proof of A. R. Smith (1969), to which the reader should refer for more details.

Let $T$ be a Turing machine, utilizing $m$ symbols and $n$ states. We can construct an ABC $\Gamma_T$ that will carry out the computation performed by $T$ on any given input tape, such that $\Gamma_T$ uses max $(n + 1, m + 1)$ states, an infinite two-dimensional Cartesian grid, and the neighborhood corresponding to the generator set

$$G^{\circ} = \{(0,1), (1,0), (-1,0), (-1,-1), (0,-1), (1,-1)\}$$

(See Fig. 8–9.)

Each cell of $g$ has a set of $M + \max(m + 1, n + 1)$ states, $Q = \{0,1,\ldots,M - 1\}$, which, "depending on context," are used to represent

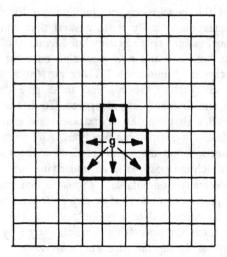

Figure 8–9. The $\Gamma_T$ neighborhood about a cell $g$.

either the states or the symbols of the Turing machine $T$. The state 0 is the quiescent state of $\Gamma_T$, however, and is not used to represent either a state or a symbol of $T$. The blank symbol $b$ of $T$ is to be represented in $\Gamma_T$ by the state 1. In general, the state $q_i$ of $T$ is to be represented in $\Gamma_T$ by the state $i + 1$, and the symbol $x_j$ of $T$ is to be represented in $\Gamma_T$ by the state $j + 1$; that is, $b = x_0$.

To simulate the computation of $T$ for a given finite input string $i$, that string is embedded in a row of the space of $\Gamma_T$, each symbol of the string $i$ being represented by a corresponding cell state in the row; the control and tapehead of $T$ are represented by the single cell above the leftmost end of the row, being placed in state 1, corresponding to the initial state $q_o$ of $T$ (see Fig. 8–10). All other cells in $\Gamma_T$ are initially given state 0. At any subsequent time $t$, the configuration of the Turing machine $T$ will be represented by a finite row of cells in nonzero state, above which there is a single cell in nonzero state.

Figure 8–11 then gives the basic design of the transition function $f$ used by the cells of $\Gamma_T$, corresponding to the next-move function $P$ of $T$. Nonzero states in the table are represented by the dummy symbol $s$ or by explicit variables: $i + 1$ represents state $q_i$ and $j + 1$ represents symbol $x_j$, etc. Figure 8–11 shows what will happen for all the various cells of $\Gamma_T$ during any unit time-interval, provided $T$ is in state $q_i$ scanning symbol $x_j$ and the next-move function $P$ contains the quintuple

$$q_i x_j x_k X q_l$$

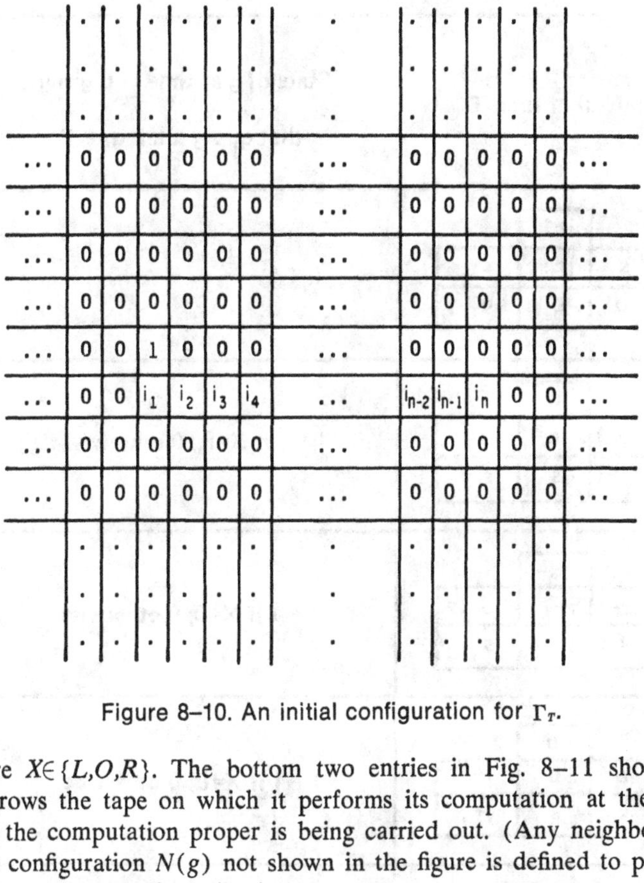

| ... | 0 | 0 | 0 | 0 | 0 | 0 | ... | 0 | 0 | 0 | 0 | 0 | ... |
|---|---|---|---|---|---|---|---|---|---|---|---|---|---|
| ... | 0 | 0 | 0 | 0 | 0 | 0 | ... | 0 | 0 | 0 | 0 | 0 | ... |
| ... | 0 | 0 | 0 | 0 | 0 | 0 | ... | 0 | 0 | 0 | 0 | 0 | ... |
| ... | 0 | 0 | 0 | 0 | 0 | 0 | ... | 0 | 0 | 0 | 0 | 0 | ... |
| ... | 0 | 0 | 1 | 0 | 0 | 0 | ... | 0 | 0 | 0 | 0 | 0 | ... |
| ... | 0 | 0 | $i_1$ | $i_2$ | $i_3$ | $i_4$ | ... | $i_{n-2}$ | $i_{n-1}$ | $i_n$ | 0 | 0 | ... |
| ... | 0 | 0 | 0 | 0 | 0 | 0 | ... | 0 | 0 | 0 | 0 | 0 | ... |
| ... | 0 | 0 | 0 | 0 | 0 | 0 | ... | 0 | 0 | 0 | 0 | 0 | ... |

Figure 8–10. An initial configuration for $\Gamma_T$.

where $X \in \{L,O,R\}$. The bottom two entries in Fig. 8–11 show that $\Gamma_T$ grows the tape on which it performs its computation at the same time the computation proper is being carried out. (Any neighborhood state configuration $N(g)$ not shown in the figure is defined to produce no change in state for cell $g$.)

The conclusion, again, is that the operation of any given AMS can be computed to any finite extent by a suitable ABC. However, in general, a universal Turing machine $U$ given an input tape $(d_T, i)$ requires longer to compute the result $(T(i))$ than does the machine $T$, given the input tape $i$. So, this suggests another question, that of whether the operation of any given AMS can be computed completely (at a constant speed ratio to that of the AMS itself) by a suitable ABC. Such an ABC would constitute a "complete simulation" for the AMS. The answer to this question is no, subject to our current definition of finite-state automata within an ABC; that is, we have so far required that all cells within an ABC operate simultaneously within unit time-intervals. It is not possible, in particular, for a given cell to operate instantaneously at the beginning of a unit time-interval and thus pass information with "zero delay"

| N(g) at time T | State of g at time t+1, given that $q_i \times j \times k \times q_i \in P$ |
|---|---|
| <table><tr><td></td><td>i+1</td><td></td></tr><tr><td>s</td><td>j+1</td><td>s</td></tr><tr><td>o</td><td>o</td><td>o</td></tr></table> | k + 1 |
| <table><tr><td></td><td>o</td><td></td></tr><tr><td>i+1</td><td>o</td><td>o</td></tr><tr><td>j+1</td><td>s</td><td>s</td></tr></table> | l + 1 if X=R; 0 otherwise |
| <table><tr><td></td><td>o</td><td></td></tr><tr><td>o</td><td>i+1</td><td>o</td></tr><tr><td>s</td><td>j+1</td><td>s</td></tr></table> | l + 1 if X=0; 0 otherwise |
| <table><tr><td></td><td>o</td><td></td></tr><tr><td>o</td><td>o</td><td>i+1</td></tr><tr><td>s</td><td>s</td><td>j+1</td></tr></table> | l + 1 if X=L; 0 otherwise |
| <table><tr><td></td><td>o</td><td></td></tr><tr><td>s</td><td>o</td><td>o</td></tr><tr><td>o</td><td>o</td><td>o</td></tr></table> | 1 |
| <table><tr><td></td><td>o</td><td></td></tr><tr><td>o</td><td>o</td><td>s</td></tr><tr><td>o</td><td>o</td><td>o</td></tr></table> | 1 |

Figure 8–11. Basic design of transition function f.

between unconnected cells. In other words, there is a limit to the speed at which information concerning one part of the space of an ABC can be carried by its cells to another part.

This limit to the speed of information transfer in an ABC can be used to show (Holland, 1970) that the ABC's are not *composition-universal:* There does not exist an ABC that can be used to compute, at a constant rate, the sequence of configurations of *any* AMS, or even of any ABC. If an ABC is being used to reproduce the successive configurations of another ABC or AMS, it must in some cases require an increasingly longer and longer amount of time to do so, even for finite configurations; there is no ABC that can simulate all ABC's in "real time," or even at a slower but still constant rate.

The AMS's are composition-universal (and thus cannot be completely simulated by the ABC's) because the tapeheads of a given AMS cell are allowed to transmit information with zero delay across varying distances of the underlying space. One can also modify the formalization for cellular automata to yield ABC's that are composition-universal: The modification consists precisely in allowing some cells to carry out their transition functions instantaneously, whenever they are in certain states, at the beginning of the unit time-intervals that occupy the operations of the other cells. Such instantaneously acting cells ("Mealy automata") are said to form *zero-delay gates* for information transfer (note 8–6).

In summary, the two notions of simulation, referred to here as *effective* and *complete,* correspond to two types of universality: computation universality and composition universality. Both concepts of universality are of relevance to the study of self-affecting systems. We shall find that computation universality in a given ABC implies the ability of that automaton to hold a self-reproducing configuration, which is itself equivalent to a universal Turing machine; also, it seems very likely that the composition-universal spaces are those best suited to modeling evolutionary systems.

## SELF-AFFECTING SYSTEMS: SELF-REPRODUCTION

A mathematical system that "affects itself" is typically composed of at least two parts, $A$ and $B$, which bear the relation that

$A$ affects $B$

and

$B$ affects $A$

The entire system $(A,B)$ is then called self-affecting if the actions of any part affect the other parts, which in turn affect the original part (note 8–7).

Equivalently, the study of self-affecting systems is the study of machines that produce and accept feedback to themselves. This viewpoint of self-affecting systems lends itself to a study of continuously self-affecting systems, via analytic function theory, a direction of research presented in N. Wiener (1948) and Formby (1965). For our own purposes it is adequate to stick to the descriptions of discrete, self-affecting processes provided by automata theory.

Many types of self-affecting systems can be studied within the contexts of cellular automata theory and the theory of Abelian machine spaces (see Exercise 8–2). Those that seem to be of particular importance to the field of artificial intelligence are the *self-diagnosing* and *self-repairing* systems, the *self-reproducing* systems, the *self-organizing* systems, and the *evolutionary* systems. Some of the basic qualities of self-diagnosing and self-repairing systems are illustrated by the Exercises at the end of the chapter; for thorough discussions on the current uses of such systems, see Carter (1971) and Randell (1971). Self-organizing and evolutionary systems will be discussed in the next section, and this section will concentrate on the nature of self-reproducing systems.

The study of self-reproducing systems can be approached from many different angles. After the discussion of a few such approaches, the reader should investigate the vast literature for himself: von Neumann (1966) was the first to investigate it extensively, using cellular automata theory, and most of the subsequent approaches are due to his influence. A semi-intuitive argument of von Neumann's provides the best introduction to the nature of self-reproducing machines. Let us assume that there exists a machine $A$, which is a *universal constructor* in the sense that if $A$ is given a finite input tape describing a given machine $X$, $A$ will eventually construct $X$. This is denoted by

$$A: d_X \to X \tag{8–1}$$

where $d_X$ is an input tape describing $X$. It should be noted that

$$A: d_A \to A \tag{8–2}$$

is *not* an example of self-reproduction, since after the process 8–2 is complete, there exist two $A$ machines and only a single tape $d_A$, which is not specified as being given as input to either of the two $A$ machines. Rather, the need is for an equation of the form

$$Y: d_Y \to Y: d_Y \tag{8–3}$$

which indicates that there is a single machine $Z = (Y:d_Y)$ such that $Z \rightarrow Z \rightarrow Z \rightarrow \ldots$ . To obtain this, we need two other machines, both simpler in design than $A$. The first of these is a machine $B$, which is capable of copying any input tape

$$B: i \rightarrow i \qquad (8\text{–}4)$$

(After process 8–4 is complete, there will exist *two* input tapes *i*.) The other machine, $C$, is to be capable of coordinating the actions of $A$ and $B$ so that the ensemble of machines $A + B + C$, given an input tape $d_X$, will operate as follows:

$$(A + B + C): d_X \rightarrow (B + C): (A:d_X) \rightarrow$$
$$(C + A + X): (B:d_X) \rightarrow X:d_X \qquad (8\text{–}5)$$

That is, $C$ first submits $d_X$ to $A$, causing $A$ to produce a copy of $X$; then $C$ submits $d_X$ to $B$, causing $B$ to copy $d_X$; then $C$ submits the copy of $d_X$ to $X$ and allows $X: d_X$ to operate on its own. Let us then denote the machine $(A + B + C)$ by the symbol $D$; the result follows immediately that

$$D: d_D \rightarrow D: d_D \qquad (8\text{–}6)$$

Thus, the machine $E = (D: d_D)$ is self-reproducing. The reader should note that there are no logic problems with this argument, and that the result follows directly from the assumption that the three machines $A$, $B$, and $C$ are each finitely describable.

Of these three machines, the only one that has not been given an effective description within the argument is $A$, the universal constructor; that is, the argument describes $A$, but not sufficiently to guarantee that it can actually be built. At the suggestion of S. Ulam, von Neumann (1966) made the first investigations in cellular automata theory in an attempt to prove the existence of a universal constructor. Although he died before he could finish his work, he did prove the existence of a universal constructor (note 8–8), using a two-dimensional ABC of 29 states. The constructor itself was effectively described and shown to occupy about 200,000 squares of the space. Since then, others have shown that the size and number of states required for a universal constructor can be considerably reduced (see Codd, 1968).

It is relatively easy to show that there are ABC's in which certain configurations of states will reproduce. A very simple example, due to E. Fredkin, uses two states—$Q = \{0, 1\}$ for each cell—the ("von Neumann") neighborhood corresponding to the generator set

$$G^\circ = \{(1,0),(0,1),(-1,0),(0,-1)\}$$

for the infinite two-dimensional Cartesian grid, and the following transition function $f$:

1. If at time $t$, $g$ is connected (by $G°$, under vector addition) to an even number of cells in state 1, then at time $t + 1$, $g$ will be in state 0.
2. If at time $t$, $g$ is connected to an odd number of cells in state 1, then at time $t + 1$, $g$ will be in state 1.

It is not difficult to prove that *any* finite initial configuration of 1's will reproduce itself endlessly in this ABC. Figure 8–12 shows a sequence of self-reproductions of a "right tromino."

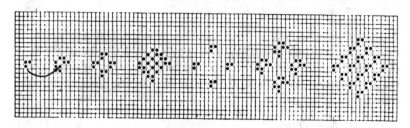

Figure 8–12. The self-replications of a "right tromino." (From "Mathematical Games" by Martin Gardner. Copyright © 1971 by Scientific American, Inc. All rights reserved.)

From the standpoint of automata theory (and artificial intelligence), it is important to search for a more general type of self-reproduction. The need is for an ABC in which there is a configuration that reproduces and which can also carry on some type of universal processing activity. Thus, we have another reason for von Neumann's motivation to show the existence of a universal constructor. (Fredkin's ABC mentioned above is not capable of holding a universal-constructor or universal-computer configuration.) Rather than reproduce a lengthy universal-construction proof, it is sufficient merely to summarize A. R. Smith's (1969) proof that there exist ABC's that can hold self-reproducing, computation-universal configurations.

The preceding section showed that there exist ABC's that are computation-universal. Such an ABC, given an initial finite configuration corresponding to the machine-tape pair $(d_T, i)$, will carry out the computation of $T$, given the input tape $i$. Suppose that $T$ given $i$ yields the (finite) output string $j$, which is denoted[5]

$$i \underset{T}{\rightarrow} j \tag{8-7}$$

---

[5] The notation of formulas 8–7 through 8–11 is similar to, but not to be confused with, that of formulas 8–1 through 8–6.

Similarly, if $\Gamma$ is a universal ABC, we denote its operation on $(d_T,i)$ by

$$(d_T,i) \underset{\Gamma}{\rightarrow} j \tag{8–8}$$

Finally, if the tapehead-control cell of $\Gamma$ is scanning the leftmost square of a finite row $x$, we write

$$\underset{\uparrow}{x} \tag{8–9}$$

To establish our result, we shall need to use the "fixed point" recursion theorem.

> THEOREM. For any total recursive function $h$ mapping programs into programs, there exists a program $P$ such that $h(P) = P$.

(A function is said to be total if it is defined for all elements of its domain; a function is recursive if it is expressible as a Turing machine program.)

> LEMMA 8–1. For any arbitrary encoding function $d$ and for any arbitrary partial recursive function $g$, there exists a program $P$ such that
>
> $$\underset{\uparrow}{i} \underset{P}{\rightarrow} (\underset{\uparrow}{(d_P, i)}, j, (d_P, i)) \underset{P}{\rightarrow} (\underset{\uparrow}{(d_P, i)}, j, (d_P, i), j, (d_P, i)) \underset{P}{\rightarrow} \ldots \tag{8–10}$$
>
> if $g(i) = j$ is defined. ($P$ is said to be *self-describing*.)

*Sketch of Proof.* We can define a function $h$ from programs to programs such that

(a)      $\underset{\uparrow}{i} \underset{h(Q)}{\rightarrow} ((d_Q, i), j, (\underset{\uparrow}{d_Q}, i))$

(b)      $(\underset{\uparrow}{d_Q}, i) \underset{h(Q)}{\rightarrow} ((d_Q, i), j, (\underset{\uparrow}{d_Q}, i))$

This can be done because, given that $h(Q)$ is in its initial state scanning the leftmost symbol of a string $x$, it can always decide whether $x$ is of the form $(d_Q, i)$ for some $i$ (it knows the function $Q$ and $d$; therefore it can compute $d_Q$, compare it to the leftmost part of $x$, etc.). If $x$ is not of this form, $h$ can be designed to perform step (a), which consists basically of setting $i = x$, computing $(d_Q,i)$, computing $g(i) = j$, copying $(d_Q,i)$, and going into its halt state (=its initial state) at the proper place. Step (b) requires all but the first two parts of step (a). However, the fact that $d$ is a total recursive function (which follows from the nature of an encoding function; see Chapter 2) implies that $h$ is total recursive. So, by the recursion theorem stated above, we know there must exist a program $P$ such that $h(P) = P$ and

$$\underset{\uparrow}{i} \underset{P}{\rightarrow} (\underset{\uparrow}{(d_P, i)}, j, (d_P, i)) \underset{P}{\rightarrow} (\underset{\uparrow}{(d_P, i)}, j, (d_P, i), j, (d_P, i)) \underset{P}{\rightarrow} \ldots$$

thus concluding our proof.

THEOREM 8–1. Let $\Gamma$ be a computation-universal ABC. There exists a finite configuration $c_0$ of $\Gamma$ which is self-reproducing and computes an arbitrary partial recursive function $g$.

*Proof.* Let $c_0$ be $(d_p,i)$. Then

$$(d_P, i) \underset{\uparrow}{\overrightarrow{\Gamma}} (d_P, i), j, (d_P, i)) \overrightarrow{\Gamma} \cdots \qquad (8\text{–}11)$$

if $j = g(i)$ is defined.

COROLLARY 8–1. If $\Gamma$ is a computation-universal ABC, then there exists a finite configuration $c_0$ of $\Gamma$ which is self-reproducing and computation-universal.

*Proof.* Let $g$ be the universal Turing machine function.

The existence of self-describing machines is more than a theoretical result of automata theory. Thatcher (1963) gave an explicit 2532-instruction program for a self-describing machine; thus, it is possible for programs to reproduce themselves inside a computer.

Myhill (1964) investigated self-reproducing machines from a recursion-theoretic viewpoint also, although his results were not concerned specifically with cellular automata. The principal result of his studies was that there exists a sequence of *self-improving* machines $M_0, M_1, M_2, \ldots$, such that each machine constructs the next one. The machines are improvements over each other in the following respect: The first machine effectively proves all decidable propositions in a given recursive axiomatization of arithmetic; the second machine uses an expanded recursive axiomatization of arithmetic and effectively proves all the decidable propositions in its own axiomatization, including some that are undecidable in the axiomatization of the first; the third machine does the same for the second, and so on (see Chapter 6).

The self-improvement of these machines is, however, not fully effective: It can be shown that some propositions of arithmetic are undecidable for every machine $M_i$ in the sequence; there is no (mathematically describable) sequence of consistent machines which effectively decides the truth or falsity of every proposition of arithmetic. Still, Myhill's results did show that machines cannot only reproduce themselves but, in a sense, also develop themselves.

The reader may have noted by this time that, except for Fredkin's ABC, all self-reproducing systems so far discussed have operated in a highly serial manner, despite orientation of the discussion (at least in the case of von Neumann and Smith) toward cellular automata and parallel processes. This is in fact the current state of affairs for the study of self-reproducing configurations in cellular automata. So far as the present author is aware, no one has as yet demonstrated an ABC configuration that is a universal computer and which self-reproduces in a highly parallel manner.

The problem seems to be much easier to deal with in the Abelian machine spaces, where we can obtain parallel universal computation rather trivially, merely by requiring that all polycephalic cells of the AMS be universal computers. This also enables them to generate programs for each other and to program each other. We might then specify that all cells of the AMS initially have blank input tapes, except for the one cell $C_0$, whose input tape contains the description $d_P$ for a self-describing universal program $P$, which contains within it a description $d_\Gamma$ for a finite cellular automaton $\Gamma$; and contains within it a description for an "activated portion" of $d_\Gamma$.

The nature of $C_0$, given $d_P$, is that $C_0$ will print "subactivations" of $d_P$ on the input tapes of two cells (say, $C_1$ and $C_2$) and erase the input tape of $C_0$. By a subactivation of $d_P$ is meant a new description $d'_P$, which is identical with $d_P$ except for its reference to an "activitated portion" of $d_\Gamma$; the activated portion of $d_\Gamma$ described in $d'_P$ should be contained within the activated portion of $d_\Gamma$ described within $d_P$. The process is to be carried in a similar manner down the levels of activation allowed in $d_\Gamma$, with the end result being that instead of $C_0$ (the "fertilized egg"), there will be a set of cells $\{C_i\}$, each "activated" to be a single cell of $\Gamma$, and all connected together within the space of the AMS by their tapeheads so as to form the cellular automaton $\Gamma$. The construction is complete if we design $\Gamma$ to be able to program the original $d_P$ (specifying complete activation of $d_\Gamma$) into a cell of the AMS. Then $\Gamma$ will be a finite automaton that reproduces in a highly parallel manner; giving it universal-computing ability would probably not be too difficult.

Again, this construction has not been rigorously formalized; however, there is no essential mathematical difficulty in proving the existence of a $d_P$ that will behave in the manner indicated above, activating different portions of $d_\Gamma$ as required and programming itself into some of the unprogrammed cells of the AMS. The most difficult problem in achieving such a self-reproducing automaton is probably that of attaining the proper coordination into a single, universal automaton of the cells that descend from $C_0$. This author suspects that even this can be solved in a relatively simple way.

The final sections of this chapter leave formalities aside and merely speculate on the usefulness of such results to artificial intelligence.

# HIERARCHICAL, SELF-ORGANIZING, AND EVOLUTIONARY SYSTEMS

## Conditions

This section briefly describes some of the ways in which large, multiprocessing systems may eventually be used in AI research. The emphasis is particularly on "hierarchical," "self-organizing," and "evolutionary" systems. Before beginning, the reader should be warned that there are still no comprehensive theories or definitions for the nature of these systems (especially for the latter two). Rather, there are a number of partial results and guidelines, primarily concerned with hierarchical systems. And the little experience so far obtained with self-organizing and evolutionary programs has been largely disenchanting.

Nevertheless, it is the present author's belief that these systems may eventually be very valuable to AI researchers, provided two conditions can be satisfied:

1. First, there is a hardware requirement. These systems may involve rather sizable complexes of computers; and it would be good if they were inexpensive.

2. Second, we must overcome the misconception that these systems are essentially incompatible with the "reasoning-program" approach (see Chapter 3), and begin to investigate the possibilities of "hybrid" (hierarchical, self-organizing, evolutionary and reasoning) programs.

While it is not the purpose of this book to discuss hardware, there are encouraging signs in that field of computer science. For example, Culver and Mehran (1971) suggested that the use of laser technology may eventually allow a computer to perform a logic operation in a time span on the order of picoseconds ($10^{-12}$ seconds); holographic storage techniques (again, "laser technology") may eventually make it possible for computer memories to store millions of bits of information per square inch (Hunt, Elser, and Wolf, 1970). At any rate, we can ignore the hardware condition and try to assume within reasonable bounds that it can be met successfully. This section is intended to show that the second condition can be met (that is, to suggest some "hybrid" programs that AI researchers may eventually investigate with profit), and possibly to restore some enchantment to the study of self-organizing and evolutionary systems.

## Hierarchical Systems

Many types of "hierarchical" systems have been encountered throughout this book. In particular, the reader may recall the discussion of PLANNER (Chapter 6), the "hierarchy of visual perception systems" described in Chapter 5, "hierarchies of languages" discussed in Chapter 7, and the "economy of invention" hierarchy suggested in Chapter 3. In general, a *hierarchical system* is an ordered collection of machines (systems, programs, procedures, processes, etc.). We may speak of the type of "order" involved as determining the "form" and the "nature" of the hierarchy, which may be different for different hierarchies. The *form* of most systems that are considered to be hierarchical corresponds to either a string, a tree, a lattice (see Fig. 8–13) or perhaps to some cyclical variation on these forms. The *nature* of a hierarchical system corresponds to the physical meaning of the ordering between its machines, the factors of which may include time, energy, composition, construction, information, and control. These factors may

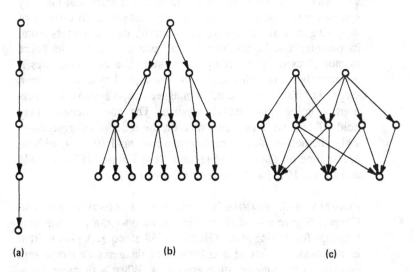

(a)           (b)           (c)

Figure 8–13. (a) String, (b) tree, (c) lattice.

be explained by noting that machines operate in time, transform energy, may be made up of other machines, may make other machines, may process information and send it to other machines or people, and may control the behavior of other machines (by programming them, altering their environments, starting them, unplugging them, etc.).

Following are two brief examples of ways the "hierarchical systems" concept is of use to computer science and artificial intelligence.

EXAMPLE 8–2. MEMORY SYSTEMS. As mentioned in Chapter 2, a memory system is a means of storing and retrieving information (data structures), and may typically be described by reference to its qualities of *size* (number of bits of information it can store) and *access time* (time necessary to determine the bits held at a particular place of storage in the memory). For a given memory system these qualities are directly related: The larger the memory size, the greater is the access time. One of the earliest hierarchical systems investigated by computer scientists (it was suggested by von Neumann) is the *hierarchical memory system*. Its value results from the fact that the utility of a given data structure varies with time: When a data structure is being used by (or as) a program, it has high utility, whereas otherwise its utility is very low, corresponding to the probability with which it may be used in the future. A hierarchical memory system consists of a lattice of memory systems, each capable of supplying data structures to, or accepting data structures from, its parents: The highest member of such a system is the "core memory," used by the computer to store the data structures it is currently using; other members of a typical system may be a magnetic "disk" or "drum," a magnetic tape system, and perhaps a holographic storage system. The core memory may hold $10^6$ bits, with an access time on the order of microseconds, while the holographic storage system may hold $10^{12}$ bits, with an access time on the order of seconds (see Katzan, 1971; Gentile and Lucas, 1971; and Arora and Gallo, 1971).

EXAMPLE 8–3. PLANNER'S HIERARCHICAL CONTROL SYSTEM. Chapter 6 gave a brief description of PLANNER, a programming language for writing plans (Hewitt, 1968 et seq.). A plan written in PLANNER consists of a collection of theorems that represent procedures for manipulating assertions. When a theorem (procedure) is used, it may affect other procedures (create them or manipulate them) or it may "call" another procedure (i.e., cause that procedure to be used). When a theorem is being used, we may think of it as having "control" of the actions currently being taken by the computer, and when it calls another theorem, we may think of it as *transferring control* to that theorem. The implementation of a plan starts by calling one of its theorems and continues as theorems manipulate data structures, transfer control, etc. Theorems in PLANNER are *goal-directed* procedures: Their purpose is generally to establish something as a fact. Consequently, the way in which they transfer control may be "conditional": Theorem $A$ may transfer control to theorem $B$; if theorem $B$ (and those theorems to which it transfers control) fails to achieve its goal, control automatically *backs up* to theorem $A$ so that it can (hopefully) do something else. PLANNER's *hierarchical control structure* enables it to keep track of the hierarchy of theorems being called and transferring control caused by the implementation of a plan.

As Holland (1970) pointed out, the chief value of the "hierarchical systems" idea is that it gives a way of describing large systems that is far more practical than the "state-transition" function approach of automata theory. A large system (e.g., the human brain) may have $10^{10}$ components; if each component has two states, the system will have $2^{10^{10}}$ possible states, and an explicit description of the state-transition function for a system of this size is not possible. Yet it may be that the components of the system are organized into a hierarchy of, say, 11 levels of "blocks," in which each block is divided into 10 lower-level blocks (a tree with a branching factor of 10 and a depth of 11), the lowest-level blocks being the "components" of the system. Should this be the case, then:

> Even for a device with as many as $10^{10}$ components, one need only make a selection at each of 10 levels to uniquely locate any given component. And, assuming a relevant functional division, much will be learned of the effect of that component by observing the use or function of the blocks involved . . . For devices of this complexity, hierarchical descriptions offer almost the only avenue to detailed understanding. (Holland, 1970.)

The reader who wishes to pursue the subject of hierarchical descriptions is also encouraged to see van Emden (1970) and Pratt (1969a,b). Mesarovic, Macko, and Takahara (1970) presented a rather comprehensive treatment of general hierarchical systems; Miller, Galanter, and Pribram (1960) presented an early, but still not superseded, discussion of the hierarchical nature of human plans; Gertz (1970) discussed hierarchical associative memory systems for parallel processing; Fikes and Nilsson (1971) discussed the hierarchical nature of the systems used in the robot STRIPS.

# Self-Organizing Systems

A collection of organs performing functions in relation to each other is called an *organism;* similarly, a collection of people who solve problems or perform functions in relation to each other is often called an *organization,* especially if the relations and functions involved seem to be relatively unchanging.

By a *self-organizing system* (SOS) is meant a collection of machines capable of solving problems by forming into (perhaps temporary) organizations. Essential to the operation of any self-organizing system is an *environment,* or collection of problems to be solved and patterns to be recognized (see Chapter 3). Two paradigms for the concept of "self-organizing system" are suggested.

The first is the standard paradigm, which received a great deal of investigation in the early 1960s (see Yovits and Cameron, 1960; Yovits, Jacobi, and Goldstein, 1962). The viewpoint of this paradigm is to see a self-organizing system as made up of parts, one of which is a *control* and governs the organization of the other parts (often called generators, characteristic functions, predicates, parametric functions, etc.); these parts begin with some initial (possibly empty) set of relations to each other. The function of control is to modify that set with experience, to make a structure of the parts which reflects its current knowledge and inference about the environment. (Note that "control" might be an "imaginary part" in reality distributed among the other parts.) Thus, for example, a Perceptron begins with a structure of predicates, and it alters that structure by changing the "weight relation" between its predicates (see Chapter 5). In general, the structure of parts produced by the control of an SOS at any given moment is to be interpreted as the SOS's current strategy for solving the problems presented by its environment. Thus, we have already encountered several simple types of self-organizing system: Pandemonium, the Samuel Checkers Player, and the Waterman Poker Player are all basic examples of programs designed to modify structures (i.e., "organize themselves") so as to solve a problem.

It is probably fair to say that this paradigm has gone out of vogue.[6]

---

[6] If such a vogueless expression is permissible.

The reason is that it gives few guidelines for designing self-organizing systems with an aptitude for specific real-world environments, leaving the burden of finding suitable predicates and controls to the human designer. The paradigm is somewhat tautological and, as a consequence, it is still being used, but many researchers no longer bother to refer to it. (Even so, the collections edited by Yovits et al. (1962) contain many insightful papers, and are worth reading.) Rather, it has been replaced by a general interest in the question of "machine induction" (learning of evaluation functions, pattern recognition, etc.) and been pursued in three directions: statistical decision theory (Duda and Hart, 1973), automatic program writing (Chapter 6), and grammatical inference (Chapter 7).

The present author will not pretend to lift the burden or to eliminate the tautology. However, it is possible that another paradigm, which emphasizes a different aspect of self-organization, may stimulate renewed interest in the subject and eventually lead to more results. The paradigm suggested here is to see a self-organizing system as having the following characteristics: First, it will consist of parts, each part being a problem-solving device, and "control" will be distributed throughout each of the parts; second, the parts of a self-organizing system will share a language capability for some language. Each part will be able to communicate with other parts of the system, and the actions of each part will be influenced by the messages it receives. The design of a self-organizing system should focus on two aspects: the nature of the individual components and the language they use to communicate with each other. Let us give an example of a way in which this kind of self-organizing system might be useful in the real world.

EXAMPLE 8–4. COMPUTER-DRIVEN VEHICLES. R. A. Schmidt (1971) presented convincing arguments that if an automated system for the transportation of people by automobile is developed, each automobile in the system should be an artificial intelligence, with its own computer and visual perception system. His basic reasons are as follows: First, the automobile (call it an "automatic car" or a "robot chauffeur") should be capable of being introduced into the existing transportation system without requiring an extensive (and expensive) road-rebuilding project. Thus, the robot chauffeur should be capable of traveling over ordinary roads and highways, just as people-driven cars do. Second, use of the robot chauffeur should be as safe as the use of ordinary cars (preferably safer). Finally, there is good reason why an automated transportation system should still be (at least partially) based on the use of automobiles; namely, the use of automobiles gives people a greater "freedom of mobility" than seems achievable otherwise. Other systems require the use of terminals, spaced relatively far apart, at which people may enter and leave the transportation system; usually the people involved may expect the terminals to be some distance from their own destinations. "This fact, along with the nuisance of scheduling present in other systems, is what induces most people to use automobiles, in spite of parking problems, congestion, delay and the host of other problems involved" (Schmidt, 1971, p. 139).

Given these reasons, Schmidt noted that the difficulty involved in automobile transportation systems is that danger areas ("incidents") are highly localized and variable. They may be caused by anything, ranging from a child running across a street, to another car with a flat tire or an erratic driver, or a hubcap lying in the road; and they may appear and disappear quickly. Given that we do not embark on an extensive project of building new, automated roads (and such a road system would still be susceptible to incidents), the information necessary to discover and avoid incidents must be obtained visually, and each car must contain its own robot chauffeur.[7]

The present author's suggestion (to illustrate the "communication" paradigm for self-organizing systems) is that it would be desirable for each of the automatic cars to be capable of communicating with the others in its area (say, on a special communications band) in a language specifically designed for expressing information about danger areas, incidents, roads, automobiles, etc. Certain kinds of automobiles would have the ability to make use of special sentences in the language. Thus, an ambulance might tell other cars it is heading along a certain road, at a certain speed and location, toward a certain destination; a car stalled on the road ahead might be able to reply with a warning to slow down. Again, suppose an accident occurs on lane 1 of a highway at point $A$ (Fig. 8–14); automobiles traveling on lane 2 should be able to relay a warning message back to automobiles at point $B$ in lane 1, telling them to slow down. As it is currently done, the warning message that is relayed back consists of the brakelight signals emitted by the cars

---

[7] Note that Schmidt's own assessment of the likelihood of achieving safe robot chauffers is pessimistic. He concludes that driving a car requires the full intellectual abilities of judgment and learning possessed by a human being: ". . . future research in computer control would be more profitable in areas such as industrial automation, remote exploration, or man-machine systems. . . ."

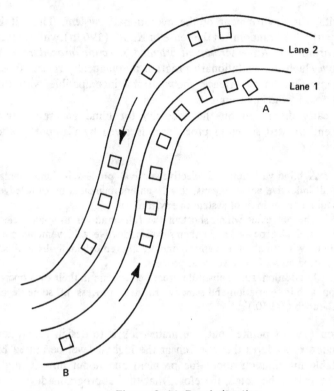

Figure 8–14. Road situation.

slowing down in lane 1; the signal does not travel as fast (by the present author's reasoning; it would be interesting to perform tests, using, of course, human drivers) to point *B* as it would if it were also carried by the cars in lane 2. (This idea could, of course, be implemented without the use of robot chauffeurs, using lights instead of electric signals.)

## Evolutionary Systems

An *evolutionary system* is a machine (program, procedure, etc.) that develops submachines according to their ability to perform tasks (solve problems, recognize patterns) in an environment produced by the real world. Generally, an evolutionary system is considered to make use of a "blind generation procedure" and an "environment-oriented selection procedure." A *blind generation procedure* is a method of creating new submachines that is partially independent of, and possibly inconsistent with, the environment of the evolutionary system. Thus, it is (at least partially) "random" in the sense of Knuth (1969b) with respect to its environment. An *environment-oriented selection procedure* is a method by which the evolutionary program automatically rejects those subprograms that its experience shows to be incompatible with the tasks required by the environment.

An early discussion on the necessity for blind generation and environment-oriented selection procedures is given by Campbell, who concluded:

> A blind-variation-and-selective-survival process is fundamental to all inductive achievements, to all genuine increases in knowledge, to all increases in fit of system to environment.
>
> The processes which shortcut the full blind-variation-and-selective-survival process are in themselves inductive achievements containing wisdom about the environment achieved originally by a blind-variation-and-selective-survival process.
>
> In addition, such substitute processes contain in their own operation a blind-variation-and-selective-survival process at some level. (Campbell, 1960.)

As Nilsson (1971) pointed out, the main trick is to design generation and selection procedures that "search at the highest level permitted by the available information about the problem and about how it might be solved." It is to be noted, therefore, that the subprograms developed by an evolutionary system may vary in the "blindness" they display. Thus a really intelligent system might first develop programs for symbolic integration similar to Slagle's (1963), later develop programs similar to Moses' (1967), and finally (eons later?) develop programs embodying the Risch (1969) algorithm (see Chapter 3). Also, note that it is possible for the evolutionary system to vary its *own* "blindness," to change the "level" at which it conducts its search (see Holland, 1960 et seq.). Even so, the present author agrees with Campbell that it is necessary for such a system to preserve some blindness because (as stated in Chapter 3) "a real-world environment has no known, complete finite description or prediction."

To the present author's knowledge, there have been only two fully general attempts to program evolutionary systems that would possess artificial intelligence. These were the attempts of Friedberg et al. (1958, 1959) and Fogel et al. (1966). Neither attempt had any success comparable to that obtained by other, nonevolutionary approaches to artificial intelligence, and we might therefore categorize them as "instructive failures." Friedberg's program attempted to develop subprograms written in machine language for a very simple computer. The generation procedure produced random (64–instruction) subprograms from those that had been previously produced. The selection procedure was used to assign success-or-failure credit to individual instructions used in these subprograms, and the credit given to an instruction was used to determine the likelihood that the generation procedure would use it in developing future subprograms. Similarly, Fogel's program was designed to develop subprograms corresponding to the state-transition diagrams of relatively simple finite-state machines (all machines developed had less than 30 states and input-output alphabets of no more than 8 symbols). These subprograms were used to make predictions of variously chosen sequences, and success-or-failure ratings were assigned to each subprogram. The generation process consisted of "mutating" a given subprogram to produce a new subprogram denoting a finite-state machine differing from its "parent" by an output symbol, a state transition, the number of states, or the initial state. "Parent" and "offspring" subprograms would then have their predictions compared for the same sequences and the subprogram with the best predictive capability would be retained (selected) and used in future mutation processes while the other would be rejected.

Besides the fact that these evolutionary systems produced only very small subprograms, they shared an essential limitation in method; namely: The generation and selection procedures used by these methods were restricted to taking very small steps through the space of possible subprograms. A change in a single machine instruction or state of a finite-state machine will only very rarely make any significant, desirable change in the behavior of the machine (subprogram); the likelihood of its doing so decreases with the size of the subprogram being mutated. We may expect any machine with a general artificial intelligence to have a huge number of states (say, greater than $10^{1,000}$) and its description in an ordinary machine language to involve a huge number of instructions ($10^7 \ldots$?). A further complication is the phenomenon of *evolutionary stagnation* (Bremmerman, 1962), also called the "Mesa phenomenon" (Minsky, 1963). It may be the case that a given subprogram could be mutated to form a better subprogram, but that such a mutation would require several "submutations" of which any partial combination would only produce a worse subprogram. If such a mutation were to occur, it would be necessary for all of its submutations to occur simultaneously. The probability that this would happen is the product of the probabilities that each of the submutations would happen. Thus, the given subprogram may tend to "stagnate" where it is.

Holland (1960 et seq.) presented a detailed scheme for the implementation of an evolutionary system that would not be so limited, using "iterative circuit computers." In particular, he suggested that the evolutionary system describe its submachines hierarchically and that the generation (mutation) procedure used be performed on the hierarchical descriptions (perhaps in LISPish notation) for the submachines, not on the submachines themselves. In addition, he suggested that the generation and selection procedures used by the evolutionary system should themselves be hierarchically described and (recursively) evolved by the system. Minsky (1963, pp. 434–435) made similar recommendations: "No scheme for learning, or for pattern recognition, can have very general utility unless there are provisions for recursive, or at least hierarchical, use of previous results." Again citing McCarthy (1956): "The enumeration of partial recursive functions should give an early place to compositions of functions that have already appeared."

The implementation of Holland-like evolutionary systems must await the satisfaction of the hardware condition cited at the beginning of this section. (Should these systems ever be implemented, it might be desirable to "prime" them with subprograms and generation and selection procedures that were already somewhat sophisticated.) However, two types of evolutionary systems may be of more immediate interest to AI researchers.

*Variable-Valued Reasoning Programs.* Chapter 6 discussed the possibility that reasoning-programs might change their rules of inference and logical calculi. Whether this is desirable is hard to say: Certainly it would seem that if it were done, the reasoning program should do it "reasonably." Still, keeping the preceding comments in mind, a certain amount of "blind" variation in the values of the reasoning program might be good. At any rate, it would be desirable that the reasoning program not sacrifice the generality of its phenomena language, however much it changed its efficiency at expressing certain concepts. Our rationale for giving reasoning programs such capabilities is the following: the program has to be able to form beliefs about its environment and recognize errors in these beliefs. It should also be able to correct the source of these errors, whenever possible. Often the source will be another belief, but in some cases it might be an inference rule, therefore it should be able to change these too. Again, the ultimate intelligent program should be able to understand that it is a program and understand the purposes of its subprograms. The program should be able to debug, rewrite, and extend itself, in order to adapt to its environment. And it should be able to perceive itself as a part of that environment.

*Networks of Question Answerers.* The possibility that networks of question answerers and protocol analyzers might demonstrate an ability to improve their intelligence was discussed in the last section of Chapter 7.

One objection to the utility of evolutionary programs, even should they be successful in producing general artificial intelligence, concerns the issues of *understanding* and *control*. To quote McCarthy and Hayes:

> . . . (the evolutionary approach) has had no substantial success so far, perhaps due to inadequate models of the world and of the evolutionary process, but it might succeed. It seems dangerous since a program that was intelligent in a way its designer didn't understand might get out of control. (McCarthy and Hayes, 1968.)

The present author agrees in general with this criticism, but thinks it can be equally well applied to the strict reasoning-program approach. It should be stressed that for any intelligent, programmed machine (whether of the strict-reasoning or evolutionary type), its designer will always be capable of examining the complete printout of all programs and other data in that machine, at least up to the moment the machine "gets out of control." Thus, he will be able to follow the development of any subprograms in an evolutionary machine, or the proofs of any theorems by a reasoning machine, to whatever extent he desires. In either case the amount of data involved might be enormous, so he might need to make use of other machines to examine it ("reasoning checkers"; see Chapter 9). It does not seem that the "reasoning patterns" of reasoning programs will necesarily be more perspicuous than the "evolutionary patterns" of evolutionary programs. And, if either type of machine is allowed to interact with a real-world environment, the designer will not be able to control precisely the information that will come into it. Thus, we do not expect that he will be able to control completely the actions of either type of machine. The only question is whether the designer will be able to foresee that his intelligent machine is going out of control, before it actually does so, and there does not seem to be a guarantee that he can have such an ability, for either machine. The best he can do is to predesign the machine so that its "freedom of will" (which is basically "freedom of action") can never exceed certain bounds. We expect that this can be done, within limits, for either the reasoning-machine or the evolutionary machine.

# SUMMARY

In this chapter we have seen that the abilities of machines to do things people would normally say require intelligence is complemented by abilities to do things people would normally say require 'life," namely: self-reproduce, evolve, self-organize, self-diagnose, and self-repair. These abilities may in the future be highly relevant to the search for artificial intelligence and to the development of future industries and technologies.

## NOTES

**8–1.** A *graph* G is an ordered pair (N,R), where N is a set of *nodes* and R is a binary relation on N. If $R(x,y)$ holds for a given x and y in N, then we say that "x is connected to y," or "x connects to y," under the relation R. If for all x and y in N, $R(x,y)$ implies $R(y,x)$, then we say the graph is *bidirectional* or *bidirected* (see Chapter 3).

**8–2.** A *group* G is an ordered *pair* (E,•), where E is a set of *elements* (or nodes, etc.) and "•" is the *group operation*, a function on $E \times E$ to E which is such that (a) "•" is associative, i.e., $x•(y•z) = (x•y)•z$ for all x,y, and z in E; (b) there exists an *identity element* e in E which is such that, for any given x in E, $x•e = e•x = x;$ (c) for each x in E there is an *inverse element*, denoted $x^{-1}$, in E such that $x • x^{-1} = e$. If A is a subset of E, then by $A^*$ we denote the set of all elements x such that

$$x = y_1•(y_2•( \cdots (y_{n-1}•y_n) \cdots))$$

for some finite n, and for some choice of the $y_i$ such that each $y_i$ is an element of A. (Cf. the equivalent notion of sets of strings, given in Chapter 6.) For any group, $E^* = E$. (Why?)

A group is said to be finitely generated if there is a finite set $G°$, called a generator set for G, such that $(G°)^* = E$. Finally, a group G is said to be *Abelian*, or *commutative*, iff $x•y = y•x$ for all x,y in E.

A given group may possess more than one generator set; for example, the Abelian group corresponding to the two-dimensional Cartesian (square) grid is generated by the set

$$\{(1,0), (0,1), (-1,0), (0,-1)\}$$

and by the set

$$\{(1, 1), (0,-1), (-1,0)\}$$

as may be seen by comparing the following two diagrams:

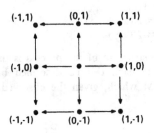

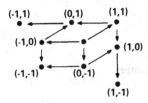

(The group operation is vector addition.)

**8–3.** Most work has been done with cellular automata whose spaces are described by Abelian groups; the main results have all been extended to the (slightly) more general case of spaces equivalent to "homogenous tessellations" (e.g., A. R. Smith, 1969). Not much work has been devoted to spaces described by nonhomogenous tessellations, and very little has been done on spaces equivalent to graphs in general, although certain results have been obtained; for example, any *finite* cellular automaton (i.e., containing a finite number of cells) is describable by a single finite-state automaton.

**8–4.** The space itself of the process can properly be said to be "self-affecting." This raises some intriguing possibilities. For example, here is a self-reproducing space: Let the process make use of a countably infinite set of cells, each cell referred to by a unique pair of integers $\langle i,j \rangle$ (positive or negative), every such pair being used to refer to a cell; let the structure, or space of the process always be described by means of four relations ($L$, $A$, $B$, $R$) between certain cells; let the initial structure $\sigma_0$ at time $t_0$ be such that

$$r(x,y) \text{ iff } x \cdot g_i = y$$

for some $g_i$ in the set

$$G^o = \{(1,0), (0,1), (-1,0), (0,-1)\}$$

where ($\cdot$) signifies the operation of vector addition; in particular let

$$y = x \cdot (1,0) \Rightarrow R(x,y)$$
$$y = x \cdot (0,1) \Rightarrow A(x,y)$$
$$y = x \cdot (-1,0) \Rightarrow L(x,y)$$
$$y = x \cdot (0,-1) \Rightarrow B(x,y)$$

Thus, the initial space structure of the process is the two-dimensional Cartesian grid. Finally, let each cell be described by the same "structure-processing" machine M which, given the observed neighborhood-structure

replaces it by the structure

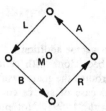

The figure below then shows the initial structure space $\sigma_0$ of the process at time $t_0$ and the subsequent structure space $\sigma_1$ produced at time $t_1$, all cells having operated simultaneously during the unit time-interval $[t_0 t_1]$, their actions being superposed in a logical manner. However, $\sigma_1$ is equivalent to two Cartesian grids, $\sigma_2$ will be equivalent to four, and so on, as the reader can easily prove. The space structure of this process is therefore self-reproducing.

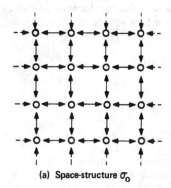

(a) Space-structure $\sigma_0$

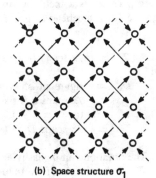

(b) Space structure $\sigma_1$

**8–5.** How to prove this fact for an AMS is briefly sketched, and the reader can fill in the details. The basic givens are as follows: There may be an infinite number of cells in an AMS, but each cell must be described by one of a finite number of (polychephalic) Turing machines; each cell has a finite number of tapeheads, whose initial locations are effectively computable; each square initially has a finite set of tapeheads scanning it. It is also required that this set be computable, both as to the number and as to the nature (to which cell(s) they are attached) of the tapeheads scanning any given square.

An inductive argument is necessary. These conditions imply that one can compute the initial configuration of tapeheads and symbols in any finite set of squares of the space. We will assume that one can compute the configuration of tapeheads and symbols in any finite set of squares of the space after a finite elapsed time $t = k$ and describe how to show that it can consequently be done for $t = k + 1$.

Let $D$ be a finite set of squares in the space of the AMS; let $a,b,c, \ldots$ be variables that range over the squares of the AMS, and let $x,y,z, \ldots$ be variables that range over its cells. As a (descriptive, but essentially valid) notation, let us then use

$D^* =$ assignment of symbols to squares of $D$ at time $t = k$
$T^* = \{x/x \text{ scans a square of } D^* \text{ (at time } t = k)\}$

The hypothesis should be that both $D^*$ and $T^*$ are computable for all $t \leq$ k. Also, let

$D^{k,j}$ = assignment of symbols to squares of
$$D \cup \{a/\rho(a,D) \leq j\}$$
at time $t = k - j$

where $\rho(a,D)$ is a metric giving the distance (shortest number of moves required for a tapehead to go) from square $a$ to a square of $D$. In general, let

$$T^{k,j} = \{x/x \text{ scans } D^{k,j} \text{ (at time } t = k - j)\}$$

and

$D_1^{k,j}$ = assignment of symbols to
$\{a/x \text{ scans } a \text{ for some } x \text{ in } T^{k,j}\}$
at time $t = k - j$.
$$T_1^{k,j} = \{x/x \text{ scans } D_1^{k,j} \text{ (at time } k - j)\}$$

etc. The notation can be easily extended for use in a complete proof.

To make the proof by induction, the major thing to note is that $D^{k+1}$ and $T^{k+1}$ depend (in a computable way) on $D^{k+1,1}$, $T^{k+1,1}$ and $D^{k+1,1}$, and that our induction hypothesis guarantees that each of these things can be computed (they are all finite, if $D$ is finite, and they are all defined for time $t = k + 1 - 1 = k$).

**8–6.** The permission of zero delay in a machine is perhaps unrealistic; automata theorists have also investigated the "middle ground" where zero delay is allowed but is limited in the amount of space it can cover. Thus, we might require (see Wagner, 1964, 1965) that all cells of the AMS be spider *automata* (i.e., polycephalic Tm's) whose tapeheads can never be greater than a fixed distance from each other. Similarly, the Holland (1960) iterative circuit computers are restricted to having no more than a fixed number of Mealy automata in any given chain of zero-delay gates. These restrictions do not, however, destroy composition universality in the AMS (see Holland, 1970, pp. 341–343).

**8–7.** From the philosopher's standpoint, the concept of *self* can be understood in at least two senses: First, self as the essence of consciousness; second, self as the image that consciousness has of self. Machines such as those we describe as "self-affecting" are composed of submachines that each operate with respect to an "image" of the other machines. The simultaneous operation of all machines is capable of changing the image that an outside observer might have of their (momentary) totality—thus the name "self-affecting." The true "self" of the machine (if there is one) presumably does not change.

**8–8.** It should be noted that there is a difference in the definition of "construction universality" (as used by von Neumann) and "composition universality" as used by Holland. A universal constructor in an ABC is defined to be able to produce any finite configuration of elements from a certain finite set of finite configurations of finite-state machines. These elements are the "parts" both of the universal constructor itself and of all the machines it can build. A universal constructor is not defined as being able to construct or simulate the construction of any finite configuration of elements from the set of *all* finite configurations of finite-state machines: As the present author understands it, this is Holland's concept of composition universality.

## EXERCISES

*8-1.* Prove that an "infinite conflict-of-print-commands" problem cannot arise in an AMS, given that each square of space initially has only a finite number of tapeheads scanning it.

*8-2.* Let $A$ and $B$ be two machines, each engaged in performing some never-ending task, with the additional feature that $A$ is able to scan $B$, recognize whenever $B$ is not performing correctly, stop $B$, repair $B$, and then start $B$ again, and that $B$ is able to do the same for $A$. Assume that $A$ and $B$ operate simultaneously during discrete time intervals, and that each machine is able to detect and repair a malfunction in the other machine during the single time interval in which it occurs, unless the repairing machine breaks down itself during that time interval, in which case neither will be repaired. If either machine is working correctly at time $t$, then the probability is $p$ that it will break down during the interval until time $t + 1$; if neither machine is working correctly at time $t$, then it is certain that they will not be repaired at time $t + 1$. (a) What is the probability that both machines will break down during the same time interval? (b) What is the mathematically expected number of time intervals that one machine would survive alone? (c) What is the mathematically expected number of time intervals that the two machines will survive together?

*8-3.* Define "nondeterministic cellular automata." Show that Checkers, Chess, and GO can be represented by nondeterministic ABCs.

*8-4.* Design some simple self-replicating machines.

# 9

# THE HARVEST OF ARTIFICIAL INTELLIGENCE

## INTRODUCTION

This chapter summarizes the preceding chapters with a brief review of current research on robots and a look at the possible future uses of artificial intelligence.

## ROBOTS

A *robot* is a mechanical intelligence capable of operating in our own real-world environment (see Chapter 3). The successful construction of a robot entails the integration of most, if not all, of the techniques discussed in previous chapters; in addition, it requires that a host of new problems be solved. Thus, a robot must have some sort of sensing and perception system that allows it to detect pattern examples in the environment; it must have some way of reasoning about its environment (and its relations to that environment); and it must also have some way of acting upon its environment. In our particular real-world environment a robot must confront the fact that no description of its current situation (the current state of the environment) can possibly be complete, in the sense of removing all uncertainties about the situation or enabling the robot to answer all questions about it. There seem to be three basic reasons for this (Munson, 1971):

 I. Sensing and perception systems are subject to accuracy limitations (e.g., Heisenberg's Uncertainty Principle) and also to gross failures (e.g., misinterpretation of lines).

 II. Many real-world objects may not be completely described, simply because of their complexity (e.g., a human or a complex piece of equipment).

 III. Any actions taken by the robot in its environment are subject to inaccuracies, or failures, and may introduce uncertainties rather than remove them.

As noted in previous chapters, AI researchers have been so far only partially successful in giving computers sensing, perceiving, and reasoning abilities. It will therefore come as no surprise to the reader that they have also been only partially successful in enabling computers to perform actions in the real world and in integrating these abilities to make the complete robot. Still, there has been some success.

Research on robots is currently being undertaken in the United States, Great Britain, and Japan.[1] Citations to this research will be found in Aida et al. (1971), J. D. Becker (1969), Coles (1969, 1970a,b), Doran (1969, 1970), Ejiri et al. (1971), H. A. Ernst (1962), J. Feldman (1967), Fikes (1971), Friedman (1969), Hart et al. (1971), Hayes (1971), Hewitt (1971a), Kinoshita et al (1971), McCarthy (1964a, 1968), Munson (1970a,b; 1971), Nilsson (1969, 1970), Paul (1971), Pingle et al. (1968), Popplestone (1969), Raphael et al. (1971), C. A. Rosen (1970), and Sutro and Kilmer (1969). The robot sensing and perception systems that have been investigated have been primarily *vision* and *touch,* with some attention to hearing; see Astrahan (1970), Bobrow and Klatt (1968), Coles (1969), Raj Reddy (1966). Robot-reasoning procedures that have been used include heuristic tree search and theorem proving (both resolution-based and PLANNER-based; also see Hart, Nilsson, and Robinson, 1971). The only robot *effector systems* (for acting on the environment) that have yet been developed are *mechanical hands* (and arms) and *locomotion* systems. As yet there is no robot that successfully combines all these systems in its performance.

The basic nature and the current limitations of robot vision systems are described in Chapter 5 of this book. Robots are still "effectively blind," at least when compared to humans. In addition to requiring special lighting conditions, robots are unable to "see" moving

---

[1] Undoubtedly, the Soviet Union also conducts robot research.

objects, even when they are moving quite slowly,[2] and they have only very limited ability to make use of color and texture.

Tactile sensing and perception systems are still at a rudimentary stage of development; however, it has been possible (in a way, somewhat natural) to integrate these systems into the operation of effector systems. Kinoshita, Aida, and Mori (1971) described a computer-controlled mechanical hand that has a very simple tactile sensing and perception system. Figure 9–1 shows the Kinoshita hand as well as the (more versatile, but less sensitive) computer-controlled arm and hand in use at the Stanford Artificial Intelligence Project (Paul, 1971). The bump-detector "whiskers" of the robot Shakey (Fig. 9–2) are integrated into use of locomotion (Coles, 1970a). Finally, Aida, Cordella, and Ivacevic (1971) present an approach to the integration of visual and tactile perception systems.[3]

To date, the real-world problems that robots have been able to solve have been of the "toy problem" variety, at the level of difficulty of the Monkey-and-Bananas Problem discussed in Chapter 6. Ejiri et al. (1971) presented a robot with the ability to solve problems that involve stacking blocks into simple configurations. Coles (1970a) described how Shakey makes use of resolution-based theorem proving to solve a problem that involves deciding to use a ramp as a tool to climb up on a platform and push a box off the platform (see Fig. 9–3). More recently, the Shakey-STRIPS configuration has been used to solve more "difficult" problems (see Chapter 6). Finally, Feldman et al. (1971) described how a computer can, with the use of an eye system and a mechanical hand ("known affectionately as Butterfingers"), solve the Instant Insanity puzzle by physically stacking up the blocks in a desired configuration.[4]

Although these tasks are relatively trivial, it is to be expected that robot research will make significant progress in the next few years, following the implementation of the PLANNER-like programming languages (see Chapters 6 and 7).

---

[2] However, when the type of motion is restricted and predefined, robots can be quite successful. Thus, Hieserman (1971) describes an optical-electronic scanning system that accurately records the type, owner, and registration number of every freight car in a train moving at speeds up to 80 miles per hour.

[3] The magazine *Electronic Design* (Nov. 11, 1971, p. 30) reported the development of a highly sophisticated mechanical arm and hand, designed to be remote-controlled by a human being. The device, called the Naval Anthropomorphic Teleoperator (NAT) has a tactile sensing system that supplies force-feedback to the human operator, and is so sensitive that it can be used to thread needles.

[4] The problem-solving program was designed specifically for the Instant Insanity Puzzle, and was not a "general problem solver" (see Chapter 3).

Figure 9–1. (a) Kinoshita's versatile computer-controlled hand; (b) the sensitive computer-controlled arm and hand used at Stanford AI Project. (Kinoshita, Aida, Mori, 1971, Paul, 1971, reprinted with permission;)

Figure 9–2. Shakey. (Coles, 1970, reprinted with permission.)

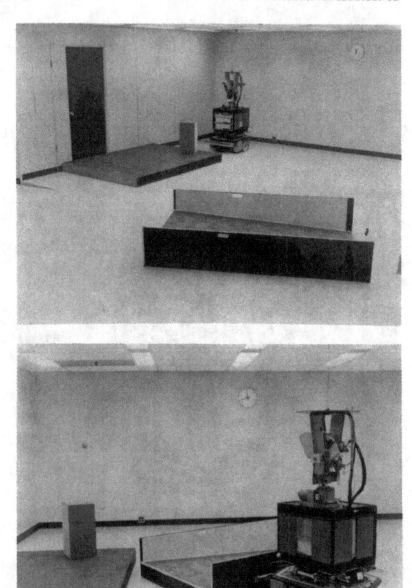

Figure 9–3. The robot and the box. (Coles, 1970, reprinted with permission.)

Figure 9-3 (Continued)

# A LOOK AT POSSIBILITIES

## Tools and People

The utility of artificial intelligence research forms an important part of a larger question, that of the relation of technology to society and nature. The urgency of this larger question is readily demonstrated in scientific terms, for technology is inherently concerned with the development of tools that affect our environment. Yet scientists and laymen have only recently begun to consider this question in detail.

Technology is, simply speaking, both the tools and the ability to make tools, which people have developed. Technology has existed since people first used shovels and spears, but it changes its form with each new invention and discovery. Moreover, a change in technology may either increase our abilities or limit them. Technology has made it possible for a man to breathe under water if he is wearing scuba tanks, and difficult for him to breathe in a large city if he is not. Changes in technology may help bring about changes in society. Examples are the invention and introduction to civilization of cannon warfare, the printing press, the telegraph, the assembly-line factory, the automobile, and the airplane. Two central facts about technology are that it gives people tools that enable them to do things they could not have done previously, and that the provision and use of these tools may cause either beneficial or detrimental consequences.

Artificial intelligence and the development of computer science in general represent a change in technology of the first magnitude, comparable to that of the discovery and development of atomic power sources. This change in technology has given us the ability to make tools (in fact, industries) that can be self-directing, that can operate more and more freely of our control. We have seen that AI researchers can give machines the ability to perform many tasks that previously could be performed only by people. In particular, such machines can solve problems, "reason" about actions and their consequences, display "learning" abilities, perceive patterns, control mechanical hands, repair themselves, and converse with us in our own languages. It is important for us to ask what consequences artificial intelligence will have.

Although it is too early to say with certainty what changes in environment and society will result from AI research, we can point to some possibilities.[5] If the research is unsuccessful at producing a general

---

[5] For the reader who is interested in comparing estimates, we cite the references in the Bibliography to Armer, Asbell, Asimov, Barth, Berkeley, Borko, Burck, Burke, Clarke, Demczynski, Diebold, Eastland et al., Eber, Foreign

artificial intelligence, over a period of more than a hundred years, then its failure may raise some serious doubt among many scientists as to the finite describability of man and his universe. However, the evidence presented in this book makes it seem likely that artificial intelligence research will be successful, that a technology will be developed which is capable of producing machines that can demonstrate most, if not all, of the mental abilities of human beings. Let us therefore assume that this will happen, and imagine two worlds that might result.

## Overmechanization of the World: The Machine as Dictator

In the author's research for this book and his conversations with people over the past several years, one of the dominant viewpoints he has encountered with respect to artificial intelligence research is the fear that it will produce a world with "everyone living in a machine." This is the viewpoint that Asimov calls the "Frankenstein complex." It is a particularly disturbing viewpoint because it is highly plausible. We should not pretend that AI techniques can be (or are being) used only for peaceful, desirable purposes. Rather, we should note that artificial intelligence can also be used to simulate those aspects of human "intelligence" which are warlike or otherwise undesirable,[6] and that such mechanical intelligences could be given perception and effector systems that would allow their "simulation" to take place in real time, in our own evironment. It is not difficult to envision actualities in which an aritificial intelligence would exert control over human beings, yet be out of their control.

Given that intelligent machines are to be used, the question of their control and noncontrol must be answered. If a machine is programmed to seek certain ends, how are we to insure that the means it

Policy Assoc., Forrester, Fromm, Galbraith, Gordon, Greenberger, Hamming, Hatt, Hilton, Johannesson, Kochen, Landers, Licklider, J. Martin and Norman, A. R. Miller, Mumford, McCarthy, Newell, Parsons and Williams, Philipson, Pierce, Polak, Pylyshyn, Reichardt, Rezler, Sackman, Samuel, Silberman, Simon, Skinner et al., Slagle, Sprague, Taviss, Toffler, Toynbee, Vonnegut, Westin, and Wiener.

[6] For example, Colby, Weber, and Hilf (1971) developed a computer program (known as PARRY) that possesses "artificial paranoia." It should be understood that no criticism of Colby and his coworkers is intended (their research was directed toward a better understanding of human paranoia); they have performed a valuable service by showing that computers can behave in a way that many people would generally think impossible, "2001" and HAL aside. Our question now is whether computers could behave this way even if we did not want them to.

chooses to employ are agreeable to people? A preliminary solution to the problem is given by the fact that we can specify state-space problems to require that their solution paths shall not pass through certain states (see Chapter 3). However, the task of giving machines more sophisticated value systems, and especially of making them "ethical," has not yet been deeply investigated by AI researchers; probably for the next few years it will not be necessary. Asimov (1950) presented a suggestion of three rules that might be included in an ethical robot, and which have become known as *Asimov's Three Laws of Robotics:*

1. A robot may not injure a human being or, through inaction, allow a human being to come to harm.
2. A robot must obey the orders given it by human beings except where such orders would conflict with the first law.
3. A robot must protect its own existence as long as such protection does not conflict with the first or second law.

Although these rules have not yet been implemented, there seems to be no a priori reason why they could not be.

The question of control should be coupled with the "lack of understanding" question; that is, the possibility exists that intelligent machines might be too complicated for us to understand in situations that require real-time analyses (see the discussion of evolutionary programs in Chapter 8). We could conceivably always demand that a machine give a complete output of its reasoning on a problem; nevertheless that reasoning might not be effectively understandable to us if the problem itself were to determine a time limit for producing a solution. In such a case, if we were to act rationally, we might have to follow the machine's advice without understanding its "motives." A good solution would be to use other machines ("reasoning checkers") to corroborate the reasoning and advice of the first machine; in such a case it would be essential to understand any "side effects" that the first machine's reasoning might have on the reasoning checker.

It has been suggested that an intelligent machine might arise accidentally, without our knowledge, through some fortuitous interconnection of smaller machines (see Heinlein, 1966). If the smaller machines each helped to control some aspect of our economy or defense, the accidental intelligence might well act as a dictator. (A similar theme has been stated in Reich, 1971, and Eiseley, 1970, and by many others, who have suggested that humanity and its machines are evolving into an autonomous meta-organism.) It seems highly unlikely that this will happen, especially if we devote sufficient time to studying the non-accidental systems we implement.

A more significant danger is that artificial intelligence might be used to further the interests of *human* dictators. A limited supply of intelligent machines in the hands of a human dictator might greatly increase his power over other human beings, perhaps to the extent of giving him complete censorship and supervision of the public. The vision of Big Brother in Orwell's *1984* could become a mechanical reality. Recent work on an "electronic battlefield" and the development of computerized data files containing individual information about many American citizens (see A. R. Miller, 1971; Sprague, 1969) illustrate this possibility. In this regard, it should be noted that most AI research in the United States is funded by the Department of Defense (specifically, by the Advanced Research Projects Agency, or ARPA). Indeed, in this country most scientific research of any sort is funded by the Department of Defense. The AI research supported by the ARPA Office of Information Processing Techniques is unclassified, available to the public, and not intended for any Big Brothers. Still, science should be supported in its own light or else our real problems may never be solved, namely—we must replace our inefficient technology before the finite resources of our planet are wasted.

Finally, industrial automation is currently foreseen as one of the chief applications of artificial intelligence. It is possible that a factory could be completely operated by machines. Indeed, it is possible that most of the physical and paper-work drudgery necessary to sustain an entire economy could be performed by intelligent machines. If this were to happen too quickly, some people might not be able to adjust to their increased leisure time (see Vonnegut, 1952).

## The Well-Natured Machine

Let us now paint another, more positive picture of the world that might result from artificial intelligence research. Hints about this world have appeared throughout this book, and it is time to look more clearly at some of its qualities. It is a world in which man and his machines have reached a state of *symbiosis,* both with each other and with the rest of the environment taken as a whole (see Landers, 1966). "Symbiosis" is a word from biology, referring to a relationship between two or more species, which is mutually beneficial to the members of each. It is an apt word because automation and artificial intelligence research are concerned with the development of a "species of machines" that will simulate at least two of the major abilities of living organisms: self-repair and intelligence (and perhaps self-reproduction).

The benefits humanity might gain from achieving such a symbiosis are enormous. As mentioned above, it may be possible for artificial intelligence to greatly reduce the amount of human labor necessary to operate the economy of the world (Simon, 1962; McCarthy, 1971). Computers and AI research may play an important part in helping to overcome the food, population, housing, and other crises that currently grip the earth, and in helping us to understand the dynamics of the relevant "world system" (Forrester, 1971). They may also play a significant role in the planning of technological systems, the decentralization of cities, and in the discovery and prevention of ecological disasters (see Graves, Pingry, and Whinston, 1971; Langlois, 1971; Olsen, 1971; Ross, 1971). Artificial intelligence may eventually be used to contain the "information explosion" (see Kochen, 1967; Licklider, 1965) and, perhaps, to partially automate the development of science itself (Buchanan, Fiegenbaum, and Lederberg, 1971; Hearn, 1971; Paton, 1970). As Toffler (1970) suggested, a sophisticated, computer-controlled economic system could give us a *greater* diversity of choices for housing and other goods than we have ever had in the past. The future may well see the development of an "information utility" (Sprague, 1969; Armer, 1968) that will enable each individual of the general public to have a computer terminal in his home that will give him access to the current "public information" of the world, as well as to the abilities of a general problem-solving artificial intelligence, for a price roughly comparable to that he currently pays for electricity or water. Perhaps artificial intelligence will someday be used in automatic teachers that will be as good at teaching as people are, and perhaps mechanical translators will someday be developed which will fluently translate human languages. And (*very* perhaps) the day may eventually come when the "household robot" and the "robot chauffeur" will be a reality (see Schmidt, 1971).

There is no space to allow discussion of these possibilities in detail, but the Bibliography references will enable the reader to make a good start at studying them himself, if he likes. In some ways it is reassuring that progress in artificial intelligence research is proceeding at a relatively slow but regular pace. It should be at least a decade before any of these possibilities becomes an actuality, which will give us some time to consider in more detail the issues involved.

Even if all these advances should occur, there will still be work for people to do, and people will still be able to feel pain or be unhappy: There will, however, be more time for people to do other things than work, and less cause for them to suffer needlessly. So, the harvest of artificial intelligence may be for the good of humanity.

# TO ☺ Tic -Tac- toe

```
10   Forward 90
20   PENUP
30   Right 90
40   Forward 30
50   Right 90
60   pen down
70   Forward 90
80   pen up
90   LEFT 135
95   Forward 90
100  Left 135
110  pen down
120  Forward 90
130  pen up
140  Right 90
150  Forward 30
160  Right 90
170  Pen down
190  Forward 90
end
```

Figure 9–4. Program written by a child. (See Papert, 1971.)

# BIBLIOGRAPHY

## ABBREVIATIONS

| | | | |
|---|---|---|---|
| *Acad.* | Academy | *Nat.* | National |
| *Amer.* | American | *Patt.* | Pattern |
| *Art.* | Artificial | *Phil.* | Philosophy |
| *Assoc.* | Association | *Prob.* | Probability |
| *Aut.* | Automata | *Proc.* | Proceedings |
| *Behav.* | Behavioral | *Proj.* | Project |
| *Bull.* | Bulletin | *Recog.* | Recognition |
| *Comm.* | Communications | *Rept.* | Report |
| *Comp.* | Computer(s) | *Res.* | Research |
| *Conf.* | Conference | *Rev.* | Review |
| *Dept.* | Department | *Sci.* | Science(s) |
| *Dev.* | Development | *Soc.* | Society |
| *Elect.* | Electronic(s) | *St.* | State |
| *Infor.* | Information | *Stat.* | Statistics |
| *Inst.* | Institute | *Symb.* | Symbolic |
| *Int.* | International | *Symp.* | Symposium |
| *Intell.* | Intelligence | *Sys.* | System(s) |
| *J.* | Journal | *Tech.* | Technical |
| *Lab.* | Laboratory | *Trans.* | Transactions |
| *Math.* | Mathematical, mathematics | *Univ.* | University |

# MNEMONICS FOR SYMPOSIA, PROCEEDINGS, AND SPECIAL COLLECTIONS

*ACM*, Association for Computing Machinery

*AFIPS*, American Federation of Information Processing Societies

*AIHP*, Findler, N. V., and Meltzer, B., eds. (1971). *Artificial Intelligence and Heuristic Programming.* New York: American Elsevier

*CACM, Communications of the Association for Computing Machinery.* New York: ACM

*CT*, Feigenbaum, E. A., and Feldman, J., eds. (1963). *Computers and Thought.* New York: McGraw-Hill

*ECA*, Burks, A. W., ed. (1970). *Essays on Cellular Automata.* Chicago: Univ. Ill. Press

*FJCC, Proceedings of the Fall Joint Computer Conference of the American Federation of Information Processing Societies.* Montvale, N.J.: AFIPS Press

*IC, Information and Control.* New York: Academic Press

*IEEE*, Institute of Electrical and Electronics Engineers

*IJCAI*, Walker, D. E., and Norton, L. M., eds. (1969). *Proceedings of the International Joint Conference on Artificial Intelligence.* Washington, D.C.: MITRE Corp.

*IJCAI2*, British Computer Society (1971). *Proceedings of the Second International Joint Conference on Artificial Intelligence.* London: British Computer Society

*IRE*, Institute of Radio Engineers

*JACM, Journal of the Association for Computing Machinery.* New York: ACM

*MI1*, Collins, N. L., and Michie, D., eds. (1967). *Machine Intelligence 1.* New York: American Elsevier

*MI2*, Dale, E., and Michie, D., eds. (1968). *Machine Intelligence 2.* New York: American Elsevier

*MI3*, Michie, D., ed. (1968). *Machine Intelligence 3.* New York: American Elsevier

*MI4*, Meltzer, B., and Michie, D., eds. (1969). *Machine Intelligence 4.* New York: American Elsevier

*MI5*, Meltzer, B., and Michie, D., eds. (1970). *Machine Intelligence 5.* New York: American Elsevier

*MI6*, Meltzer, B., and Michie, D., eds. (1971). *Machine Intelligence 6.* New York: American Elsevier

*PCR*, Pylyshyn, Z. W., ed. (1970). *Perspectives on the Computer Revolution* Englewood Cliffs, N.J.: Prentice-Hall

*RM*, Simon, H. A., and Siklossy, L., eds. (1972). *Representation and Meaning: Experiments with Information Processing Systems.* Englewood Cliffs: Prentice-Hall

*SIP*, Minsky, M., ed. (1968). *Semantic Information Processing*. Cambridge: M.I.T. Press

*SJCC, Proceedings of the Spring Joint Computer Conference of the American Federation of Information Processing Societies*. Montvale, N.J.: AFIPS Press

*TAPR*, Banerji, R., and Mesarovic, M. D., eds. (1970). *Theoretical Approaches to Non-Numerical Problem Solving*. New York: Springer-Verlag

# BIBLIOGRAPHY

Abelson, R. P., and Carroll, J. D. (1965). Computer simulation of individual belief systems. *Amer. Behav. Sci.*, vol. 9, pp. 24–30.

Abelson, R. P., and Reich, C. M. (1969). Implicational molecules: A method for extracting meaning from input sentences. *IJCAI*, pp. 641–647.

Abend, K. (1968). Compound decision procedures for unknown distributions and for dependent states of nature. In *Pattern Recognition*, ed. L. N. Kanal, pp. 207–249. Washington, D.C.: Thompson.

Abrahams, G. (1960). *The Chess Mind*. Baltimore: Penguin.

Ackermann, R. (1967). *An Introduction to Many-Valued Logics*. New York: Dover.

Aho, A. V., and Ullman, J. D. (1968). The theory of languages. *Math. Sys. Theory*, vol. 2, no. 2, pp. 97–125.

Aida, S., Cordella, L., and Ivacevic, N. (1971). Visual-tactile symbiotic system for stereometric pattern recognition. *IJCAI2*, pp. 365–375.

——— See also Kinoshita, Aida, and Mori.

Aiken, H. (1937). Proposed automatic calculating machine. *PCR*, pp. 29–36.

Ajdukiewicz, K., ed. (1965). *The Foundation of Statements and Decisions*. Warsaw: Panstwowe Wydawnictwo Naukowe.

Alekhine, A. (1964). *My Best Games of Chess*. (3 vols.) London: G. Bell.

Aleksander, I. (1971). Artificial intelligence and all that. *Wireless World* (*Britain*), vol. 77, no. 1432, pp. 494–495.

Amarel, S. (1967). An approach to heuristic problem solving and theorem proving in the propositional calculus. In *Systems and Computer Science*, eds. J. Hart and S. Takasu. Toronto: Univ. Toronto Press.

——— (1968). On representations of problems of reasoning about actions. *MI3*, pp. 131–171.

——— (1969). On the representation of problems and goal directed procedures for computers. *Comm. Amer. Soc. Cybernetics*, vol. 1, no. 2.

——— (1970). On the representation of problems and goal-directed procedures for computers. *TAPR*, pp. 179–244.

Anderson, A. R., ed. (1964). *Minds and Machines*. Englewood Cliffs, N.J.: Prentice-Hall.

Andrews, H. C., Pratt, W. K., and Caspari, K. (1970). *Computer Techniques in Image Processing*. New York: Academic Press.

Andrews, P. B. (1968). Resolution with merging. *JACM*, vol. 15, pp. 367–381.

Apter, M. J. (1966). *Cybernetics and Development*. Oxford: Pergamon.

Arbib, M. A. (1964). *Brains, Machines, and Mathematics*. New York: McGraw-Hill.

———, ed. (1968). The Algebraic Theory of Machines, Languages, and Semigroups. New York: Academic Press.

——— (1969). *Theories of Abstract Automata*. Englewood Cliffs, N.J.: Prentice-Hall.

——— (1972). Toward an automata theory of brains. *CACM*, vol. 15, no. 7, pp. 521–527.

——— (1972). *The Metaphorical Brain: An Introduction to Cybernetics as Artificial Intelligence and Brain Theory*. New York: Wiley.

Arbib, M. A., and Blum, M. (1965). Machine dependence of degrees of difficulty. *Proc. Amer. Math. Soc.*, vol. 16, pp. 568–570.

Arkadev, A. G., and Braverman, E. M. (1967). *Computers and Pattern Recognition*. Washington, D.C.: Thompson.

Armer, P. (1963). Attitudes toward intelligent machines. *CT*, pp. 389–405.

——— (1968). Social implications of the computer utility. In *Computers and Communications—Toward a Computer Utility*, ed. F. Gruenberger, pp. 191–198. Englewood Cliffs, N.J.: Prentice-Hall.

Arnheim, R. (1971). *Visual Thinking*. Berkeley: Univ. California Press.

Arora, S. R., and Gallo, A. (1971). Optimal sizing, landing and reloading in a multi-level memory hierarchy system. *SJCC*, vol. 38, pp. 337–344.

Asbell, B. (1963). *The New Improved American*. New York: McGraw-Hill.

Ashcroft, E., and Manna, Z. (1971). The translation of "go to" programs to "while" programs. AIM-138. Stanford: Stanford Art. Intell. Proj.

Ashcroft, E., and Pneuli, A. (1971). Decidable properties of monadic functional schemas. AIM-148. Stanford: Stanford Art. Intell. Proj. *Also in Theory of Machines and Computations*, eds. Z. Kohavi and A. Paz, pp. 3–18. New York: Academic Press.

Asimov, I. (1950). *I, Robot*. Greenwich, Conn.: Fawcett.

——— (1964). *Eight Stories from the Rest of the Robots*. New York: Doubleday.

——— *See also* Skinner et al.

Astrahan, M. M. (1970). Speech analysis by clustering, or the hyperphoneme method. AIM-124. Stanford: Stanford Art. Intell. Proj.

Atkinson, S. M. (1971). The control of short term memory. *Scientific American*, August, vol. 225, no. 2, pp. 82–90.

Auerbach Publishers (1971). *Auerbach on Optical Character Recognition*. Philadelphia: Auerbach.

Averbakh, Y. (1966). *Chess Endings: Essential Knowledge*. New York: Pergamon.

Avizienis, A., Gilley, G. C., Mathur, F. P., Rennels, D. A., Rohr, J. A., and Rubin, D. K. (1971). The STAR (self-testing and repairing) computer: An investigation of the theory and practice of fault-tolerant computer design. *IEEE Trans. Comp.*, vol. C-20, no. 11, pp. 1312–1321.

Babbage, C. (1864). Of the analytical engine. *PCR*, pp. 16–28.

———— *See also* Morrison and Morrison.

Backus, J. W. (1959). The syntax and semantics of the proposed international algebraic language of the Zurich ACM-GAMM Conference. *Proc. Int. Conf. Information Processing*, pp. 125–132. Paris: UNESCO.

Baer, J.-L. E., and Estrin, G. (1969). Bounds for maximum parallelism in a bilogic graph model of computations. *IEEE Trans. Comp.*, vol. C-18, no. 11, pp. 1012–1014.

Bajcsy, R. (1972). Computer identification of textured visual scenes. AIM-180. Stanford: Stanford Art. Intell. Proj.

Baker, R. (1972). An iterative array which can represent the rotation and translation of objects. *International J. Man-Machine Studies*, vol. 4, no. 2, pp. 85–103.

Balzer, R. M. (1969). Search for a solution: a case study. *IJCAI*, pp. 21–31.

———— (1972). Automatic programming. Inst. Tech. Memorandum. Marina Del Rey: Infor. Sci. Inst., Univ. Southern California.

Banerji, R. B. (1969). *Theory of Problem Solving: An Approach to Artificial Intelligence*. New York: Elsevier.

———— (1970). Game playing programs: An approach and overview. *TAPR*, pp. 21–58.

———— (1971). Some linguistic and statistical problems in pattern recognition. *Patt. Recog.*, vol. 3, pp. 409–419.

Banerji, R. B., and Ernst, G. W. (1971). Changes in representation which preserve strategies in games. *IJCAI2*, pp. 651–658.

Barbizet, J. (1970). *Human Memory and Its Pathology*. San Francisco: Freeman.

Bar-Hillel, Y. (1964). *Language and Information*. Reading, Mass.: Addison-Wesley.

Bar-Hillel, Y., Perles, M., and Shamir, E. (1961). On formal properties of simple phrase-structure grammars. *Z.f. Phonetik Sprachwissenschaft und Communikationsforschung*, vol. 15, pp. 143–172.

———— *See also* Pierce.

Barondes, S. H. (1970). Brain glycomacromolecules interneuronal recognition. In *The Neurosciences: Second Study Program*, ed. F. O. Schmidt, pp. 747–760. New York: Rockefeller Univ. Press.

———— (1970). Multiple steps in the biology of memory. In *The Neurosciences: A Second Study Program*, ed. F. O. Schmitt, pp. 272–278. New York: Rockefeller Univ. Press.

Barrow, H. G., Ambler, A. P., and Burstall, R. M. (1972). Some techniques for recognizing structures in pictures. In *Frontiers of Pattern Recognition*, ed. S. Watanabe, pp. 1–29. New York: Academic.

Bartee, T. C. (1966). *Digital Computer Fundamentals.* New York: Mc-Graw-Hill.

Barth, J. (1966). *Giles Goat-Boy, or the Revised, New Syllabus.* New York: Doubleday.

Bartley, W. W. (1972). Lewis Carroll's lost book on logic. *Scientific American,* July, pp. 38–46.

Bauer, F. L., and Wössner, H. (1972). The "Plankalkül" of Konrad Zuse: A forerunner of today's programming languages. *CACM,* vol. 15, no. 7, pp. 678–685.

Baum, L. F. (1907). *Ozma of Oz.* Reprinted, 1968. New York: Rand McNally.

Baumgart, B. G. (1972). Micro-planner alternate reference manual. SAILON No. 67. Stanford: Stanford Art. Intell. Proj.

Becker, J. D. (1969). The modelling of simple analogic and inductive processes in a semantic memory system. *IJCAI,* pp. 655–668.

Becker, P. W. (1968). Recognition of patterns using the frequencies of occurrence of binary words. Lyngby, Denmark: Elect. Lab., Technical Univ. Denmark.

Beckwith, J. (1964). *Early Medieval Art.* New York: Praeger.

Beeler, M., Gosper, R. W., and Schroeppel, R. (1972). HAKMEM. A.I. Memo 239. Cambridge: M.I.T. Art. Intell. Lab.

Beizer, B. (1971). *The Architecture and Engineering of Digital Computer Complexes.* New York: Plenum.

Bell, C. G., and Newell, A., eds. (1971). *Computer Structures: Readings and Examples.* New York: McGraw-Hill.

Bell, C. G., et al. (1971). C.ai: A computing environment for AI research. Pittsburgh: Comp. Sci. Dept., Carnegie-Mellon Univ.

Benacerraf, P., and Putnam, H., eds. (1964). *Philosophy of Mathematics.* Englewood Cliffs, N.J.: Prentice-Hall.

Berkeley, E. C. (1962). The social responsibilities of computer people. *PCR,* pp. 461–471.

Berliner, H. J. (1970). Experiences gained in constructing and testing a chess program. Presented at IEEE Symp. Sys. Sci. and Cybernetics, Pittsburgh, Pa., October 1970.

Bernstein, J. (1964). *The Analytical Engine: Computers—Past, Present and Future.* New York: Random House.

Bertoni, A. (1972). Complexity problems related to the approximation of probabilistic languages and events by deterministic machines. Milan: Gruppo di Elettronica e Cibernetica, Universita' Degli Studi di Milano.

Biss, K. O., Chien, R. T., and Stahl, F. A. (1972). A data structure for cognitive information retrieval. *Int. J. Comp. and Infor. Sci.,* vol. 1, no. 1, pp. 17–27.

Black, F. (1964). A deductive question-answering system. *SIP,* pp. 354–402.

Bledsoe, W. W. (1961). A basic limitation on the speed of digital computers. *IRE Trans. Elect. Comp.,* vol. EC-10, p. 530.

———— (1971). Splitting and reduction heuristics in automated theorem proving. *Art. Intell.*, vol. 2, pp. 55–77.

Bledsoe, W. W., Boyer, S., and Henneman, W. H. (1971). Computer proofs of limit theorems. *IJCAI2*, pp. 586–600.

Blum, M. (1964). Measures on the computation speed of partial-recursive functions. *Quarterly Progress Report 72*. Cambridge: M.I.T. Res. Lab. Electronics.

———— (1967). A machine-independent theory of the complexity of recursive functions. *JACM*, vol. 15, pp. 332–336.

———— *See also* Arbib and Blum.

Bobrow, D. G. (1964). Natural language input for a computer problem-solving system. Rept. TR-1. Cambridge: M.I.T. Art. Intell. Lab. Also in *SIP*, pp. 146–226.

————, ed. (1968). *Symbol Manipulation Languages and Techniques*. Amsterdam: North-Holland.

———— (1972). Requirements for advanced programming systems for List processing. *CACM*, vol. 15, no. 7, pp. 618–627.

Bobrow, D. G., and Fraser, B. (1969). An augmented state transition network analysis procedure. *IJCAI*, pp. 557–567.

Bobrow, D. G., and Klatt, D. H. (1968). A limited speech recognition system. *FJCC*, vol. 33, part 1, pp. 305–318.

Bobrow, D. G., and Wegbreit, B. (1972). A model and stack implementation of multiple environments. BBN Rept. No. 2334. Cambridge: Bolt Beranek and Newman.

Bobrow, D. G., et al. *See also* Teitelman, Bobrow, et al.

Bochvar, D. A. (1972). Two papers on partial predicate calculus. AIM-165. Stanford: Stanford Art. Intell. Proj.

Boeke, K. (1957). *Cosmic View*. New York: John Day.

Boole, G. (1854). *An Investigation of the Laws of Thought*. New York: Dover.

Borko, H. (1968). National and international information networks in science and technology. *FJCC*, vol. 33, part 2, pp. 1469–1472.

Botvinnik, M. M. (1970). *Computers, Chess and Long-Range Planning*. New York: Springer-Verlag.

Boulding, K. E. (1964). *The Meaning of the 20th Century: The Great Transition*. New York: Harper & Row.

Brady, J. V. (1951). The effect of electroconvulsive shock on a conditioned emotional response: The permanence of the effect. *J. Comparative and Physiological Psychology*, vol. 41, pp. 507–511.

Brand, S. (1972). Spacewar: Fanatic life and symbolic death among the computer bums. *Rolling Stone*, Dec. 7, no. 123, pp. 50–58.

Bredt, T. H. (1970a). A survey of models for parallel computing. SU-SEL-70-044. Stanford: Stanford Elect. Lab.

———— (1970a). Analysis and synthesis of concurrent sequential programs. SU-SEL-70-024. Stanford: Stanford Elect. Lab.

Bredt, T. H., and McCluskey, E. J. (1970). A model for parallel computer systems. SU-SEL-70-014. Stanford: Stanford Elect. Lab.

Bremermann, H. J. (1962). Optimization through evolution and recombination. In *Self-Organizing Systems*, eds. M. C. Yovits, G. T. Jacobi, and G. O. Goldstein, pp. 93–106. Washington, D.C.: Spartan.

———— (1967). Quantum noise and information. *Proc. 5th Berkeley Symp. on Math. Stat. and Prob.*, vol. 4, pp. 15–20. Berkeley: Univ. Calif. Press.

Brice, C. R., and Fennema, C. L. (1970). Scene analysis using regions. *Art. Intell.*, vol. 1, pp. 205–226.

Brodatz, P. (1966). *Textures*. New York: Dover.

Brown, S., Gries, D., and Szymanski, T. (1972). Program schemes with pushdown stores. TR 72–126. New York: Cornell Univ. Dept. Comp. Sci.

Buchanan, B. G., Feigenbaum, E. A., and Lederberg, J. (1971). A heuristic programming study of theory formation in science. *IJCAI2*, pp. 40–50.

Bull, G. M., and Packham, S. F. G. (1971). *Time-Sharing Systems*. London: McGraw-Hill.

Buneman, P. (1970). A grammar for the topological analysis of plane figures. *MI5*, pp. 383–394.

Burck, G., et al. (1965). *The Computer Age and Its Potential for Management*. New York: Harper & Row.

Burke, J. G., ed. (1967). *The New Technology and Human Values*. Belmont, Calif.: Wadsworth.

Burks, A. W., ed. (1970). *Essays on Cellular Automata*. Chicago: Univ. Ill. Press.

Burks, A. W., Goldstine, H. H., and von Neumann, J. (1946). Preliminary discussion of the logical design of an electronic computing instrument. *PCR*, pp. 37–46.

Burstall, R. M., Collins, J. S., and Popplestone, R. J. (1971). *Programming in POP-2*. Edinburgh: Edinburgh Univ. Press.

Bush, V. (1945). As we may think. *PCR*, pp. 47–59.

Cade, C. M. (1966). *Other Worlds Than Ours*. New York: Taplinger.

Cadiou, J. M. (1972). Recursive definitions of partial functions and their computations. AIM-163. Stanford: Stanford Art. Intell. Proj.

Cameron, A. G. W., ed. (1963). *Interstellar Communication*. New York: Benjamin.

Campbell, D. (1960). Blind variation and selective survival as a general strategy in knowledge-processes. In *Self-Organizing Systems*, eds. M. Yovits and S. Cameron, pp. 205–231. New York: Pergamon.

Carnap, R. (1947). *Meaning and Necessity, A Study in Semantics and Modal Logic*. Chicago: Univ. Chicago Press.

———— (1950). *Logical Foundations of Probability*. Chicago: Univ. Chicago Press.

Carroll, J. B. (1964). *Language and Thought*. Englewood Cliffs, N.J.: Prentice-Hall.

Carroll, L. (1932). Photography extraordinary. In *The Rectory Umbrella and Mischmasch*. New York: Dover.

—— (1956). What the Tortoise said to Achilles and other riddles. In *The World of Mathematics*, ed. J. R. Newman, vol. 4, pp. 2403–2409. New York: Simon and Schuster.

—— (1958). *Symbolic Logic and the Game of Logic*. New York: Dover.

—— *See also* Bartley; Collingwood.

Carter, W. C. (1971). Diagnostic procedures to find component faults in self-repairing computers. *Mexico Int. IEEE Conf. Sys., Networks, Comp.*, vol. 2, pp. 954–958.

Casey, R., and Nagy, G. (1966). Recognition of printed Chinese characters. *IEEE Trans. Elect. Comp.*, vol. EC-15, pp. 91–101.

Castaneda, C. (1972). Journey to Ixtlan. *Psychology Today*, December, vol. 6, no. 7, pp. 102–109.

Chaitin, G. J. (1966). On the length of programs for computing finite binary sequences. *JACM*, vol. 13, no. 4, pp. 547–569.

—— (1969). On the length of programs for computing finite binary sequences. *JACM*, vol. 16, no. 1, pp. 145–159.

Chamberlin, D. C. (1971). The "single-assignment" approach to parallel processing. *FJCC*, vol. 39, pp. 263–269.

Chambers, K. L., ed. (1970). *Biochemical Coevolution*. Corvallis: Oregon St. Univ. Press.

Chandra, A. K., and Manna, Z. (1973). On the power of programming features. AIM-185. Stanford: Stanford Art. Intell. Proj.

Chang, C. C. (1958). Algebraic analysis of many-valued logics. *Trans. Amer. Math. Soc.*, vol. 88, pp. 467–490.

Chang, C. L. (1968). Fuzzy topological spaces. *J. Math. Analysis and Applications*, vol. 24, no. 1, pp. 182–190.

Chang, C. L., and Slagle, J. R. (1971). An admissible and optimal algorithm for searching AND/OR graphs. *Art. Intell.*, vol. 2, no. 2, pp. 117–128.

Chang, C. L., et al. *See also* Slagle, Chang, and Lee.

Chang, S.-K. (1971). Fuzzy programs. *Proc. Brooklyn Polytechnical Inst. Symp. Comp. Aut.*, vol. 1, no. 21.

Chapin, N. (1957). *An Introduction to Automatic Computers*. Princeton, N.J.: Van Nostrand.

—— (1971a). *Computers: A Systems Approach*. New York: Van Nostrand-Reinhold.

—— (1971b). *Flowcharts*. Princeton, N.J.: Auerbach.

Charniak, E. (1969a). CARPS, a program which solves calculus word problems. MAC-TR-51. Cambridge: M.I.T. Art. Intell. Lab.

—— (1969b) Computer solution of calculus word problems. *IJCAI*, pp. 303–316.

—— (1972). Toward a model of children's story comprehension. AI-TR-266. Cambridge, Mass.: M.I.T. Art. Intell. Lab.

Chartrand, R. L. (1971). *Systems Technology Applied to Social and Community Problems*. New York: Spartan.

Chen, T. C. (1971). Unconventional superspeed computer systems. *SJCC*, vol. 38, pp. 365–372.

Cherry, C. (1957). *On Human Communication*. Cambridge: M.I.T. Press.

Childs, D. L. (1968). Description of a set-theoretic data structure. *FJCC*, vol. 33, part 1, pp. 557–564.

Chomsky, N. (1959). On certain formal properties of grammars. *IC*, vol. 2, pp. 137–167.

——— (1966). *Cartesian Linguistics: A Chapter in the History of Rationalist Thought*. New York: Harper & Row.

——— (1968). *Language and Mind*. New York: Harcourt.

Chomsky, N., and Miller, G. A. (1958). Finite state languages. *IC*, vol. 1, pp. 91–112.

Chow, Y. S., Robbins, H., and Siegmund, D. (1971). *Great Expectations: The Theory of Optimal Stopping*. Boston: Houghton-Mifflin.

Church, A. (1941). *The Calculi of Lambda-Conversion*. Princeton, N.J.: Princeton Univ. Press.

——— (1956). *Introduction to Mathematical Logic*. Princeton, N.J.: Princeton Univ. Press.

Churchman, C. W. (1938). On finite and infinite modal systems. *J. Symb. Logic*, vol. 3, pp. 77–82.

——— (1968). Real time systems and public information. *FJCC*, vol. 33, part 2, pp. 1467–1468.

——— (1970). The role of Weltanschauung in problem solving and inquiry. *TAPR*, pp. 141–160.

Citrenbaum, R. L. (1972). Strategic pattern generation: a solution technique for a class of games. *Patt. Recog.*, vol. 4, pp. 317–329.

Clark, R., and Miller, W. (1966). Computer based data analysis systems at Argonne. In *Methods in Computational Physics*, ed. F. Alt, vol. 5, pp. 47–98. New York: Academic Press.

Clarke, A. C. (1962). *Profiles of the Future: An Inquiry into the Limits of the Possible*. New York: Harper & Row.

——— (1968). The mind of the machine. *Playboy*, December, pp. 116–119, 122,293–294.

Clarkson, G. P. E. (1963). A model of the trust investment process. *CT*, pp. 347–371.

Cleave, J. P. (1963). A hierarchy of primitive recursive functions. *Z. F. Math. Logik und Grundlagen d. Math.*, vol. 9, pp. 331–345.

Clowes, M. B. (1970a). Picture syntax. In *Picture Language Machines*, ed. S. Kaneff, pp. 119–149. New York: Academic Press.

——— (1970b). On the description of board games. In *Picture Language Machines*, ed. S. Kaneff, pp. 397–420. New York: Academic Press.

——— (1971). On seeing things. *Art. Intell.*, vol. 2, no. 1, pp. 79–116.

Cobham, H. A. (1964). The intrinsic computational difficulty of functions. *Proc. Int. Congress for Logic, Methodology, and Phil. Sci.*, pp. 24–30. Amsterdam: North-Holland.

Codd, E. F. (1968). *Cellular Automata.* New York: Academic Press.

Colby, K. M. (1967). Computer-aided language development in nonspeaking mentally disturbed children. Tech. Rept. CS 85. Stanford: Stanford Univ. Comp. Sci. Dept.

—— (1970). Mind and brain, again. AIM-116. Stanford: Stanford Art. Intell. Proj.

Colby, K. M., and Enea, H. (1967a). Machine utilization of the natural language word 'good'. Tech. Rept. CS 78. Stanford: Stanford Comp. Sci. Dept.

—— (1967b). Heuristic methods for computer understanding of natural language in context-restricted on-line dialogues. *Math. Biosciences,* vol. 1, pp. 1–25.

—— (1968). Inductive inference by intelligent machines. *Scientia,* vol. 103, pp. 669–720.

Colby, K. M., and Smith, D. C. (1969). Dialogues between humans and an artificial belief system. *IJCAI,* pp. 319–324.

Colby, K. M., Enea, H., and Tesler, L. (1969). Experiments with a search algorithm for the data-base of a human belief system. *IJCAI,* pp. 649–654.

Colby, K. M., Weber, S., and Hilf, F. D. (1971). Artificial paranoia. *Art. Intell.,* vol. 2, no. 1, pp. 1–25.

Colby, K. M., Weber, S., Hilf, F. D., and Kraemer, H. C. (1971). A resemblance test for the validation of a computer simulation of paranoid processes. AIM-156. Stanford: Stanford Art. Intell. Proj.

Coles, L. S. (1969). Talking with a robot in English. *IJCAI,* pp. 587–596.

—— (1970a). An experiment in robot tool using. Tech. Note 41. Menlo Park: Art. Intell. Center, Stanford Research Institute.

—— (1970b). Bibliography of literature in the field of robots. Tech. Note 23. Menlo Park: Art. Intell. Center, Stanford Research Institute.

—— (1971). The application of theorem proving to information retrieval. Tech. Note 51. Menlo Park: Art. Intell. Center, Stanford Research Institute.

—— (1972). Syntax directed interpretation of natural language. *RM,* pp. 211–287.

Coles, L. S., and Green, C. C. (1969). Chemistry question-answering. Tech. Note 9. Menlo Park: Art. Intell. Center, Stanford Research Institute.

Collingwood, S. D., ed. (1961). *Diversions and Digressions of Lewis Carroll.* New York: Dover.

Colodny, R. G., ed. (1966). *Mind and Cosmos: Essays in Contemporary Science and Philosophy.* Pittsburgh: Univ. Pittsburgh Press.

Conklin, G., ed. (1954). *Science-Fiction Thinking Machines: Robots, Androids, Computers.* New York: Vanguard.

Cooper, D. C. (1966). The equivalence of certain computations. *Comp. J.,* vol. 9, no. 1, pp. 45–52.

Cover, T. M. (1965). Geometrical and statistical properties of systems of linear inequalities with applications in pattern recognition. *IEEE Trans. Elect. Comp.*, vol. EC-14, pp. 326–334.

Cragg, B. G. (1967). The density of synapses and neurons in the motor and visual areas of the cerebral cortex. *J. Anatomy*, vol. 101, no. 4, pp. 639–654.

Crespi-Reghizzi, S. (1971). Reduction of enumeration in grammar acquisition. *IJCAI2*, pp. 546–552.

CRM Books, Inc. (1970). *Psychology Today, An Introduction*. Del Mar, Calif.: CRM Books.

Crosson, F. J., ed. (1970). *Human and Artificial Intelligence*. New York: Appleton.

Culbertson, J. T. (1963). *The Minds of Robots*. Urbana: Univ. Illinois Press.

Culver, W. H., and Mehran, F. (1971). Coherent optical logic. RC 3533. Yorktown Heights, N.Y.: IBM Watson Research Center.

Damron, S., et al. (1968). A random access terabit magnetic memory. *FJCC*, vol. 33, part 2, pp. 1381–1387.

Darden, S. C. (1969). A contextual recognition system for formal languages. *IJCAI*, pp. 597–607.

Darlington, J. (1964). Translating ordinary English into symbolic logic. MAC-M-149. Cambridge: M.I.T. Proj. MAC.

———— (1968). Automatic theorem proving with equality substitutions and mathematical induction. *MI3*, pp. 113–127.

Davis, M. (1958). *Computability and Unsolvability*. New York: McGraw-Hill.

———— ed. (1965). *The Undecidable*. Hewlett, N.Y.: Raven.

Davis, M., and Putnam, H. (1960). A computing procedure for quantification theory. *JACM*, vol. 7, no. 3, pp. 201–215.

de Bono, E. (1969). *The Mechanism of Mind*. New York: Simon and Schuster.

de Bruijn, N. G. (1972). A solitaire game and its relation to a finite field. *J. Recreational Math.*, vol. 5, no. 2, pp. 133–137.

De Groot, A. D. (1965). *Thought and Choice in Chess*. Paris: Mouton.

Delgado, J. M. R. (1969). *Physical Control of the Mind: Toward a Psychocivilized Society*. New York: Harper & Row.

Demczynski, S. (1964). *Automation and the Future of Man*. London: George Allen and Unwin.

Denes, P. B., and Mathews, M. V. (1968). Computer models for speech and music appreciation. *FJCC*, vol. 33, part 1, pp. 319–327.

Derksen, J. A., Rulifson, J. F., and Waldinger, R. J. (1972). The QA4 language applied to robot planning. *FJCC*, vol. 41, part 2, pp. 1181–1192.

Deutsch, J. A. (1969). The physiological basis of memory. *Annual Rev. Psychology*, vol. 20, pp. 85–104.

Deutsch, D. (1972). Music and memory. *Psychology Today*, December, vol. 6, no. 7, pp. 86–89, 119.

Deutsch, E. S., and Hayes, K. C., Jr. (1972). A heuristic solution to the tangram puzzle. TR-177, GJ-754. College Park, Md.: Univ. Maryland Comp. Sci. Center.

Diebold, J. (1969). *Man and the Computer: Technology as an Agent of Social Change.* New York: Praeger.

Dijkstra, E. W. (1965). Cooperating sequential processes. Rept. EWD 123. Endhoven, The Netherlands: Math. Dept., Technological Univ. *Also* in *Programming Languages,* ed. F. Genuys, pp. 43–112. New York: Academic Press.

———— (1968). The structure of "THE" multiprogramming system. *CACM,* vol. 11, no. 5, pp. 341–346.

———— (1971). Hierarchical ordering of sequential processes. *Acta Informatica,* vol. 1, no. 2, pp. 115–138.

Doran, J. E. (1969). Planning and generalization in an automaton automaton/environment system. *MI4,* pp. 435–454.

———— (1970). Planning and robots. *MI5,* pp. 519–532.

Dreben, B. (1952). On the completeness of quantification theory. *Proc. Nat. Academy Sci.,* vol. 38, pp. 1047–1052.

Dreyfus, H. L. (1965). Alchemy and artificial intelligence. RAND P3244. Santa Monica, Calif.: RAND Corp.

———— (1972). *What Computers Can't Do: A Critique of Artificial Reason.* New York: Harper & Row.

Dreyfus, H. L., et al. *See also* Pierce.

Duda, R. O., and Hart, P. E. (1968). Experiments in the recognition of hand-printed text: part II—context analysis. *FJCC,* vol. 33, part 2, pp. 1139–1149.

———— (1973). *Pattern Classification and Scene Analysis.* New York: Wiley.

Dudeney, H. (1958). *The Canterbury Puzzles.* New York: Dover.

———— (1967). *536 Puzzles and Curious Problems,* ed. M. Gardner. New York: Dover.

Dunnington, G. W. (1955). *Carl Friedrich Gauss: Titan of Science.* New York: Hafner.

Dutton, J. M., and Starbuck, W. H., eds. (1971). *Computer Simulation of Human Behavior.* New York: Wiley.

Earley, J., and Sturgis, H. (1970). A formalism for translator interactions. *CACM,* vol. 13, no. 10, pp. 607–617.

Earnest, L. (1967). Choosing an eye for a computer. Memo No. 51. Stanford: Stanford Art. Intell. Proj.

Eastland, J. O., Long, E. V., et al. (1967). *Computer Privacy: Hearings before the Subcommittee on Administrative Practice and Procedure of the Committee of the Judiciary, United States Senate.* Washington, D.C.: U.S. Government Printing Office.

Eastman, C. M. (1971). Heuristic algorithms for automated space planning. *IJCAI2,* pp. 27–39.

Eber, D. (1969). *The Computer Centre Party.* Montreal: Tundra Books.

The Ecologist (1972). A blueprint for survival. *The Ecologist*, January, vol. 2, no. 1.

Eiseley, L. (1970). *The Invisible Pyramid*. New York: Scribner.

Ejiri, M., Uno, T., Yoda, H., Goto, T., Takeyasu, K. (1971). An intelligent robot with cognition and decision-making ability. *IJCAI2*, pp. 350–358.

Elcock, E. W. (1971). Problem-solving compilers. *AIHP*, pp. 37–50.

Elcock, E. W., Foster, J. M., Gray, P. M. D., McGregor, J. J., and Murray, A. M. (1971). ABSET: A programming language based on sets; motivation and examples. *MI6*, pp. 467–492.

Electronic Design (1971). Mechanical arm can even thread needles. *Electronic Design*, vol. 23, November 11, p. 30.

Elithorn, A., and Jones, D., eds. (1972). *Artificial and Human Thinking*. San Francisco: J. Bass.

Elsasser, W. M. (1969). Acausal phenomena in physics and biology: A case for reconstruction. *American Scientist*, vol. 57, no. 4, pp. 502–516.

Enea, H. (1968). MLISP. Tech. Rept. CS 92. Stanford: Stanford Univ. Comp. Sci. Dept.

Ernst, G. W. (1970). GPS and decision-making: An overview. *TAPR*, pp. 59–107.

Ernst, G. W., and Newell, A. (1967). Some issues of representation in a general problem solver. *AFIPS Conf. Proc.*, vol. 30, pp. 583–600.

——— (1969). *GPS: A Case Study in Generality and Problem Solving*. New York: Academic Press.

Ernst, G. W., et al. *See also* Banerji and Ernst.

Ernst, H. A. (1962). MH-1: A computer-operated mechanical hand. Ph.D. Thesis. Cambridge: M.I.T. Dept. Electrical Engineering.

Ershov, A. P. (1971). Parallel Programming. AIM-146. Stanford: Stanford Art. Intell. Proj.

——— (1972). Aesthetics and the human factor in programming. *CACM*, vol. 15, no. 7, pp. 501–505.

Evans, T. G. (1963). A heuristic program to solve geometric analogy problems. Ph.D. Thesis. Cambridge: M.I.T. *Also* in *SIP*, pp. 271–353.

——— (1971). Grammatical inference techniques in pattern analysis. *Software Engineering*, vol. 2, pp. 183–202.

Fang, I. (1972). FOLDS, a declarative formal language definition system. STAN-CS-72-329. Stanford: Stanford Univ. Comp. Sci. Dept.

Feigenbaum, E. A. (1963). The simulation of verbal learning behavior. *CT*, pp. 297–309.

——— (1969). Artificial intelligence: Themes in the second decade. In *Information Processing 68*, ed. A. J. H. Morrell, vol. 2, pp. 1008–1022. Amsterdam: North-Holland.

Feigenbaum, E. A., Buchanan, B. G., and Lederberg, J. (1971). Generality and problem solving: A case study using the DENDRAL program. *MI6*, pp. 165–190.

Fein, L. (1968). Impotence principles for machine intelligence. In *Pattern Recognition*, ed. L. N. Kanal, pp. 442–447. Washington, D.C.: Thompson.

Feldman, J. A. (1967). First thoughts on grammatical inference. AI Memo No. 55. Stanford: Stanford Art. Intell. Proj.

—— (1972). Automatic programming. AIM-160. Stanford: Stanford Art. Intell. Proj.

Feldman, J. A., and Gries, D. (1968). Translator writing systems. *CACM*, vol. 11, pp. 77–113.

Feldman, J. A., and Shields, P. C. (1972). Total complexity and the inference of best programs. AIM-159. Stanford: Stanford Art. Intell. Proj.

Feldman, J. A., Gips, J., Horning, J. J., and Reder, S. (1969). Grammatical complexity and inference. AI-89. Stanford: Stanford Art. Intell. Proj.

Feldman, J. A., Pingle, K., Binford, T., Falk, G., Kay, A., Paul, R., Sproull, R., and Tenenbaum, J. (1971). The use of vision and manipulation to solve the "Instant Insanity" puzzle. *IJCAI2*, pp. 359–364.

Feldman, J. A., Low, J. R., Swinehart, D. C., and Taylor, R. H. (1972). Recent developments in SAIL—an ALGOL-based language for artificial intelligence. *FJCC*, vol. 41, part 2, pp. 1193–1202.

Feys, R. (1965). *Modal Logics*. Louvain: E. Nauwelaerts.

Fialkowski, K. R. (1971). The evolutionary process of randomly growing mutated digital structures as a model of evolution of the first living organisms. *IJCAI2*, pp. 148–158.

Fikes, R. E. (1968). A heuristic program for solving problems stated as nondeterministic procedures. Ph.D. Thesis. Pittsburgh: Carnegie-Mellon Univ.

—— (1970). REF-ARF: A system for solving problems stated as procedures. *Art. Intell.*, vol. 1, pp. 27–120.

—— (1971). Monitored execution of robot plans produced by STRIPS. Tech. Note 55. Menlo Park: Art. Intell. Center, Stanford Research Institute.

Fikes, R. E., and Nilsson, N. J. (1971). STRIPS: A new approach to the application of theorem proving to problem solving. *IJCAI2*, pp. 608–620. Also in *Art. Intell.*, vol. 2, pp. 189–208.

Findler, N. V. (1961). Computer model of gambling and bluffing. *IRE Trans. Elect. Comp.*, vol. EC-10, pp. 97–98.

—— (1970). On the role of exact and non-exact associative memories in human and machine information processing. In *Software Engineering*, ed. J. T. Tou, vol. 2, pp. 141–153. New York: Academic Press.

Findler, N. V., and Chen, D. (1971). On the problems of time, retrieval of temporal relations, causality, and co-existence. *IJCAI2*, pp. 531–544.

Findler, N. V., and McKinsie, W. R. (1969). Computer simulation of a self-preserving and learning organism. *Bull. Math. Biophysics*, vol. 31, pp. 247–253.

Firschein, O., and Fischler, M. A. (1971). A study in descriptive representation of pictorial data. *IJCAI2*, pp. 258–269.

Fischer, M. J. (1972). Lambda calculus schemata. *SIGPLAN Notices*, vol. 7, no. 1, pp. 104–109. New York: ACM.

Floyd, R. W. (1967a). Assigning meaning to programs. *Proc. Symp. Applied Math.*, vol. 19, pp. 19–32.

Floyd, R. W. (1967b). Nondeterministic algorithms. *JACM,* vol. 14, no. 4, pp. 636–644.

Fogel, L. J., Owens, A. J., and Walsh, M. J. (1966). *Artificial Intelligence Through Simulated Evolution.* New York: Wiley.

Foreign Policy Association, eds. (1968). *Toward the Year 2018.* New York: Cowles Education Corp.

Formby, J. (1965). *An Introduction to the Mathematical Formulation of Self-Organizing Systems.* New York: Van Nostrand.

Forrester, J. W. (1971). *World Dynamics.* Cambridge: Wright-Allen.

——— (1971). Alternatives to catastrophe, understanding the counter-intuitive behavior of social systems. *The Ecologist,* vol. 1, no. 14, pp. 4–9; vol. 1, no. 15, pp. 16–23.

Foster, J. M. (1968). Self-improvement in query languages. *MI2,* pp. 195–204.

Frayn, M. (1965). *The Tin Men.* New York: Ace.

Freeman, H., and Gardner, L. (1964). Apictorial jigsaw puzzles: The computer solution of a problem in pattern-recognition. *IEEE Trans. Elect. Comp.,* vol. EC-13, pp. 118–127.

Freeman, P., and Newell, A. (1971). A model for functional reasoning in design. *IJCAI2,* pp. 621–633.

Freidberg, R. M. (1958). A learning machine, part I. *IBM J. Res. Dev.,* vol. 2, pp. 2–13.

Freidberg, R. M., Dunham, B., and North, J. H. (1959). A learning machine, part II. *IBM J. Res. Dev.,* vol. 3, pp. 282–287.

Friedman, L. (1969). Robot control strategy. *IJCAI,* pp. 527–540.

Freudenthal, H. (1960). *Lincos: Design of a Language for Cosmic Intercourse.* Amsterdam: North-Holland.

Fromm, E. (1968). *The Revolution of Hope: Toward a Humanized Technology.* New York: Harper & Row.

——— (1971). Humanizing a technological society. In *Information Technology in a Democracy,* ed. A. F. Westin, pp. 198–206. Cambridge, Mass.: Harvard Univ. Press.

Fukunaga, K. (1972). *Introduction to Statistical Pattern Recognition.* New York: Academic Press.

Galbraith, J. K. (1967). *The New Industrial State.* Boston: Houghton Mifflin.

Gardner, D., and Kandel, E. R. (1972). Diphasic post-synaptic potential: a chemical synapse capable of mediating conjoint excitation and inhibition. *Science,* vol. 176, pp. 675–678.

Gardner, M. (1958), *Logic Machines, Diagrams, and Boolean Algebras.* New York: Dover.

——— (1970). The fantastic combinations of John Conway's new solitaire game 'Life'. *Scientific American,* October, pp. 120–123.

——— (1971). On cellular automata, self-reproduction, the Garden of Eden, and the game 'Life'. *Scientific American,* February, pp. 112–117.

Gardner, M. (ed.). *See also* Dudeney; Loyd.

Gardner, R. A., and Gardner, B. T. (1969). Teaching sign language to a chimpanzee. *Science*, vol. 165, pp. 664–672.

Gazzaniga, M. S. (1967). The split brain in man. *Scientific American*, August, vol. 217, pp. 24–29.

Gelb, J. P. (1971a). The computer solution of English probability problems. Ph.D. Thesis. Troy, N.Y.: Rensselaer Polytechnic Inst.

———— (1971b). Experiments with a natural language problem solving system. *IJCAI2*, pp. 455–462.

Gelernter, H. (1959). Realization of a geometry-theorem proving machine. *CT*, pp. 134–152.

Gentile, R. B., and Lucas, J. R. (1971). The TABLON mass storage network. *SJCC*, vol. 38, pp. 345–356.

Gertz, J. L. (1970). Hierarchical associative memories for parallel computation. MAC TR-69. Cambridge, Mass.: M.I.T. Proj. MAC.

Geschwind, N. (1972). Language and the brain. *Scientific American*, April, vol. 226, pp. 76–83.

Gibbons, G. D. (1972). Beyond REF-ARF: toward an intelligent processor for a nondeterministic programming language. Pittsburgh: Carnegie-Mellon Univ. Comp. Sci. Dept.

Gibson, J. J. (1966). *The Senses Considered as Perceptual Systems*. New York: Houghton Mifflin.

Gillogly, J. J. (1971). The Technology chess program. CMU-CS-71-109. Pittsburgh: Dept. Comp. Sci., Carnegie-Mellon Univ.

Gilmore, P. C. (1959). A program for the production from axioms, of proofs for theorems derivable within the first order predicate calculus. *Proc. Int. Conf. Infor. Processing*, pp. 265–272. Paris: UNESCO.

———— (1960). A proof method for quantification theory. *IBM J. Res. Dev.*, vol. 4, pp. 28–35.

———— (1970) An examination of the Geometry Theorem Machine. *Art. Intell.*, vol. 1, pp. 171–187.

Glaser, D., McCarthy, J., and Minsky, M. (1964). The automated biological laboratory. Report of the subgroup on ABL summer study in exobiology sponsored by the Space Science Board of the National Academy of Sciences.

Godel, K. (1931). On formally undecidable propositions of Principia Mathematica and related systems. *Monatshefte fur Mathematik und Physic*, pp. 173–198. Translated into English by B. Meltzer, and published by Oliver and Boyd, London, 1962.

Good, I. J. (1962). The mind-body problem, or could an android feel pain? In *Theories of the Mind*, ed. J. M. Scher, pp. 490–518. New York: The Free Press.

———— (1968). A five-year plan for automatic chess. *MI2*, pp. 89–118.

Gordon, T. J. (1965). *The Future*. New York: St. Martin's.

Graham, W. R. (1970). The parallel and the pipeline computers. *Datamation*, April, pp. 68–71.

Graves, G., Pingry, D., and Whinston, A. (1971). Application of a large scale nonlinear programming problem to pollution control. *FJCC*, vol. 39, pp. 123–134.

Green, C. (1969a). Application of theorem-proving to problem-solving. *IJCAI*, pp. 219–239.

———— (1969b). Theorem-proving by resolution as a basis for question-answering systems. *MI4*, pp. 183–205.

———— (1969c). The application of theorem-proving techniques to question-answering systems. AIM-96. Stanford: Stanford Art. Intell. Proj.

Green, C., et al. *See also* Coles and Green.

Greenberger, M., ed. (1962). *Computers and the World of the Future*. Cambridge, Mass.: M.I.T. Press.

Greenblatt, R., Eastlake, D., and Crocker, S. (1967). The Greenblatt chess program. *FJCC*, vol. 31, pp. 801–810.

Gregory, R. L. (1967). Will seeing machines have illusions? *MI1*, pp. 169–180.

———— (1970). *The Intelligent Eye*. New York: McGraw-Hill.

Griffith, A. K. (1970). Computer recognition of prismatic solids. MAC TR-73. Cambridge: M.I.T. Proj. MAC.

Gruenberger, F., and Jaffray, G. (1966). *Problems for Computer Solution*. New York: Wiley.

Guard, J. R., Bennett, J. H., Easton, W. B., and Settle, L. G. (1967). CRT-aided semi-automated mathematics. AFCRL-67-0167. Cambridge, Mass.: Air Force Cambridge Res. Lab.

Guzman, A. (1968a). Computer recognition of three-dimensional objects in a visual scene. MAC-TR-59. Cambridge: M.I.T. Proj. MAC.

———— (1968b). Decomposition of a visual scene into three-dimensional bodies. *FJCC*, vol. 33, part 1, pp. 291–304.

———— (1971). Analysis of curved line drawings using context and global information. *MI6*, pp. 325–375.

Halliday, M. A. K. (1961). Categories of the theory of grammar. *Word*, vol. 17.

———— (1966). Some notes on "deep" grammar. *J. Linguistics*, vol. 2.

———— (1967). Notes on transivity and theme in English. *J. Linguistics*, vol. 3.

———— (1970). Functional diversity in language as seen from a consideration of modality and mood in English. *Foundations of Language*, vol. 6, pp. 322–361.

Halmos, P. R. (1960). *Naive Set Theory*. New York: Van Nostrand-Reinhold.

Hamming, R. W. (1963). Intellectual implications of the computer revolution. *Amer. Math. Monthly*, vol. 70, no. 1, pp. 4–11.

Hannan, J. F., and Robbins, H. (1955). Asymptotic solutions of the compound decision theory problem for two completely specified distributions. *Annals Math. Stat.* vol. 26, no. 1, pp. 37–51.

Harrison, M. A. (1965). On the error correcting capacity of finite automata. *IC*, vol. 8, pp. 430–450.

Hart, P., Nilsson, N., and Raphael, B. (1968). A formal basis for the heuristic determination of minimum cost paths. *IEEE Trans. Sys. Sci. and Cybernetics*, vol. SSC-4, no. 2, pp. 100–107.

Hart, P., Nilsson, N., and Robinson, A. E. (1971). A causality representation for enriched robot task domains. Tech. Rept., Proj. 1187. Menlo Park: Art. Intell. Center, Stanford Research Institute.

Hart, P., et al. *See also* Duda and Hart.

Hartmanis, J., and Stearns, R. E. (1965). On the computational complexity of algorithms. *Trans. Amer. Math. Soc.*, vol. 117, pp. 285–306.

Hatt, H. E. (1968). *Cybernetics and the Image of Man: A Study of Freedom and Responsibility in Man and Machine.* Nashville: Abingdon.

Hayes, K. C., and Rosenfeld, A. (1972). Efficient edge detectors and applications. Tech. Rept. TR-207. College Park, Maryland: Univ. Maryland Comp. Sci. Center.

Hayes, P. J. (1971). A logic of actions. *MI6*, pp. 495–520.

Hayes, P. I., et al. *See also* McCarthy and Hayes; Kowalski and Hayes.

Hearn, A. C. (1969). Standard LISP. Memo No. AI-90. Stanford: Stanford Art. Intell. Proj.

——— (1971). Applications of symbol manipulation in theoretical physics. *CACM*, vol. 14, no. 8, pp. 511–516.

Heinlein, R. A. (1966). *The Moon Is a Harsh Mistress.* New York: Berkeley.

Herbrand, J. (1930). Recherches sur la théoria de la démonstration. *Travaux de la Société des Sciences et des Lettres de Varsovie*, Classe III science mathematiques et physiques, no. 33.

Herksovits, A. (1970). On boundary detection. Ph.D. Thesis. Cambridge, Mass.: M.I.T.

Hewitt, C. (1968). Functional abstraction in LISP and PLANNER. AI Memo 151. Cambridge: M.I.T. Art. Intell. Lab.

——— (1969). PLANNER: A language for proving theorems in robots. *IJCAI*, pp. 295–301.

——— (1970a). Teaching procedures in humans and robots. AI Memo 208. Cambridge: M.I.T. Art. Intell. Lab.

——— (1970b). More comparative schematology. AI Memo 207. Cambridge: M.I.T. Art. Intell. Lab.

——— (1971). Procedural embedding of knowledge in PLANNER. *IJCAI2*, pp. 167–182.

——— (1972). Description and theoretical analysis (using schemata) of PLANNER: A language for proving theorems and manipulating models in a robot. AI TR-258. Cambridge. Mass.: M.I.T. Art. Intell. Lab.

Hierserman, D. L. (1971). Automatic railroad-car identification. *Elect. World*, vol. 86, no. 3, pp. 46–47.

Hilbert, D., and Ackerman, W. (1928). *Principles of Mathematical Logic.* New York: Chelsea.

Hill, D. R. (1967). Automatic speech recognition: A problem for machine intelligence. *MI1*, pp. 199–228.

420      INTRODUCTION TO ARTIFICIAL INTELLIGENCE

Hill, W. A. (1966). The impact of EDP systems on office employees: Some empirical conclusions. *Academy of Management,* vol. 9, pp. 9–19.

Hilton, A. M., ed. (1966). *The Evolving Society.* New York: Institute for Cybercultural Research Press.

Hintikka, J. (1962). *Knowledge and Belief: An Introduction to the Logic of the Two Notions.* Ithaca, N.Y.: Cornell Univ. Press.

—— (1969). *Models for Modalities.* Dordrecht, Netherlands: D. Reidel.

Hoare, C. A. R., and Allison, D. C. S. (1972). Incomputability. *Computing Surveys,* vol. 4, no. 3, pp. 169–177.

Hodes, L. (1966). Programming languages, logic, and cooperative games. Abstract in *CACM,* vol. 9, pp. 549.

—— (1971). Solving problems by formula manipulation in logic and linear inequalities. *IJCAI2,* pp. 553–559.

Holland, J. H. (1960). Iterative circuit computers. *Proc. Joint Comp. Conf.,* vol. 17, pp. 259–265.

—— (1962). Outline for a logical theory of adaptive systems. *JACM,* vol. 9, no. 3, pp. 297–314. *Also in ECA,* pp. 297–319.

—— (1970). Hierarchical descriptions, universal spaces, and adaptive systems. *ECA,* pp. 320–353.

Holt, A. W. (1968). What was promised—what we have—and what is being promised in character recognition. *FJCC,* vol. 33, part 2, pp. 1451–1458.

Holt, R. C. (1972). Some deadlock properties of computer systems. *Computing Surveys,* vol. 4, no. 3, pp. 179–196.

Hooper, P. K. (1966). The undecidability of the Turing machine immortality problem. *J. Symb. Logic,* vol. 31, pp. 219–234.

Hopcroft, J. E., and Ullman, J. D. (1969). *Formal Languages and Their Relation to Automata.* Reading, Mass.: Addison-Wesley.

Horning, J. J. (1969). A study of grammatical inference. AIM-98. Stanford: Stanford Art. Intell. Proj.

Horning, J. J., and Randell, B. (1972). Process structuring. Tech. Rept. Series, no. 31. Newcastle upon Tyne, England: Computing Lab., Univ. Newcastle upon Tyne.

Hübel, D. H., and Wiesel, T. N. (1962). Receptive fields, binocular interaction, and functional architecture in the cat's visual cortex. *J. Physiology,* vol. 160, pp. 106–154.

—— (1963). Shape and arrangement of columns in cat's striate cortex. *J. Physiology,* vol. 165, pp. 559–568.

Hueckel, M. H. (1969). An operator which locates edges in digitized pictures. AIM-105. Stanford: Stanford Art. Intell. Proj.

—— (1971). An operator which locates edges in digitized pictures. *JACM,* vol. 18, no. 1, pp. 113–125.

Huffman, D. A. (1971). Impossible objects as nonsense sentences. *MI6,* pp. 295–323.

Hunt, E. B., Marin, J., and Stone, F. (1966). *Experiments in Induction.* New York: Academic Press.

Hunt, R. P., Elser, T., and Wolf, I. W. (1970). The future role of magneto-optical memory systems. *Datamation,* April, pp. 97–101.

Hyden, H. (1969). Biochemical aspects of learning and memory. In *On the Biology of Learning,* ed. K. H. Probram, pp. 95–125. New York: Harcourt.

Igarashi, S. (1972). Admissibility of fixed-point induction in first-order logic of typed theories. AIM-168. Stanford: Stanford Art. Intell. Proj.

Irons, E. T. (1970). Experience with an extensible language. *CACM,* vol. 13, no. 1, pp. 31–40.

Itkin, V. E., and Zwinogrodzki, Z. (1972). On program schemata equivalence. *J Comp. Sys. Sci.,* vol. 6, no. 1, pp. 88–101.

Jacobs, W. W. (1971). A structure for systems that plan abstractly. *SJCC,* vol. 38, pp. 357–364.

Jaki, S. L. (1969). *Brain, Mind and Computers.* New York: Herder.

Jarvik, M. E. (1972). Effects of chemical and physical treatments on learning and memory. *Annual Rev. Psychology,* vol. 23, pp. 457–486.

Jensen, D. (1972). A syntactic model for resolution theorem proving in type theory. Res. Rept. CSRR 2055. Waterloo, Ontario: Dept. Comp. Sci., Univ. Waterloo.

Jensen, D., et al. *See also* Pietrzykowski and Jensen.

Johannesson, O. (1966). *The Tale of the Big Computer: A Vision.* Translated by N. Walford. New York: Coward-McCann Inc.

Kac, M., and Ulam, S. (1968). *Mathematics and Logic: Retrospect and Prospect.* New York: Praeger.

Kahn, M. E. (1969). The near-minimum-time control of open-loop articulated kinematic chains. AIM-106. Stanford: Stanford Art. Intell. Proj.

Kain, R. Y. (1972). *Automata Theory: Machines and Languages.* New York: McGraw-Hill.

Kamman, A. B. (1971). Development of computer applications in emerging nations. *FJCC,* vol. 39, pp. 17–26.

Kanal, L. N., ed. (1968). *Pattern Recognition.* Washington, D.C.: Thompson.

Kandel, E. R. (1970). Nerve cells and behavior. *Scientific American,* vol. 223, July, pp. 57–70.

Kaneff, S., ed. (1970). *Picture Language Machines.* New York: Academic Press.

Kaplan, R. M. (1971). Augmented transition networks as psychological models of sentence comprehension. *IJCA12,* pp. 429–443.

——— (1972). Augmented transition networks as psychological models of sentence comprehension. *Art. Intell.,* vol. 3, pp. 77–100.

Karp, C. R. (1964). *Languages with Expressions of Infinite Length.* Amsterdam: North-Holland.

Kasner, E., and Newman, J. R. (1956). Pastimes of past and present times. In *The World of Mathematics,* ed. J. R. Newman, vol. 4, pp. 2416–2438. New York: Simon and Schuster.

Katzan, H. (1971). Storage hierarchy systems. *SJCC,* vol. 38, pp. 325–336.

Kay, A. C. (1968). FLEX—a flexible extendable language. Tech. Rept. 4–7. Salt Lake City: Comp. Sci. Dept., Univ. Utah.

Kelly, M. D. (1970a). Edge detection in pictures by computer using planning. AIM-108. Stanford: Stanford Art. Intell. Proj.

——— (1970b). Visual identification of people by computer. AIM-130. Stanford: Stanford Art. Intell. Proj.

Kinoshita, G., Aida, S., and Mori, M. (1971). Pattern recognition by an artificial tactile sense. *IJCAI2*, pp. 376–384.

Kirkpatrick, P., and Pattee, H. H. (1953). *Advances in Biological and Medical Physics*. New York: Academic Press.

Kirsch, R. A. (1964). Computer interpretation of English text and picture patterns. *IEEE Trans. Elect. Comp.*, vol. EC-13, pp. 363–381.

Kleene, S. C. (1967). *Mathematical Logic*. New York: Wiley.

——— (1969). The new logic. *Amer. Scientist*, vol. 57, no. 3, pp. 333–347.

——— (1971). *Introduction to Metamathematics*. New York: Elsevier.

Kling, R. E. (1971a). Reasoning by analogy with applications to heuristic problem solving: A case study. AIM-147. Stanford: Stanford Art. Intell. Proj.

——— (1971b). A paradigm for reasoning by analogy. *IJCAI2*, pp. 568–585.

——— (1971c). A paradigm for reasoning by analogy. *Art. Intell.*, vol. 2, no. 2, pp. 147–178.

Knast, R. (1969). Continuous-time probabilistic automata. *IC*, vol. 15, pp. 335–352.

Knuth, D. E. (1965). On the translation of languages from left to right. *IC*, vol. 8, no. 6, pp. 607–639.

——— (1967). A characterization of parenthesis languages. *IC*, vol. 11; no. 3, pp. 269–289.

——— (1968). Semantics of context-free languages. *Math. Sys. Theory*, vol. 2, no. 2, pp. 127–145.

——— (1969a). *Fundamental Algorithms*. Reading, Mass.: Addison-Wesley.

——— (1969b). *Seminumerical Algorithms*. Reading, Mass: Addison-Wesley.

——— (1970). Von Neumann's first computer program. *Computing Surveys*, vol. 2, pp. 247–260.

——— (1972). Ancient Babylonian algorithms. *CACM*, vol. 15, no. 7, pp. 671–677.

Knuth, D. E., and Bigelow, R. (1967). Programming languages for automata. *JACM*, vol. 14, no. 4, pp. 615–635.

Kochen, M. (1965). *Some Problems in Information Science*. New York: Scarecrow.

———, ed. (1967). *The Growth of Knowledge: Readings on Organization and Retrieval of Information*. New York: Wiley.

Kochen, M., et al. *See also* Uhr and Kochen.

Koffman, E. G. (1967). Learning through pattern recognition applied to a class of games. Rept. No. SRC 107-A-67-45. Cleveland: Case Western Reserve Univ.

Kohavi, Z., and Paz, A., eds. (1971). *Theory of Machines and Computations.* New York: Academic Press.

Kolers, P. A., and Eden, M., eds. (1968). *Recognizing Patterns: Studies in Living and Automatic Systems.* Cambridge, Mass.: M.I.T. Press.

Kolmogorov, A. N. (1968). Logical basis for information theory and probability theory. *IEEE Trans. Infor. Theory,* vol. IT-14, no. 5, pp. 662–664.

Koontz, W. L. G., and Fukunaga, K. (1971). A nonparametric valley-seeking technique for cluster analysis. *IJCAI2,* pp. 411–417.

Kowalski, R. (1970a). Search strategies for theorem-proving. *MI5,* pp. 181–201.

––––––– (1970b). Studies in the completeness and efficiency of theorem-proving by resolution. Ph.D. Thesis. Edinburgh: Univ. Edinburgh.

Kowalski, R., and Hayes, P. (1969). Semantic trees in automatic theorem-proving. *MI4,* pp. 87–101.

Kuhn, T. (1962). *The Structure of Scientific Revolutions.* Chicago: Univ. Chicago Press.

Kuno, S. (1965). The predictive analyzer and a path elimination technique. *CACM,* vol. 8, no. 7, pp. 453–462.

––––––– (1972). Mathematical linguistics and automatic translation. Rept. No. NSF-28. Cambridge, Mass.: Harvard Univ. Comp. Lab.

Kurki-Suonio, R. (1971). *A Programmer's Introduction to Computability and Formal Languages.* Philadelphia: Auerbach.

Laffal, J. (1965). *Pathological and Normal Language.* New York: Atherton.

Laing, R. D. (1968). *The Politics of Experience.* New York: Ballantine.

Landers, R. R. (1966). *Man's Place in the Dybosphere.* Englewood Cliffs, N.J.: Prentice-Hall.

Landin, P. J. (1963). The mechanical evaluation of expressions. *Comp. J.,* vol. 6, pp. 308–320.

––––––– (1965). A correspondence between Algol and Church's λ-calculus. *CACM,* vol. 8, pp. 89–101, 158–165.

––––––– (1966). The next 700 programming languages. *CACM,* vol. 9, no. 3, pp. 157–164.

Langlois, W. E. (1971). Digital simulation of the general atmosphere circulation using a very dense grid. *FJCC,* vol. 39, pp. 97–103.

Lanham, U. (1962). *The Fishes.* New York: Columbia Univ. Press.

Laotzu (600 B.C.?). *The Way of Life.* Translated by W. Bynner. New York: Capricorn.

Laszlo, E. (1969). *System, Structure, and Experience: Toward a Scientific Theory of Mind.* New York: Gordon and Breach.

Lee, C. H. (1963). A Turing machine which prints its own code script. *Proc. Symp. Math. Theory Automata,* pp. 155–164. Brooklyn: Polytechnic.

Lee, J. A. N. (1972). *Computer Semantics: Studies of Algorithms, Processors, and Languages.* New York: Van Nostrand-Reinhold.

Lee, R. C. T. (1971). Fuzzy logic and the resolution principle. *IJCAI2*, pp. 560–567.

Lee, R. C. T., et al. *See also* Slagle, Chang, and Lee; Waldinger and Lee.

Lee, S. E. (1964), *A History of Far Eastern Art.* Englewood Cliffs, N.J.: Prentice-Hall.

Leovy, C. B., Smith, B. A., Young, A. T., and Leighton, R. B. (1971). Mariner Mars 1969: Atmospheric results. *J. Geophysical Res.*, vol. 76, no. 2, pp. 297–312.

Lewis, C. I. (1918). *A Survey of Symbolic Logic.* Berkeley: Univ. California Press.

Lewis, C. I., and Langford, C. H. (1959). *Symbolic Logic.* New York: Dover.

Licklider, J. C. R. (1965). Libraries of the Future. Cambridge, Mass.: M.I.T. Press.

Licklider, J. C. R., et al. *See also* Pierce, J. R.

Lilly, J. C. (1967). *The Mind of the Dolphin: A Nonhuman Intelligence.* New York: Doubleday.

Lindsay, P. H., and Norman, D. A. (1972). *Human Information Processing.* New York: Academic Press.

Lindsay, R. K. (1963). Inferential memory as the basis of machines which understand natural language. *CT*, pp. 217–233.

———— (1971). Jigsaw heuristics and a language learning model. *AIHP*, pp. 173–189.

Lohman, R. D., Mezrich, R. S., and Stewart, W. C. (1971). Holographic mass memory's promise: megabits accessible in microseconds. *Electronics*, January 18, pp. 61–66.

London, R. L. (1971). Correctness of two compilers for a LISP subset. AIM-151. Stanford: Stanford Art. Intell. Proj.

Loveland, D. W. (1968). Mechanical theorem proving by model elimination. *JACM*, vol. 15, pp. 236–251.

———— (1969). A variant of the Kolmogorov concept of complexity. *IC*, vol. 15, pp. 510–526.

Loyd, S. (1960). *Mathematical Puzzles of Sam Loyd*, M. Gardner, ed. New York: Dover.

Lucchesi, C. L. (1972). The undecidability of the unification problem for 3rd order languages. Res. Rept. CSRR-2059. Waterloo, Ontario: Dept. Comp. Sci., Univ. Waterloo.

Luckham, D. (1967). The resolution principle in theorem-proving. *MI1*, pp. 47–61.

Luckham, D., and Nilsson, N. J. (1971). Extracting information from resolution proof-trees. *Art. Intell.*, vol. 2, no. 1, pp. 27–54.

Lucky, R. W. (1969). Structure in error-correcting codes. In *Hierarchical Structures*, ed. L. L. Whyte et al., pp. 253–278. New York: Elsevier.

Luconi, F. L. (1968). Asynchronous computational structures. MAC-TR-49. Cambridge, Mass.: M.I.T. Proj. MAC.

Lukasiewicz, J. (1963). *Elements of Mathematical Logic.* New York: Macmillan.

—— (1970). Logical foundations of probability theory. In *Jan Lukasiewicz, Selected Works,* ed. L. Berkowski, pp. 16–43. Amsterdam: North-Holland.

Luria, A. R. (1970). The functional organization of the brain. *Scientific American,* March, vol. 222, pp. 66–78.

MacKay, D. M. (1969). *Information, Mechanism, and Meaning.* Cambridge, Mass.: M.I.T. Press.

Manna, Z. (1968a). Termination of algorithms. Ph.D. thesis. Pittsburgh: Comp. Sci. Dept., Carnegie-Mellon Univ.

—— (1968b). Formalization of properties of programs. AIM-64. Stanford: Stanford Art. Intell. Proj.

—— (1969). The correctness of programs. *J. Comp. Sys. Sci.,* vol. 3, no. 2, pp. 119–127.

—— (1970a). Second order mathematical theory of computation. *Proc. ACM Symp. Theory Computing,* pp. 158–168. New York: Assoc. Computing Machinery.

—— (1970b). The correctness of nondeterministic programs. *Art. Intell.,* vol. 1, no. ½, pp. 1–26.

—— (1971). Mathematical theory of partial correctness. AIM-139. Stanford: Stanford Art. Intell. Proj.

Manna, Z., and Vuillemin, J. (1972). Fixpoint approach to the theory of computation. *CACM,* vol. 15, no. 7, pp. 528–536.

Manna, Z., and Waldinger, R. J. (1970). Towards automatic program synthesis. AIM-127. Stanford: Stanford Art. Intell. Proj.

Manna, Z., Ness, S., and Vuillemin, J. (1971). Inductive methods for proving properties of programs. AIM-154. Stanford: Stanford Art. Intell. Proj.

Manna, Z., et al. *See also* Ashcroft and Manna.

Margenau, H. (1939). Probability, many-valued logics, and physics. *Phil. Sci.,* vol. 6, pp. 65–87.

Marinos, P. N. (1969). Fuzzy logic and its application to switching systems. *IEEE Trans. Comp.,* vol. C-18, no. 4, pp. 343–348.

Marschak, J., and Radner, R. (1969). *The Economic Theory of Teams.* Cowles Foundation Monograph. New Haven, Conn.: Yale Univ.

Martin, D. F., and Estrin, G. (1969). Path length computations on graph models of computations. *IEEE Trans. Comp.,* vol. C-18, no. 6, pp. 530–536.

Martin, J., and Norman, A. D. (1970). *The Computerized Society.* Englewood Cliffs, N.J.: Prentice-Hall.

Martin-Lof, P. (1966). The definition of random sequences. *IC,* vol. 9, pp. 602–619.

426        INTRODUCTION TO ARTIFICIAL INTELLIGENCE

Maruyama, K. (1971). An approximation method for solving the sofa problem. UIUCDS-R-71-489. Urbana-Champaign: Dept. Comp. Sci., Univ. Illinois at Urbana-Champaign.

Masterman, M. (1970). Semantic language games or philosophy by computer. Paper No. 210. Cambridge, England: Cambridge Language Research Unit.

Masuda, K. (1971). ISUPPOSEW—a computer program that finds regions in the plan model of a visual scene. Tech. Note 54. Menlo Park, Calif.: Art. Intell. Group, Stanford Research Institute.

Maurer, H. A., and Williams, M. R. (1972). *A Collection of Programming Problems and Techniques.* Englewood Cliffs, N.J.: Prentice-Hall.

McCarthy, J. (1956). The inversion of functions defined by Turing machines. In *Automata Studies*, eds. C. E. Shannon and J. McCarthy, pp. 177–181. Princeton, N.J.: Princeton Univ. Press.

——— (1959). Programs with common sense. *Proc. Symp. Mechanization of Thought Processes*, pp. 75–84. London: Her Majesty's Stationery Office.

——— (1960). Recursive functions of symbolic expressions and their computation by machine. *CACM*, vol. 3, pp. 184–195.

——— (1962). Towards a mathematical science of computation. *Proc. IFIP Congress 1962*, pp. 21–28. Amsterdam: North-Holland.

——— (1963a). Situations, actions, and causal laws. AI Memo 2. Stanford: Stanford Art. Intell. Proj.

——— (1963b). Predicate calculus with "undefined" as a truth-value. AI Memo 1. Stanford: Stanford Art. Intell. Proj.

——— (1964a). Computer control of a machine for exploring Mars. AI Memo 14. Stanford: Stanford Art. Intell. Proj.

——— (1964b). A tough nut for proof procedures. AI Memo 16. Stanford: Stanford Art. Intell. Proj.

——— (1965). Problems in the theory of computation. *Proc. IFIP Congress 1965*, pp. 219–222. Washington, D.C.: Spartan.

——— (1968). Programs with common sense. *SIP*, pp. 403–418.

——— (1971). Technology's potential. *Chaparral*, vol. 72, no. 4, pp. 5. Stanford: Storke Publications Building.

McCarthy, J., and Hayes, P. (1968). Some philosophical problems from the standpoint of artificial intelligence. AIM-73. Stanford: Stanford Art. Intell. Proj. *Also in MI4*, pp. 463–502.

McCarthy, J., and Painter, J. A. (1967). Correctness of a compiler for arithmetic expressions. *Proc. Symp. Applied Mathematics*, vol. 19, pp. 33–41. Providence, R.I.: Amer. Math. Soc.

McCarthy, J., Earnest, L., Reddy, D., and Vicens, P. (1968). A computer with hands, eyes, and ears. *FJCC*, vol. 33, part 1, pp. 329–338.

McCarthy, J., Abrahams, P. W., Edwards, D. J., Hart, T. P., and Levin, M. I. (1962). *LISP 1.5 Programmer's Manual.* Cambridge, Mass.: M.I.T. Press.

McCarthy, J., et al. *See also* Glaser, McCarthy, and Minsky; Pierce; Shannon and McCarthy.

McConlogue, K., and Simmons, R. F. (1965). Analyzing English syntax with a pattern-learning parser. *CACM*, vol. 8, no. 11, pp. 687–698.

McCormick, E. M. (1959). *Digital Computer Primer*. New York: McGraw-Hill.

McCracken, D., and Robertson, G. (1971). C.ai(P.L*)—An L* processor for C.ai. CMU-CS-71-106. Pittsburgh: Dept. Comp. Sci., Carnegie-Mellon Univ.

McGraw-Hill, Inc. (1971). *McGraw-Hill Yearbook of Science and Technology*.

McKinsey, J. C. C. (1935). On the independence of undefined ideas. *Bull. Amer. Math. Soc.*, vol. 41, pp. 291–297.

McLamore, H. (1968). Recursive product development. *J. Data Management*, vol. 6, pp. 28–31.

McWhirter, N., and McWhirter, R. (1971). *Guinness Book of World Records*. New York: Bantam.

Meltzer, B. (1968). Some notes on resolution strategies. *MI3*, pp. 71–76.

Meltzer, B., et al. *See also* Godel; Meltzer, trans.

Mendel, J. M., and Fu, K. S., eds. (1970). *Adaptive, Learning and Pattern Recognition Systems*. New York: Academic Press.

Mermelstein, P. (1969). Computer simulation of articulatory activity in speech production. *IJCAI*, pp. 447–454.

Mesarovic, M. D. (1969). On some metamathematical results as properties of general systems. *J Math. Sys. Theory*, vol. 2, pp. 357–361. New York: Springer-Verlag.

——— (1970). Systems theoretic approach to formal problem solving. *TAPR*, pp. 161–178.

——— ed. (1964). *Views on General Systems Theory*. New York: Wiley.

Mesarovic, M. D., Macko, D., and Takahara, Y. (1970). *Theory of Hierarchical, Multilevel Systems*. New York: Academic Press.

Michie, D. (1971). Formation and execution of plans by machine. *AIHP*, pp. 101–124.

——— (1972). Programmer's gambit. *New Scientist*, August 17, pp. 329–332.

——— (1971). On not seeing things. Experimental Programming Rept. No. 22. Edinburgh: Dept. Machine Intell. and Perception, Univ. Edinburgh.

Miczaika, G. R., and Sinton, W. M. (1961). *Tools of the Astronomer*. Cambridge, Mass.: Harvard Univ. Press.

Millar, P. H. (1971). On defining the intelligence of behaviour and machines. *IJCAI2*, pp. 279–286.

Miller, A. R. (1971). *The Assault on Privacy*. Ann Arbor: Univ. Michigan Press.

Miller, G. A. (1967). *The Psychology of Communication: Seven Essays*. New York: Basic Books.

Miller, G. A., Galanter, E., and Pribram, K. H. (1960). *Plans and the Structure of Behavior.* New York: Holt, Rinehart and Winston.

Miller, G. A., et al. *See also* Chomsky and Miller.

Miller, J. G. (1965). Living systems: Basic Concepts. *Behavioral Sci.*, vol. 10, pp. 193–237.

Miller, W., and Shaw, A. C. (1968). Linguistic methods in picture processing—a survey. *FJCC*, vol. 33, part 1, pp. 279–290.

Milner, R. (1971). An algebraic definition of simulation between programs. *IJCAI2*, pp. 481–489.

———— (1972). Logic for computable functions, description of a machine implementation. AIM-169. Stanford: Stanford Art. Intell. Proj.

Minsky, M. (1963). Steps toward artificial intelligence. *CT*, pp. 406–450.

———— (1966). Artificial intelligence. In *Information*, ed. G. Piel et al., pp. 193–210. San Francisco: Freeman.

———— (1967). *Computation: Finite and Infinite Machines.* Englewood Cliffs, N.J.: Prentice-Hall.

———— (1968a). Descriptive languages and problem solving. *SIP*, pp. 419–424.

———— (1968b). Matter, mind and models. *SIP*, pp. 425–432.

———— (1970). Form and content in computer science. *JACM*, vol. 17, no. 2, pp. 197–215.

Minsky, M., and Papert, S. (1969). *Perceptrons: An Introduction to Computational Geometry.* Cambridge, Mass.: M.I.T. Press.

Minsky, M., et al. *See also* Glaser, McCarthy, and Minsky; Pierce; Skinner, Woolridge, Asimov, and Minsky, et al.

Montanari, U. (1971). On the optimal detection of curves in noisy pictures. *CACM*, vol. 14, no. 5, pp. 335–345.

Moore, E. F. (1962). Machine models of self-reproduction. *ECA*, pp. 187–203.

Moore, R. C. (1973). D-SCRIPT: a computational theory of descriptions. AI Memo No. 278. Cambridge, Mass.: M.I.T. Art. Intell. Lab.

Morgan, C. G. (1971). Hypothesis generation by machine. *Art. Intell.*, vol. 2, no. 2, pp. 179–187.

Morofsky, E. L., and Wong, A. K. C. (1971). Computer perception of complex patterns. *IJCAI2*, pp. 248–257.

Morrison, P., and Morrison, E., eds. (1961). *Charles Babbage and His Calculating Engines: Selected Writings by Charles Babbage and Others.* New York: Dover.

Moses, J. (1967). Symbolic integration. MAC-TR-47. Cambridge, Mass.: M.I.T. Proj. MAC.

———— (1970). The function of FUNCTION in LISP, or Why the FUNARG problem should be called the environment problem. AI-199. Cambridge, Mass.: M.I.T. Art. Intell. Lab.

———— (1971). Symbolic integration: the stormy decade. *CACM*, vol. 14, no. 8, pp. 548–560.

———— (1972). Toward a general theory of special functions. *CACM*, vol. 15, no. 7, pp. 550–556.

Mowshowitz, A. (1967). Entropy and the complexity of graphs. Document 67-17,819. Ann Arbor, Mich.: Univ. Microfilms.

Mumford, L. (1966). *The Myth of the Machine: Technics and Human Development*. New York: Harcourt.

Munson, J. H. (1968). Experiments in the recognition of hand-printed text: Part I—character recognition. *FJCC*, vol. 33, part 2, pp. 1125–1138.

———— (1970a). The SRI intelligent automaton program. Tech. Note 19. Menlo Park, Calif.: Art. Intell. Group, Stanford Research Institute.

———— (1970b). A cost-effectiveness basis for robot problem solving and execution. Tech. Note 29. Menlo Park, Calif.: Art. Intell. Group, Stanford Research Institute.

———— (1971). Robot planning, execution, and monitoring in an uncertain environment. *IJCAI2*, pp. 338–349.

Myhill, J. (1963). The converse of Moore's Garden of Eden theorem. *ECA*, pp. 204–205.

———— (1964). The abstract theory of self-reproduction. *ECA*, pp. 206–218.

Nakano, K., and Nagumo, J. (1971). Information processing using a model of associative memory. *IJCAI2*, pp. 101–110.

Narasimhan, R. (1964). Labeling schemata and syntactic descriptions of pictures. *IC*, vol. 7, pp. 151–179.

Newell, A. (1960). On programming a highly parallel machine to be an intelligent technician. *Proc. Joint Comp. Conf.*, vol. 17, pp. 267–282. New York: ACM.

———— (1963). Learning, generality, and problem-solving. RM-3285-1-PR. Santa Monica: RAND Corp.

Newell, A., and Simon, H. A. (1961). Computer simulation of human thinking. *Science*, vol. 134, pp. 2011–2017.

———— (1972). *Human Problem Solving*. Englewood Cliffs, N.J.: Prentice-Hall.

———— (1963). GPS, a program that simulates human thought. *CT*, pp. 279–293.

Newell, A., Shaw, J. C., and Simon, H. A. (1957). Empirical explorations with the Logic Theory machine: A case study in heuristics. *CT*, pp. 109–133.

———— (1963). Chess-playing programs and the problem of complexity. *CT*, pp. 39–70.

Newell, A., et al. *See also* Bell and Newell; Ernst and Newell; Freeman and Newell; Waterman and Newell.

Nilsson, N. J. (1965). *Learning Machines: Foundations of Trainable Pattern-Classifying Systems*. New York: McGraw-Hill.

———— (1969). A mobile automaton: An application of artificial intelligence techniques. *IJCAI*, pp. 509–520.

Nilsson, N. J. (1971). *Problem-Solving Methods in Artificial Intelligence.* New York: McGraw-Hill.

Nilsson, N. J., and Hart, P. (1970). An information-processing model of operant behavior. Tech. Note 38. Menlo Park, Calif:. Art. Intell. Group, Stanford Research Institute.

Nilsson, N. J., et al. *See also* Fikes and Nilsson; Hart, Nilsson, and Raphael; Luckham and Nilsson.

Norman, D. A. (1972). Memory, knowledge, and the answering of questions. CHIP 25. La Jolla, Calif.: Center for Human Infor. Processing, Univ. Calif., San Diego.

Ogden, C. K. (1933). *Basic English: An Introduction with Rules and Grammar.* London: Kegan Paul, Trench, Trubner.

Ogden, W. (1968). A helpful result for proving inherent ambiguity. *Math. Sys. Theory,* vol. 2, pp. 191–194.

Olsen, D. J. (1971). Programming the war against water pollution. *FJCC,* vol. 39, pp. 115–121.

Papert, S. (1968). The artificial intelligence of Hubert L. Dreyfus: A budget of fallacies. AI Memo 154. Cambridge, Mass.: M.I.T. Proj. MAC.

——— (1971). A computer laboratory for elementary schools. AI Memo 246. Cambridge, Mass.: M.I.T. Art. Intell. Lab.

Parnas, D. L. (1972). On a solution to the cigarette smokers' problem (without conditional statements). Pittsburgh, Pa.: Comp. Sci. Dept., Carnegie-Mellon Univ.

Parsons and Williams, Inc. (1968). *Forecast 1968–2000 of Computer Developments and Applications.* Copenhagen: Parsons & Williams.

Paton, K. A. (1970). Conic sections in automatic chromosome analysis. *MI5,* pp. 411–434.

Paul, R. (1971). Trajectory control of a computer arm. *IJCAI2,* pp. 385–390.

Pfaltz, J. L., and Rosenfeld, A. (1969). Web grammars. *IJCAI,* pp. 609–619.

Philipson, M., ed. (1972). *Automation: Implications for the Future.* New York: Random House.

Piaget, J. (1946). *The Child's Conception of Movement and Speed.* London: Routledge and Kegan Paul.

——— (1950). *The Psychology of Intelligence.* New York: Harcourt.

——— (1952). *The Child's Conception of Number.* New York: Humanities Press.

Pierce, J. R. (1970). What computers should be doing. *PCR,* pp. 274–297. (With discussion by Bar-Hillel, Bush, Shannon, McCarthy, Minsky, Abrahams, Rosenblith, Dreyfus, and Licklider.)

Pietrzykowski, T., and Jensen, D. C. (1972). A complete mechanization of $\omega$-order logic. Res. Rept. CSRR-2060. Waterloo, Ontario: Dept. Comp. Sci., Univ. Waterloo.

Pingle, K. K., Singer, J. A., and Wichman, W. M. (1968). Computer control of a mechanical arm through visual input. In *Information Proc-*

*essing 68,* ed. A. J. H. Morrell, pp. 1563–1569. Amsterdam: North-Holland.

Pingle, K. K., and Tenenbaum, J. M. (1971). An accommodating edge follower. *IJCAI2,* pp. 1–7.

Pitrat, J. (1971). A general game-playing program. *AIHP,* pp. 125–155.

Pohl, I. (1970). First results on the effect of error in heuristic search. *MI5,* pp. 219–236.

Polak, F. L. (1961). *The Image of the Future.* Dobbs Ferry, N.Y.: Oceana.

Polanyi, M. (1958). *Personal Knowledge.* Chicago: Univ. Chicago Press.

———— (1968). Life's irreducible structure. *Science,* vol. 160, no. 3834, pp. 1308–1312.

Pollack, B. W., ed. (1972). *Compiler Techniques.* Philadelphia: Auerbach.

Polya, G. (1945). *How to Solve It: A New Aspect of Mathematical Method.* New York: Doubleday.

———— (1954a). *Induction and Analogy in Mathematics.* Princeton, N.J.: Princeton Univ. Press.

———— (1954b). *Patterns of Plausible Inference.* Princeton, N.J.: Princeton Univ. Press.

Pople, H. E. (1972). A goal-oriented language for the computer. *RM,* pp. 331–413.

Popplestone, R. J. (1969). Freddy in toyland. *MI4,* pp. 455–462.

Popplestone, R. J., et al. *See also* Burstall, Collins, and Popplestone.

Post, E. L. (1944). Recursively enumerable sets of positive integers. *Bull. Amer. Math. Soc.,* vol. 50, pp. 284–316.

Potvin, J. N. T. (1971). The Star-Ring system of loosely coupled digital devices. CSRG-7. Toronto: Comp. Sys. Res. Group, Univ. Toronto.

Pratt, T. W. (1969a). Semantic modeling by hierarchical graphs. TSN-6. Austin: Computation Center, Univ. Texas.

———— (1969b). A hierarchical graph model of the semantics of programs. *SJCC,* vol. 34, pp. 813–825.

———— (1971). Kernel equivalence of programs and proving kernel equivalence and correctness by test cases. *IJCAI2,* pp. 474–480.

Pratt, T. W., and Friedman, D. P. (1971). A language extension for graph-processing and its formal semantics. *CACM,* vol. 14, no. 7, pp. 460–467.

Prawitz, D. (1960). An improved proof procedure. *Theoria,* vol. 26, pp. 102–139.

Premack, A. J., and Premack, D. (1972). Teaching language to an ape. *Scientific American,* October, pp. 92–99.

Premack, D. (1970). The education of Sarah: A chimp learns the language. *Psychology Today,* vol. 4, pp. 55–58.

Preparata, F. P., and Yeh, R. T. (1972). Continuously valued logic. *J Comp. Sys. Sci.,* vol. 6, pp. 397–418.

Pribram, K. H. (1971). *Languages of the Brain: Experimental Paradoxes and Principles in Neurophysiology.* Englewood Cliffs, N.J.: Prentice-Hall.

Pribram, K. H., ed. (1969). *Memory Mechanisms*. Baltimore: Penguin.

Pribram, K. H., et al. *See also* Miller, Galanter, and Pribram.

Price, D. J. de S. (1959). An ancient Greek computer. *Scientific American,* June, vol. 200, no. 6, pp. 60–67.

Prior, A. N. (1957). *Time and Modality*. Oxford: Clarendon.

—— (1967). *Past, Present, and Future*. Oxford: Clarendon.

Puccetti, R. (1968). *Persons: A Study of Possible Moral Agents in the Universe*. London: Macmillan.

Pylyshyn, Z. W., ed. (1970). *Perspectives on the Computer Revolution*. Englewood Cliffs, N.J.: Prentice-Hall.

—— (1972). The problem of cognitive representation. Res. Bull. 227. London, Canada: Dept. Psychology, Univ. Western Ontario.

Quam, L. H., Liebes, S., Tucker, R. B., Hannah, M. J., and Eross, B. G. (1972). Computer interactive picture processing. AIM-166. Stanford: Stanford Art. Intell. Proj.

Quillian, M. R. (1966). Semantic memory. AFCRL-66-189. Ph.D. thesis. Pittsburgh: Carnegie-Mellon Univ. Also in *SIP*, pp. 227–270.

—— (1969). The Teachable Language Comprehender: A simulation program and theory of language. *CACM*, vol. 12, no. 8, pp. 459–476.

Quine, W. V. (1955). A proof procedure for quantification theory. *J. Symb. Logic*, vol. 20, pp. 141–149.

—— (1959). *Methods of Logic*. New York: Holt, Rinehart, and Winston.

—— (1960). *Word and Object*. Cambridge, Mass.: M.I.T. Press.

—— (1961). *Mathematical Logic*. Cambridge, Mass.: Harvard Univ. Press.

—— (1964). *From a Logical Point of View*. Cambridge, Mass.: Harvard Univ. Press.

Quinlan, J. R. (1969). A task-independent experience-gathering scheme for a problem-solver. *IJCAI*, pp. 193–198.

Quinlan, J. R., and Hunt, E. B. (1968). A formal deductive problem-solving system. *JACM*, vol. 15, no. 4, pp. 625–646.

Raj Reddy, D. (1966). An approach to computer speech recognition by direct analysis of the speech wave. Tech. Rept. CS 49. Stanford: Comp. Sci. Dept., Stanford Univ.

Ramani, S. (1971). A language based problem-solver. *IJCAI2*, pp. 463–473.

Randell, B. (1971). Highly reliable computing systems. Tech. Rept. 20. Newcastle, England: Univ. Newcastle upon Tyne.

Randell, B., et al. *See also* Horning and Randell.

Raphael, B. (1964). SIR: A computer program for semantic information retrieval. Rept. TR-2. Cambridge: M.I.T. Proj. MAC. *Also in SIP*, pp. 33–145.

—— (1971). The frame problem in problem-solving systems. *AIHP*, pp. 159–169.

Raphael, B., Duda, R. O., Fikes, R. E., Hart, P. E., Nilsson, N. J., Thorndyke, P. W., and Wilber, B. M. (1971). Research and applications—artificial intelligence. Final Rept. (1970–1971). Menlo Park, Calif.: Stanford Res. Inst. Art. Intell. Center.

Raphael, B., and Robinson, A. E. (1972). Bibliography on computer semantics. Tech. Note 72. Menlo Park, Calif.: Stanford Res. Inst. Art. Intell. Center.

Raphael, B., et al. See also Hart, Nilsson, and Raphael.

Ratliff, F. (1972). Contour and contrast. *Scientific American,* June, vol. 226, pp. 90–101.

Raviv, J. (1966). Decision making in Markov chains. RC-1672. Yorktown Heights, N.Y.: IBM Watson Research Center.

Reich, C. (1970). *The Greening of America.* New York: Random House.

Reichardt, J. (1971). *The Computer in Art.* London: Van Nostrand Reinhold.

——— (1971). *Cybernetics, Art and Ideas.* Greenwich, Conn.: New York Graphic Society.

Reid, R. J. (1971). An associative memory for auditory recall. *IJCAI2,* pp. 111–118.

Reiter, R. (1972). The use of models in automatic theorem-proving. Tech. Rept. 72–09. Vancouver: Dept. Comp. Sci., Univ. British Columbia.

Renwick, W., and Cole, A. J. (1971). *Digital Storage Systems.* London: Chapman and Hall.

Rescher, N. (1964). *Hypothetical Reasoning.* Amsterdam: North-Holland.

——— ed. (1967). *The Logic of Decision and Action.* Pittsburgh: Univ. Pittsburgh Press.

Rezler, J. (1969). *Automation and Industrial Labor.* New York: Random House.

Rice, R., and Smith, W. R. (1971). SYMBOL—a major departure from classic software dominated von Neumann computing systems. *SJCC,* vol. 38, pp. 575–587.

Richardson, D. (1968). Some unsolvable problems involving elementary functions of a real variable. *J. Symb. Logic,* vol. 33, pp. 514–520.

Richardson, D. (1972). Tessellations with local transformations. *J. Comp. and Sys. Sci.,* vol. 6, pp. 373–388.

Riley, W. B. (1970). Illiac 4 enters the home stretch. *Electronics,* June 8, pp. 123–127.

Risch, R. (1969). The problem of integration in finite terms. *Trans. Amer. Math. Soc.,* vol. 139, pp. 167–189.

Ritchie, R. W. (1963). Classes of predictably computable functions. *Trans. Amer. Math. Soc.,* vol. 106, pp. 139–173.

Roberts, L. (1963). Machine perception of three-dimensional solids. Tech. Rept. 315. Cambridge, Mass.: M.I.T. Lincoln Lab.

Robinson, G. A., and Wos, L. (1969). Paramodulation and theorem-proving in first order theories with equality. *MI4,* pp. 135–150.

Robinson, J. A. (1965). A machine-oriented logic based on the resolution principle. *JACM,* vol. 12, pp. 23–41.

——— (1969). Mechanizing higher-order logic. *MI4,* pp. 151–170.

Robinson, J. A. (1970). An overview of mechanical theorem proving. *TAPR*, pp. 2–20.

Rodriguez, J. E. (1969). A graph model for parallel computations. MAC-TR-64. Cambridge, Mass.: M.I.T. Proj. MAC.

Rogers, H, (1967). *Theory of Recursive Functions and Effective Computability*. New York: McGraw-Hill.

Rose, J. E., Malis, L. I., and Baker, C. P. (1969). Neural growth in the cerebral cortex after lesions produced by monoenergetic deutrons. In *Memory Mechanism (Brain and Behavior,* vol. 3), ed. K. H. Pribram, pp. 22–32. Baltimore: Penguin.

Rosen, C. A. (1970). An experimental mobile automaton. Tech. Note 39. Menlo Park, Calif.: Art. Intell. Group, Stanford Research Institute.

Rosen, S. (1971). *A Quarter Century View*. New York: ACM.

———— (1972). Programming systems and languages. *CACM*, vol. 15, no. 7, pp. 591–600.

Rosenberg, J. M. (1969). *The Computer Prophets*. London: Macmillan.

Rosenblatt, F. (1959a). Two theorems of statistical separability in the perceptron. *Proc. Symp. Mechanization of Thought Processes*, pp. 421–456. London: Her Majesty's Stationery Office.

———— (1959b). On the convergence of reinforcement procedures in simple perceptions. Rept. VG-1196-G-3. Buffalo, N.Y.: Aeronautical Lab., Cornell Univ.

Rosenblueth, A., Wiener, N., and Bigelow, J. (1943). Behavior, purpose, and teleology. *Phil. Sci.*, vol. 10, pp. 18–24.

Rosenfeld, A. (1969). Picture processing by computer. *Computing Surveys*, vol. 1, no. 3, pp. 147–176.

———— (1969). *Picture Processing by Computer*. New York: Academic Press.

———— (1972). Picture processing: 1972. Tech. Rept. TR-217. College Park, Maryland: Comp. Sci. Center, Univ. Maryland.

———— (1973). Non-purposive perception in computer vision. Tech. Rept. TR-219. College Park: Comp. Sci. Center, Univ. Maryland.

———— (1973). Arcs and curves in digital pictures. *JACM*, vol. 20, no. 1, pp. 81–87.

Rosenzweig, M. R., Bennet, E. L., and Diamond, M. C. (1972). Brain changes in response to experience. *Scientific American*, February, vol. 226, pp. 22–29.

Ross, L. W. (1971). Simulating the dynamics of air and water pollution. *FJCC*, vol. 39, pp. 105–113.

Rosser, J. B., and Turquette, A. R. (1952). *Many-Valued Logics*. Amsterdam: North-Holland.

Rulifson, J. F. (1971). QA4 programming concepts. Tech. Note 60. Menlo Park, Calif.: Art. Intell. Group, Stanford Research Institute.

Rulifson, J. F., Derksen, J. A., and Waldinger, R. J. (1972). QA4: a procedural calculus for intuitive reasoning. Tech. Note 73. Menlo Park, Calif.: Stanford Res. Inst. Art. Intell. Center.

Rumelhart, D. E., Lindsay, P. H., and Norman, D. A. (1971). A process model for long-term memory. CHIP 17. La Jolla, Calif.: Center for Human Infor. Processing, Univ. Calif. San Diego.

Russell, B. (1948). *Human Knowledge: Its Scope and Limits.* New York: Simon and Schuster.

Russell, R. (1964). Kalah, the game and the program. AI Memo 22. Stanford: Stanford Art. Intell. Proj.

Russell, S. W. (1972). Semantic categories of nominals for conceptual dependency analysis of natural language. AIM-172. Stanford: Stanford Art. Intell. Proj.

Russell, S. W., et al. *See also* Colby, Weber, and Hilf; Schank, Tesler, and Weber.

Ryder, J. L. (1971). Heuristic analysis of large trees as generated in the game of Go. AIM-155. Stanford: Stanford Art. Intell. Proj.

Sackman, H. (1971). *Mass Information Utilities and Social Excellence.* Philadelphia: Auerbach.

Salton, G. (1972). Dynamic document processing. *CACM*, vol. 15, no. 7, pp. 658–670.

Sammet, J. (1972). Programming languages: History and future. *CACM*, vol. 15, no. 7, pp. 601–609.

Samuel, A. L. (1959). Some studies in machine learning using the game of Checkers. *CT*, pp. 71–105.

——— (1960). Programming computers to play games. In *Advances in Computers*, ed. F. Alt, vol. 1, pp. 165–192. New York: Academic Press.

——— (1962). Some moral and technical consequences of automation—A refutation. In *Automation: Implications for the Future*, ed. M. Philipson, pp. 174–179. New York: Random House.

——— (1962). Artificial intelligence—A frontier of automation. *Annals Amer. Acad. Political and Social Sci.*, vol. 340, pp. 10–20.

——— (1967). Some studies in machine learning using the game of Checkers, II—Recent progress. *IBM J. Res. Dev.*, vol. 11, no. 6, pp. 601–617.

Sandewall, E. J. (1969). Concepts and methods for heuristic search. *IJCAI*, pp. 199–218.

——— (1971). Formal methods in the design of question-answering systems. *Art. Intell.*, vol. 2, no. 2, pp. 129–146.

Sayre, K. M. (1969). *Consciousness: A Philosophic Study of Minds and Machines.* New York: Random House.

Sayre, K. M., and Crosson, F., eds. (1963). *The Modelling of Mind.* Notre Dame, Ind.: Univ. Notre Dame Press.

Schaffner, K. F. (1969). Chemical systems and chemical evolution: The philosophy of molecular biology. *American Scientist*, vol. 57, no. 4, pp. 410–420.

Schank, R. (1970). Semantics in conceptual analysis. AIM-122. Stanford: Stanford Art. Intell. Proj.

Schank, R. (1971a). Intention, memory, and computer understanding. AIM-140. Stanford: Stanford Art. Intell. Proj.

——— (1971b). Finding the conceptual content and intention in an utterance in natural language conversation. *IJCAI2*, pp. 444–454.

——— (1972). Adverbs and belief. AIM-171. Stanford: Stanford Art. Intell. Proj.

——— (1973). The fourteen primitive actions and their inferences. AIM-183. Stanford: Stanford Art. Intell. Proj.

Schank, R., and Tesler, L. G. (1969). A conceptual parser for natural language. *IJCAI*, pp. 569–578.

Schank, R., Tesler, L. G., and Weber, S. (1970). Spinoza II: Conceptual case-based natural language analysis. AIM-109. Stanford: Stanford Art. Intell. Proj.

Schmidt, K. P., and Inger, R. F. (1962). *Living Reptiles of the World*. New York: Doubleday.

Schmidt, R. A. (1971). A study of the real-time control of a computer-driven vehicle. AIM-149. Stanford: Stanford Art. Intell. Proj.

Schoenfield, J. R. (1971). *Degrees of Unsolvability*. New York: Elsevier.

Schrandt, R. G., and Ulam, S. M. (1970). On recursively defined geometrical objects and patterns of growth. *ECA*, pp. 232–243.

Schwarcz, R. M., Burger, J. F., and Simmons, R. F. (1970). A deductive question-answerer for natural language inference. *CACM*, vol. 13, no. 3, pp. 167–183.

Scott, D. (1970). Outline of a mathematical theory of computation. Tech. Monograph PRG-2. Oxford: Programming Res. Group, Oxford Univ. Computing Lab.

——— (1971). Continuous lattices. Tech. Monograph PRG-7. Oxford: Programming Res. Group, Oxford Univ. Computing Lab.

Scriven, M. (1970). The compleat robot: A prolegomena to androidology. In *Human and Artificial Intelligence*, ed. F. Crosson, pp. 117–140. New York: Appleton.

Segoe, K. (1960). *Go Proverbs Illustrated*. Tokyo: Nihon Ki-in.

Selfridge, O. G. (1958). Pandemonium: A paradigm for learning. In *Pattern Recognition*, ed. L. Uhr, pp. 339–348. New York: Wiley.

Selfridge, O. G., and Neisser, U. (1963). Pattern recognition by machine. *CT*, pp. 237–250.

Shannon, C. E. (1948). A mathematical theory of communication. *Bell Sys. Tech. J.*, vol. 27, pp. 379–423, 623–656.

——— (1950a). Programming a digital computer for playing chess. *Philosophy Magazine*, vol. 41, pp. 356–375.

——— (1950b). Automatic chess player. *Scientific American*, February, vol. 128, p. 48.

——— (1953). Computers and automata. *PCR*, pp. 114–127.

——— (1959). Von Neumann's contributions to automata theory. *Bull. Amer. Math. Soc.*, vol. 64, no. 3, part 2, pp. 123–129.

Shannon, C. E., and McCarthy, J., eds. (1956). *Automata Studies.* Princeton, N.J.: Princeton Univ. Press.

Shannon, C. E., and Weaver, W. (1949). *The Mathematical Theory of Communication.* (Revised 1962.) Urbana: Univ. Illinois Press.

Shapiro, S. C. (1971a). A net structure for semantic information storage, deduction and retrieval. Tech. Rept. 109. Madison: Comp. Sci. Dept., Univ. Wisconsin.

—— (1971b). A net structure for semantic information storage, deduction, and retrieval. *IJCAI2*, pp. 512–523.

Shaw, A. (1968). The formal description and parsing of pictures. SLAC-84. Stanford: Stanford Linear Accelerator Center.

Sheppard, W. (1961). *Mazes and Labyrinths: A Book of Puzzles.* New York: Dover.

Shirrai, Y. (1972). A heterarchical program for recognition of polyhedra. AI Memo No. 263. Cambridge, Mass.: M.I.T. Art. Intell. Lab.

Shklovskii, I. S., and Sagan, C. (1966). *Intelligent Life in the Universe.* New York: Dell.

Shoenfield, J. R. (1967). *Mathematical Logic.* Reading, Mass.: Addison-Wesley.

Sibert, E. E. (1969). A machine-oriented logic incorporating the equality relation. *MI4*, pp. 103–134.

Siklossy, L. (1972). Natural language learning by computer. *RM*, pp. 288–328.

Siklossy, L., and Simon, H. A. (1972). Some semantic methods for language processing. *RM*, pp. 44–66.

Silberman, C. E. (1966). *The Myths of Automation.* New York: Harper & Row.

Simmons, R. F. (1965). Answering English questions by computer: A survey. *CACM*, vol. 8, no. 1, pp. 53–70.

—— (1969). Natural language question-answering systems: 1969. *CACM*, vol. 13, pp. 15–30.

Simmons, R. F., and Bruce, B. C. (1971). Some relations between predicate calculus and semantic net representations of discourse. *IJCAI2*, pp. 524–530.

Simmons, R. F., Burger, J. F., and Long, R. E. (1966). An approach towards answering English questions from text. *FJCC*, vol. 29, pp. 357–364.

Simmons, R. F., Burger, J. F., and Schwarcz, R. M. (1968). A computer model of verbal understanding. *FJCC*, vol. 33, part 1, pp. 441–456.

Simmons, R. F., et al. *See also* McConlogue and Simmons.

Simon, H. A. (1957). *Models of Man, Social and Rational.* New York: Wiley.

—— (1962). The corporation: Will it be managed by machines? In *Automation: Implications for the Future,* ed. M. Philipson, pp. 230–263. New York: Random House.

Simon, H. A. (1963). Experiments with a heuristic compiler. *JACM*, vol. 10. pp. 482–506.

———— (1965). *The Shape of Automation*. New York: Harper & Row.

———— (1966a). Scientific discovery and the psychology of problem solving. In *Mind and Cosmos*, ed. R. G. Colodny, pp. 22–40. Pittsburgh, Pa.: Univ. Pittsburgh Press.

———— (1966b). Thinking by computers. In *Mind and Cosmos*, ed. R. G. Colodny, pp. 3–21. Pittsburgh, Pa.: Univ. Pittsburgh Press.

———— (1969). *The Sciences of the Artificial*. Cambridge, Mass.: M.I.T. Press.

———— (1972). The theory of problem solving. *Information Processing 71*, pp. 261–277. Amsterdam: North-Holland.

———— (1972a). The heuristic compiler. *RM*, pp. 9–43.

———— (1972b). On reasoning about actions. *RM*, pp. 414–430.

Simon, H. A., et al. *See also* Newell and Simon.

Sirovich, F. (1972). Memory system of a problem solver generator. Pittsburgh: Dept. Comp. Sci., Carnegie-Mellon Univ.

Skinner, B. F., Woolridge, D. E., Asimov, I., Bradbury, R., Minsky, M., Popoff, D., Switzgebel, R. L., Hall, M. H., and Bunch, R. (1969). Man and machine. *Psychology Today*, April, vol. 2, no. 11, pp. 20–58.

Sklansky, J. (1969). Recognizing convex blobs. *IJCAI*, pp. 107–116.

Sklansky, J., Finkelstein, M., and Russell, E. C. (1968). A formalism for program translation. *JACM*, vol. 15, no. 2, pp. 165–175.

Skolem, T. (1971). Peano's axioms and models of arithmetic. In *Mathematical Interpretation of Formal Systems*, ed. T. Skolem et al., pp. 1–14. Amsterdam: North-Holland.

Slagle, J. R. (1963). A heuristic program that solves symbolic integration problems in freshman calculus. *CT*, pp. 191–203.

———— (1965a). Experiments with a deductive question-answering program. *CACM*, vol. 8, no. 12, pp. 792–798.

———— (1965b). Automatic theorem-proving with renameable and semantic resolution. *JACM*, vol. 14, no. 4, pp. 687–697.

———— (1970a). Heuristic search programs. *TAPR*, pp. 246–273.

———— (1970b). Interpolation theorems for resolution in lower predicate calculus. *JACM*, vol. 17, pp. 535–542.

———— (1971). *Artificial Intelligence: The Heuristic Programming Approach*. New York: McGraw-Hill.

———— (1972). Automatic theorem proving with built-in theories including equality, partial ordering, and sets. *JACM*, vol. 19, no. 1, pp. 120–135.

Slagle, J. R., and Bursky, P. (1968). Experiments with a multipurpose, theorem-proving heuristic program. *JACM*, vol. 15, no. 1, pp. 85–99.

Slagle, J. R., and Farrell, C. D. (1971). Experiments in automatic learning for a multipurpose heuristic program. *CACM*, vol. 14, no. 2, pp. 91–99.

Slagle, J. R., and Lee, R. C. T. (1971). Application of game-tree searching techniques to sequential patter recognition. *CACM*, vol. 14, no. 2, pp. 103–110.

Slagle, J. R., Chang, C. L., and Lee, R. C. T. (1969). Completeness theorems for semantic resolution in consequence finding. *IJCAI*, pp. 281–286.

Slotnick, D. L. (1967). Achieving large computing capabilities through an array computer. *SJCC*, vol. 30, pp. 477–481.

———— (1971). The fastest computer. *Scientific American*, February, vol. 224, no. 2, pp. 76–87.

Smith, A. R. (1969). Cellular automata theory. SEL-70-016. Stanford: Stanford Electronics Lab.

Smith, D. C. (1970). MLISP. AIM-135. Stanford: Stanford Art. Intell. Proj.

Smith, T. M. (1970). Some perspectives on the early history of computers. *PCR*, pp. 7–15.

Smoliar, S. W. (1971). A parallel processing model of musical structures. TR-91. Cambridge, Mass.: M.I.T. Proj. MAC.

Sobel, I. (1970). Camera models and machine perception. AIM-121. Stanford: Stanford Art. Intell. Proj.

Spinelli, D. N. (1966). Visual receptive fields in the cat's retina: Complications. *Science*, vol. 152, pp. 1768–1769.

Sprague, R. E. (1969). *Information Utilities*. Englewood Cliffs, N.J.: Prentice-Hall.

Sternberg, S. (1969). Memory-scanning: Mental processes revealed by reaction-time experiments. *American Scientist*, vol. 57, no. 4, pp. 421–457.

Stoppard, T. (1967). *Rosencrantz and Guildenstern Are Dead*. New York: Grove.

Suppes, P. (1957). *Introduction to Logic*. New York: Van Nostrand.

Sussman, G. J. (1972). Teaching of procedures—Progress report. AI Memo No. 270. Cambridge, Mass.: M.I.T. Art. Intell. Lab.

———— (1972). Teaching of procedures—progress report. AI Memo No. 270. Cambridge, Mass.: M.I.T. Art. Intell. Lab.

Sussman, G. J., and McDermott, D. V. (1972a). Why conniving is better than planning. AI Memo No. 255A. Cambridge, Mass.: M.I.T. Art. Intell. Lab.

———— (1972b). The CONNIVER reference manual. AI Memo No. 259. Cambridge, Mass.: M.I.T. Art. Intell. Lab.

Sussman, G. J., and Winograd, T. (1970). Micro-planner reference manual. AI Memo 203. Cambridge: M.I.T. Art. Intell. Lab.

Sutro, I. L., and Kilmer, W. L. (1969). Assembly of computers to command and control a robot. *SJCC*, vol. 34, pp. 113–137.

Tarski, A. (1956). *Logic, Semantics, Metamathematics*. Oxford: University Press.

Taviss, I., ed. (1970). *The Computer Impact*. Englewood Cliffs, N.J.: Prentice-Hall.

Taylor, G. R. (1970). *The Doomsday Book: Can the World Survive?* New York: World Publishing.

Teale, E. W. (1962). *The Strange Lives of Familiar Insects.* New York: Dodd, Mead.

Teitelman, W. (1967). Design and implementation of FLIP, a LISP format directed list processor. BBN Rept. No. AFCRL-67-0514. Cambridge: Bolt Beranek and Newman.

——— (1969). Toward a programming laboratory. *IJCAI,* pp. 1–8.

Teitelman, W., Bobrow, D. G., Hartley, A. K., and Murphy, D. L. (1972). BBN-LISP, TENEX reference manual. Cambridge: Bolt Beranek and Newman.

Tesler, L. G., and Enea, H. J. (1968). A language design for concurrent processes. *SJCC,* vol. 32, pp. 403–408.

Tesler, L. G., et al. *See also* Colby, Enea, and Tesler; Schank and Tesler.

Thatcher, T. W. (1963). The construction of a self-describing Turing machine. *Proc. Symp. Math. Theory of Automata,* pp. 165–171. Brooklyn: Polytechnic.

Thompson, D. W. (1917). *On Growth and Form.* Cambridge, Britain: Cambridge Univ. Press.

Thompson, R. F. (1967). *Foundations of Physiological Psychology.* New York: Harper & Row.

Thorne, J. P., Bratley, P., and Dewar, H. (1968). The syntactic analysis of English by machine. *MI3,* pp. 281–309.

Thorp, E., and Walden, W. (1970). A computer-assisted study of GO on MxN boards. *TAPR,* pp. 303–343.

Titus, J. P. (1966). The computer as a threat to personal privacy. *CACM,* vol. 9, no. 11, pp. 824–826.

Toffler, A. (1970). *Future Shock.* New York: Random House.

Tonge, F. M. (1963). Summary of a heuristic line-balancing procedure. *CT,* pp. 168–190.

——— (1969). Hierarchical aspects of computer languages. In *Hierarchical Structures,* ed. L. L. Whyte et al., pp. 233–251. New York: Elsevier.

Toynbee, A. J. (1966). *Change and Habit: The Challenge of Our Time.* New York: Oxford Univ. Press.

Trakhtenbrot, B. A. (1963). *Algorithms and Automatic Computing Machines.* Boston: Heath.

Tsichritzis, D. C. (1968). Partially solvable and almost solvable problems. Ph.D. Thesis. Princeton Univ. Electrical Engineering Dept. *Also:* Document 69–10,528, Ann Arbor: Univ. Microfilms.

——— (1970). The equivalence problem of simple programs. *JACM,* vol. 17, no. 4, pp. 729–738.

——— (1971). IFF programs. In *Theory of Machines and Computations,* eds. Z. Kohavi and A. Paz, pp. 99–114. New York: Academic Press.

Turing, A. M. (1936). On computable numbers, with an application to the Entscheidungs-problem. *Proc. London Math. Soc.,* ser. 2, vol. 42, pp.

230–265, and vol. 43, pp. 544–546. *Also* in *The Undecidable,* ed. M. Davis, pp. 115–154. Hewlett, N.Y.: Raven.

——— (1937). Computability and λ-definability. *J. Symb. Logic,* vol. 2, pp. 153–163.

——— (1939). Systems of logic based on ordinals. *Proc. London Math. Soc.,* ser. 2, vol. 45, pp. 161–228. *Also* in *The Undecidable,* ed. M. Davis, pp. 154–222. Hewlett, N.Y.: Raven.

——— (1947). Intelligent machinery. *MI5,* pp. 3–26.

——— (1950). Computing machinery and intelligence. *Mind,* vol. 59, pp. 433–460. *Also* in *CT,* pp. 11–35.

——— (1953). Digital computers applied to games. In *Faster than Thought,* ed. B. V. Bowden, pp. 286–310. London: Pitman.

Turner, K. J. (1971). Scene analysis and recognition: a survey. Experimental Programming Rept. No. 21. Edinburgh: Dept. Machine Intell. and Perception, Univ. Edinburgh.

Uhr, L. (1960). Intelligence in computing machines: The psychology of perception in people and in machines. *Behavioral Sci.,* vol. 5, pp. 177–182.

——— (1971). Flexible linguistic pattern recognition. *Patt. Recog.,* vol. 3, pp. 363–383.

——— ed. (1967). *Pattern Recognition.* New York: Wiley

Uhr, L., and Kochen, M. (1969). MIKROKOSMS and robots. *IJCAI,* pp. 541–555.

Uhr, L., and Vossler, C. (1963). A pattern-recognition program that generates, evaluates, and adjusts its own operators. *CT,* pp. 251–268.

Ulam, S. M. (1962). On some mathematical problems connected with patterns of growth. *ECA,* pp. 219–231.

Ulam, S. M., et al. *Se also* Kac and Ulam; Schrandt and Ulam.

van Dantzig, D. (1955). Is $10^{10^{10}}$ a finite number? *Dialectica,* vol. 9, pp. 273–277.

van Emden, M. H. (1970). Hierarchical decomposition of complexity. *MI5,* pp. 361–382.

——— (1971). *An Analysis of Complexity.* Amsterdam: Mathematical Centre.

Van Heijenoort, J., ed. (1967). *From Frege to Godel: A Source Book in Mathematical Logic.* Cambridge, Mass.: Harvard Univ. Press.

Varshavsky, V. I. (1969). The organization of interaction in collectives of automata. *MI4,* pp. 285–314.

Vicens, P. (1969). Aspects of speech recognition by computer. AI Memo 85. Stanford: Stanford Art. Intell. Proj.

von Bertalanffy, L. (1967). *Robots, Men and Minds.* New York: Braziller.

——— (1968). *General System Theory: Foundations, Development, Applications.* New York: Braziller.

Vonnegut, K. (1952). *Player Piano.* New York: Holt, Rinehart and Winston.

—————— (1970). *The Sirens of Titan.* New York: Dell.

von Neumann, J. (1951). *Mathematical Foundations of Quantum Mechanics.* Princeton, N.J.: Princeton Univ. Press.

—————— (1951). The general and logical theory of automata. In *Collected Works,* ed. A. H. Taub, vol. 5, pp. 288–328. New York: Pergamon.

—————— (1956). Probabilistic logics and the synthesis of reliable organisms from unreliable components. In *Automata Studies,* eds. C. E. Shannon and J. McCarthy, pp. 43–98. Princeton, N.J.: Princeton Univ. Press.

—————— (1958). *The Computer and the Brain.* New Haven, Conn.: Yale Univ. Press.

—————— (1966). *Theory of Self-Reproducing Automata.* Edited and completed by A. W. Burks. Urbana: Univ. Illinois Press.

von Neumann, J., and Morgenstern, O. (1944). *Theory of Games and Economic Behavior.* New York: Wiley.

Vygotsky, L. S. (1962). *Thought and Language.* Cambridge, Mass.: M.I.T. Press.

Wagner, E. G. (1964). An approach to modular computers—I. spider automata and embedded automata. RC 1107. Yorktown Heights, N.Y.: IBM Watson Research Center.

—————— (1965). An approach to modular computers—II. graph theory and the interconnection of modules. RC 1414. Yorktown Heights, N.Y.: IBM Watson Research Center.

Waltz, D. L. (1972). Generating semantic descriptions from drawings of scenes with shadows. AI-TR-271. Cambridge, Mass.: M.I.T. Art. Intell. Lab.

Waksman, A. (1966). An optimum solution to the firing squad synchronization problem. *IC,* vol. 9, no. 1, pp. 66–78.

Waldinger, R. J. (1970). Robot and state variable. Tech. Note 26. Menlo Park, Calif.: Art. Intell. Group, Stanford Research Institute.

Waldinger, R. J., and Lee, R. C. T. (1969). PROW: A step toward automatic program writing. *IJCAI,* pp. 241–252.

Waldinger, R. J., et al. See also Manna and Waldinger; Rulifson, Derksen, and Waldinger.

Wand, M. (1972). A concrete approach to abstract recursive definitions. AI Memo No. 262. Cambridge, Mass.: M.I.T. Art. Intell. Lab.

Wang, H. (1957). A variant to Turing's theory of computing machines. *JACM,* vol. 4, pp. 63–92.

—————— (1960). Toward mechanical mathematics. *IBM J. Res. Dev.,* vol. 4, pp. 2–22.

—————— (1962). *A Survey of Mathematical Logic.* Amsterdam: North-Holland.

Wasserman, A. I. (1970a). Achievement of skill and generality in an artificial intelligence program. Ph.D. Thesis. Madison: Univ. Wisconsin.

—————— (1970b). Realization of a skillful bridge bidding program. *FJCC,* vol. 37, pp. 433–444.

Watanabe, S. (1969a). Modified concepts of logic, probability, and information based on generalized continuous characteristic function. *IC*, vol. 15, pp. 1–21.

────── (1969b). *Knowing and Guessing*. New York: Wiley.

────── (1971). Ungrammatical grammar in pattern recognition. *Patt. Recog.*, vol. 3, pp. 385–408.

────── ed. (1969). *Methodologies of Pattern Recognition*. New York: Academic Press.

Waterman, D. A. (1968). Machine learning of heuristics. AI Memo 74. Stanford: Stanford Art. Intell. Proj.

────── (1970). Generalization learning techniques for automating the learning of heuristics. *Art Intell.*, vol. 1, numbers 1 and 2, pp. 121–170.

Waterman, D. A., and Newell, A. (1971). Protocol analysis as a task for artificial intelligence. *IJCA12*, pp. 190–217.

────── (1972). Preliminary results with a system for automatic protocol analysis. CIP Working Paper #211. Pittsburgh, Pa.: Depts. Psych. and Comp. Sci., Carnegie-Mellon Univ.

Watson, J. D. (1970). *The Molecular Biology of the Gene*. New York: W. A. Benjamin.

Webb, E. J., Campbell, D. T., Schwartz, R. D., and Sechrest, L. (1969). *Unobtrusive Measures: Nonreactive Research in the Social Sciences*. Chicago: Rand McNally.

Wegbreit, B. (1970). Studies in extensible programming languages. ESD-TR-70-297. Hanscom Field, Mass.: Directorate of Systems Design and Development, HQ Electronic Systems Division (AFSC).

────── (1971). The ECL programming system. *FJCC*, vol. 39, pp. 253–262.

────── (1972). Multiple evaluators in an extensible programming system. *FJCC*, vol. 41, part 2, pp. 905–915.

Wegbreit, B., et al. *See also* Bobrow and Wegbreit.

Wegner, P. (1972). The Vienna Definition Language. *Computing Surveys*, vol. 4, no. 1, pp. 5–63.

Weiner, N. (1971). Regulation of norepinephrine biosynthesis. *Annual Rev. Pharmacology*, pp. 273–290.

Weir, R. H. (1962). *Language in the Crib*. The Hague, Netherlands: Mouton.

Weiskrantz, L. (1970). A long-term view of short-term memory in psychology. In *Short Term Changes in Neural Activity and Behavior*, eds. G. Horn and R. A. Hinde, pp. 63–74. London: Cambridge Univ. Press.

Weissman, C. (1967). *LISP 1.5 Primer*. Belmont, Calif.: Dickenson.

Weizenbaum, J. (1966). ELIZA—A computer program for the study of natural language communication between man and machine. *CACM*, vol. 9, no. 1, pp. 36–45.

────── (1967). Contextual understanding by computers. *CACM*, vol. 10, no. 8, pp. 474–480.

Westin, A. F., ed. (1971). *Information Technology in a Democracy.* Cambridge, Mass.: Harvard Univ. Press.

Weyl, H. (1949). *Philosophy of Mathematics and Natural Science.* Princeton, N.J.: Princeton Univ. Press.

White, L. (1969). *Machina Ex Deo.* Cambridge, Mass.: M.I.T. Press.

Whyte, L. L., Wilson, A. G., and Wilson, D., eds. (1969). *Hierarchical Structures.* New York: Elsevier.

Wiener, N. (1948). *Cybernetics.* New York: Wiley.

——— (1950). *The Human Use of Human Beings; Cybernetics and Society.* Boston: Houghton Mifflin.

——— (1958). *Nonlinear Problems in Random Theory.* New York: Wiley.

——— (1962). Some moral and technical consequences of automation. In *Automation: Implications for the Future,* ed. M. Philipson, pp. 162–173. New York: Random House.

——— (1964). *God and Golem, Inc.* Cambridge, Mass.: M.I.T. Press.

Wiener, N., et al. *See also* Rosenblueth, Wiener, and Bigelow.

Wiens, J. A., ed. (1972). *Ecosystem Structure and Function.* Corvallis: Oregon State Univ. Press.

Wilks, Y. (1971). One small head—Some remarks on the use of "model" in linguistics. CS-247. Stanford: Comp. Sci. Dept., Stanford Univ.

——— (1972a). An artificial intelligence approach to machine translation. AIM-161. Stanford: Stanford Art. Intell. Proj.

——— (1972b). Lakoff on linguistics and natural logic. AIM-170. Stanford: Stanford Art. Intell. Proj.

Williams, D. S. (1972). Computer program organization induced from problem examples. *RM,* pp. 143–206.

Williams, J. (1954). *The Compleat Strategyst, Being a Primer on the Theory of Games of Strategy.* New York: McGraw-Hill.

Williams, T. G. (1972). Some studies in game playing with a digital computer. *RM,* pp. 71–142.

Wilner, W. T. (1971). Declarative semantic definition. STAN-CS-233-71. Stanford: Comp. Sci. Dept., Stanford Univ.

Winograd, T. (1968). Linguistics and the computer analysis of tonal harmony. *J. Music Theory,* vol. 12, no. 1, pp. 2–49.

——— (1971). Procedures as a representation for data in a computer program for understanding natural language. MAC-TR-84. Cambridge, Mass.: M.I.T. Proj. MAC.

——— (1972). *Understanding Natural Language.* New York: Academic Press.

Winograd, T., et al. *See also* Sussman and Winograd.

Winston, P. H. (1970). Learning structural descriptions from examples. MAC-TR-76. Cambridge, Mass.: M.I.T. Proj. MAC.

Wittgenstein, L. (1958). *The Blue and Brown Books.* New York: Harper & Row.

Woods, W. A. (1968). Procedural semantics for a question-answering machine. *FJCC,* vol. 33, part 1, pp. 457–471.

———— (1969). Augmented transition networks for natural language analysis. Cambridge, Mass.: Computation Lab., Harvard.

———— (1970). Transition network grammars for natural language analysis. *CACM*, vol. 13, pp. 591–602.

Wos, L., Robinson, G. A., Carson, D. F., and Shalla, L. (1967). The concept of demodulation in theorem proving. *JACM*, vol. 14, pp. 698–709.

Yasuhara, A. (1971). *Recursive Function Theory and Logic.* New York: Academic Press.

Yovits, M., and Cameron, S., eds. (1960). *Self-Organizing Systems.* New York: Pergamon.

Yovits, M., Jacobi, G. T., and Goldstein, G. D., eds. (1962). *Self-Organizing Systems 1962.* Washington, D.C.: Spartan.

Zadeh, L. A. (1965). Fuzzy sets. *IC*, vol. 8, pp. 338–353.

———— (1968). Fuzzy algorithms. *IC*, vol. 12, pp. 94–102.

———— (1972). A system-theoretic view of behavior modification. ERL-M320. Berkeley: Elect. Res. Lab., Univ. Calif.

Zadeh, L. A., and Polak, E. (1969). *System Theory.* New York: McGraw-Hill.

Zobrist, A. (1969). A model of visual organization for the game of GO. *SJCC*, vol. 34, pp. 103–112.

# INDEX

447